RAND McNALLY

GOODE'S

ATLAS OF Political Geography

to accompany titles published by
John Wiley & Sons, Inc.

Howard Veregin, Ph.D., Editor

Editorial Advisory Board

Byron Augustin, D.A., Texas State University-San Marcos
Joshua Comenetz, Ph.D., University of Florida
Francis Galgano, Ph.D., United States Military Academy
Sallie A. Marston, Ph.D., University of Arizona
Virginia Thompson, Ph.D., Towson University

Abridgement of
21ST Edition

John Wiley & Sons, Inc. and **RAND MCNALLY**

Working together to bring you the best in geography education

Few publishers can claim as rich a history as John Wiley & Sons, Inc. (publishers since 1807) and Rand McNally & Company (publishers since 1856). Even fewer can claim as long-standing a commitment to geographic education.

Wiley's partnership with the geographic community began at the very beginning of the 20th century with the publication of textbooks on surveying. Rand McNally's partnership began even earlier, with the publication of the first Rand McNally maps in 1872. Since then, both companies have worked in parallel to help students visualize spatial relationships and appreciate the earth's dynamic landscapes and diverse cultures.

Now these two publishers have combined their efforts to bring you this new atlas, which represents the very best in educational resources for geography.

Based on the 21st edition of the *Goode's World Atlas*, the *Goode's Atlas of Political Geography* features:

- An emphasis on map accuracy and legibility, and the mixture of maps of different types and scales to facilitate interpretation of geographic phenomena.
- World, continental, and regional population density maps, which have been created using LandScan, a digital population database developed using satellite and computer-mapping technology.
- Graphs accompanying many of the maps, to show important statistical information, trends over time, and relationships between variables.
- Maps and graphs that have been updated, based on the most current available data in accordance with the high standards and quality that have always been a defining feature of the *Goode's World Atlas*.

Wiley and Rand McNally are currently offering seven new course-specific atlases, which can be packaged with any of Wiley's best-selling textbooks, or sold separately as stand-alones. These atlases include:

Rand McNally Goode's Atlas of Political Geography	0-471-70694-9
Rand McNally Goode's Atlas of Latin America	0-471-70697-3
Rand McNally Goode's Atlas of North America	0-471-70696-5
Rand McNally Goode's Atlas of Asia	0-471-70699-X
Rand McNally Goode's Atlas of Urban Geography	0-471-70695-7
Rand McNally Goode's Atlas of Physical Geography	0-471-70693-0
Rand McNally Goode's Atlas of Human Geography	0-471-70692-2

This book was set by GGS Book Services and printed and bound by Walsworth Press. The cover was printed by Phoenix Color.

To order books or for customer service please, call 1-800-CALL WILEY (225-5945).

ISBN 0471-70694-9

Printed in the United States

10 9 8 7 6 5 4 3 2 1

Table of Contents

Tables and Indexes

Introduction

Basic Earth Properties

The subject matter of **geography** includes people, landforms, climate, and all the other physical and human phenomena that make up the earth's environments and give unique character to different places. Geographers construct maps to visualize the **spatial distributions** of these phenomena: that is, how the phenomena vary over geographic space. Maps help geographers understand and explain phenomena and their interactions.

To better understand how maps portray geographic distributions, it is helpful to have an understanding of the basic properties of the earth.

The earth is essentially **spherical** in shape. Two basic reference points — the **North and South Poles** — mark the locations of the earth's axis of rotation. Equidistant between the two poles and encircling the earth is the **equator**. The equator divides the earth into two halves, called the **northern and southern hemispheres**. (See the figures to the right.)

Latitude and longitude are used to identify the locations of features on the earth's surface. They are measured in degrees, minutes and seconds. There are 60 minutes in a degree and 60 seconds in a minute. Latitude is the angle north or south of the equator. The symbols °, ', and " represent degrees, minutes and seconds, respectively. The N means north of the equator. For latitudes south of the equator, S is used. For example, the Rand McNally head office in Skokie, Illinois, is located at 42°1'51" N. The minimum latitude of 0° occurs at the equator. The maximum latitudes of 90° N and 90° S occur at the North and South Poles.

A **line of latitude** is a line connecting all points on the earth having the same latitude. Lines of latitude are also called **parallels**, as they run parallel to each other. Two parallels of special importance are the **Tropic of Cancer** and the **Tropic of Capricorn**, at approximately 23°30' N and S respectively. This angle coincides with the inclination of the earth's axis relative to its orbital plane around the sun. These tropics are the lines of latitude where the noon sun is directly overhead on the solstices. (See figure on page 66.) Two other important parallels are the **Arctic Circle** and the **Antarctic Circle**, at approximately 66°30' N and S respectively. These lines mark the most northerly and southerly points at which the sun can be seen on the solstices.

While latitude measures locations in a north-south direction, longitude measures them east-west. Longitude is the angle east or west of the **Prime Meridian**. A **meridian** is a line of longitude, a straight line extending from the North Pole to the South Pole. The Prime Meridian is the meridian passing through the Royal Observatory in Greenwich, England. For this reason the Prime Meridian is sometimes referred to as the **Greenwich Meridian**. This location for the Prime Meridian was adopted at the International Meridian Conference in Washington, D.C., in 1884.

Like latitude, longitude is measured in degrees, minutes, and seconds. For example, the Rand McNally head office is located at 87°43'6" W. The qualifiers E and W indicate whether a location is east or west of the Greenwich Meridian. Longitude ranges from 0° at Greenwich to 180° E or W. The meridian at 180° E is the same as the meridian at 180° W. This meridian, together with the Greenwich Meridian, divides the earth into **eastern and western hemispheres**.

Any circle that divides the earth into equal hemispheres is called a **great circle**. The equator is an example. The shortest distance between any two points on the earth is along a great circle. Other circles, including all other lines of latitude, are called **small circles**. Small circles divide the earth into two unequal pieces.

View of earth centered on 30° N, 30° W

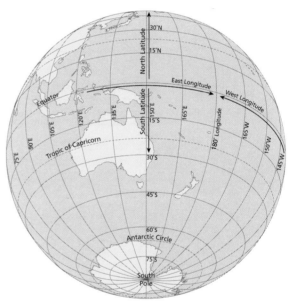

View of earth centered on 30° S, 150° E

The Geographic Grid

The grid of lines of latitude and longitude is known as the **geographic grid**. The following are some important characteristics of the grid.

All lines of longitude are equal in length and meet at the North and South Poles. These lines are called meridians.

All lines of latitude are parallel and equally spaced along meridians. These lines are called parallels.

The length of parallels increases with distance from the poles. For example, the length of the parallel at 60° latitude is one-half the length of the equator.

Meridians get closer together with increasing distance from the equator, and finally converge at the poles.

Parallels and meridians meet at right angles.

Map Scale

To use maps effectively it is important to have a basic understanding of map scale.

Map scale is defined as the ratio of distance on the map to distance on the earth's surface. For example, if a map shows two towns as separated by a distance of 1 inch, and these towns are actually 1 mile apart, then the scale of the map is 1 inch to 1 mile.

The statement "1 inch to 1 mile" is called a **verbal scale**. Verbal scales are simple and intuitive, but a drawback is that they are tied to the specific set of map and real-world units in the numerator and denominator of the ratio. This makes it difficult to compare the scales of different maps.

A more flexible way of expressing scale is as a **representative fraction**. In this case, both the numerator and denominator are converted to the same unit of measurement. For example, since there are 63,360 inches in a mile, the verbal scale "1 inch to 1 mile" can be expressed as the representative fraction 1:63,360. This means that 1 inch on the map represents 63,360 inches on the earth's surface. The advantage of the representative fraction is that it applies to any linear unit of measurement, including inches, feet, miles, meters, and kilometers.

Map scale can also be represented in graphical form. Many maps contain a **graphic scale** (or **bar scale**) showing real-world units such as miles or kilometers. The bar scale is usually subdivided to allow easy calculation of distance on the map.

Map scale has a significant effect on the amount of detail that can be portrayed on a map. This concept is illustrated here using a series of maps of the Washington, D.C., area. (See the figures to the right.) The scales of these maps range from 1:40,000,000 (top map) to 1:4,000,000 (center map) to 1:25,000 (bottom map). The top map has the **smallest scale** of the three maps, and the bottom map has the **largest scale**.

Note that as scale increases, the area of the earth's surface covered by the map decreases. The smallest-scale map covers thousands of square miles, while the largest-scale map covers only a few square miles within the city of Washington. This means that a given feature on the earth's surface will appear larger as map scale increases. On the smallest-scale map, Washington is represented by a small dot. As scale increases the dot becomes an orange shape representing the built-up area of Washington. At the largest scale Washington is so large that only a portion of it fits on the map.

Because small-scale maps cover such a large area, only the largest and most important features can be shown, such as large cities, major rivers and lakes, and international boundaries. In contrast, large-scale maps contain relatively small features, such as city streets, buildings, parks, and monuments.

Small-scale maps depict features in a more simplified manner than large-scale maps. As map scale decreases, the shapes of rivers and other features must be simplified to allow them to be depicted at a highly reduced size. This simplification process is known as **map generalization**.

Maps in *Goode's Atlas of Political Geography* have a wide range of scales. The smallest scales are used for the world thematic map series, where scales range from approximately 1:200,000,000 to 1:75,000,000. Reference map scales range from a minimum of 1:100,000,000 for world maps to a maximum of 1:1,000,000 for city maps. Most reference maps are regional views with a scale of 1:4,000,000.

1:40,000,000 scale

1:4,000,000 scale

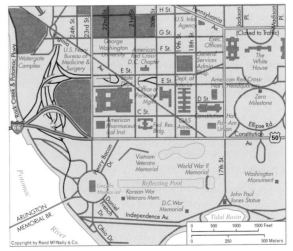

1:25,000 scale

Map Projections

Map projections influence the appearance of features on the map and the ability to interpret geographic phenomena.

A **map projection** is a geometric representation of the earth's surface on a flat or plane surface. Since the earth's surface is curved, a map projection is needed to produce any flat map, whether a page in this atlas or a computer-generated map of driving directions on www.randmcnally.com. Hundreds of projections have been developed since the dawn of mapmaking. A limitation of all projections is that they distort some geometric properties of the earth, such as shape, area, distance, or direction. However, certain properties are preserved on some projections.

If shape is preserved, the projection is called **conformal**. On conformal projections the shapes of features agree with the shapes these features have on the earth. A limitation of conformal projections is that they necessarily distort area, sometimes severely.

Equal-area projections preserve area. On equal area projections the areas of features correspond to their areas on the earth. To achieve this effect, equal-area projections distort shape.

Some projections preserve neither shape nor area, but instead balance shape and area distortion to create an aesthetically-pleasing result. These are often referred to as **compromise** projections.

Distance is preserved on **equidistant** projections, but this can only be achieved selectively, such as along specific meridians or parallels. No projection correctly preserves distance in all directions at all locations. As a result, the stated scale of a map may be accurate for only a limited set of locations. This problem is especially acute for small-scale maps covering large areas.

The projection selected for a particular map depends on the relative importance of different types of distortion, which often depends on the purpose of the map. For example, world maps showing phenomena that vary with area, such as population density or the distribution of agricultural crops, often use an equal-area projection to give an accurate depiction of the importance of each region.

Map projections are created using mathematical procedures. To illustrate the general principles of projections without using mathematics, we can view a projection as the geometric transfer of information from a globe to a flat projection surface, such as a sheet of paper. If we allow the paper to be rolled in different ways, we can derive three basic types of map projections: **cylindrical, conic,** and **azimuthal**. (See the figures to the right.)

For cylindrical projections, the sheet of paper is rolled into a tube and wrapped around the globe so that it is **tangent** (touching) along the equator. Information from the globe is transferred to the tube, and the tube is then unrolled to produce the final flat map.

Conic projections use a cone rather than a cylinder. The figure shows the cone tangent to the earth along a line of latitude with the apex of the cone over the pole. The line of tangency is called the **standard parallel** of the projection.

Azimuthal projections use a flat projection surface that is tangent to the globe at a single point, such as one of the poles.

The figures show the **normal orientation** of each type of surface relative to the globe. The **transverse orientation** is produced when the surface is rotated 90 degrees from normal. For azimuthal projections this orientation is usually called **equatorial** rather than transverse. An **oblique orientation** is created if the projection surface is oriented at an angle between normal and transverse. In general, map distortion increases with distance away from the point or line of tangency. This is why the normal orientations of the cylindrical, conic, and azimuthal projections are often used for mapping equatorial, mid-latitude, and polar regions, respectively.

The projection surface model is a visual tool useful for illustrating how information from the globe can be projected to the map. However, each of the three projection surfaces actually represents scores of individual projections. There are, for example, many projections with the term "cylindrical" in the name, each of which has the same basic rectangular shape, but different spacings of parallels and meridians. The projection surface model does not account for the numerous mathematical details that differentiate one cylindrical, conic, or azimuthal projection from another.

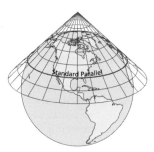
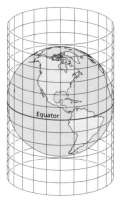

Cylindrical Projection

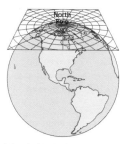
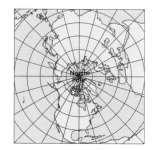

Conic Projection

Azimuthal Projection

Map Projections Used in *Goode's Atlas of Political Geography*

Of the hundreds of projections that have been developed, only a fraction are in everyday use. The main projections used in *Goode's Atlas of Political Geography* are described below.

Simple Conic

Type: Conic   **Conformal:** No   **Equal-area:** No

Notes: Shape and area distortion on the Simple Conic projection are relatively low, even though the projection is neither conformal nor equal-area. The origins of the Simple Conic can be traced back nearly two thousand years, with the modern form of the projection dating to the 18th century.

Uses in *Goode's Atlas of Political Geography*: Larger-scale reference maps of North America, Europe, Asia, and other regions.

Simple Conic Projection

Lambert Conformal Conic

Type: Conic   **Conformal:** Yes   **Equal-area:** No

Notes: On the Lambert Conformal Conic projection, spacing between parallels increases with distance away from the standard parallel, which allows the property of shape to be preserved. The projection is named after Johann Lambert, an 18th century mathematician who developed some of the most important projections in use today. It became widely used in the United States in the 20th century following its adoption for many statewide mapping programs.

Uses in *Goode's Atlas of Political Geography*: Thematic maps of the United States and Canada, and reference maps of parts of Asia.

Lambert Conformal Conic Projection

Albers Equal-Area Conic

Type: Conic   **Conformal:** No   **Equal-area:** Yes

Notes: On the Albers Equal-Area Conic projection, spacing between parallels decreases with distance away from the standard parallel, which allows the property of area to be preserved. The projection is named after Heinrich Albers, who developed it in 1805. It became widely used in the 20th century, when the United States Coast and Geodetic Survey made it a standard for equal area maps of the United States.

Uses in *Goode's Atlas of Political Geography*: Thematic maps of North America and Asia.

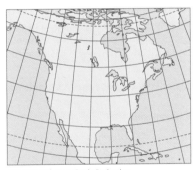
Albers Equal-Area Conic Projection

Polyconic

Type: Conic   **Conformal:** No   **Equal-area:** No

Notes: The term polyconic — literally "many-cones" — refers to the fact that this projection is an assemblage of different cones, each tangent at a different line of latitude. In contrast to many other conic projections, parallels are not concentric, and meridians are curved rather than straight. The Polyconic was first proposed by Ferdinand Hassler, who became Head of the United States Survey of the Coast (later renamed the Coast and Geodetic Survey) in 1807. The United States Geological Survey used this projection exclusively for large-scale topographic maps until the mid-20th century.

Uses in *Goode's Atlas of Political Geography*: Reference maps of North America and Asia.

Polyconic Projection

Lambert Azimuthal Equal-Area

Type: Azimuthal   **Conformal:** No   **Equal-area:** Yes

Notes: This projection (another named after Johann Lambert) is useful for mapping large regions, as area is correctly preserved while shape distortion is relatively low. All orientations — polar, equatorial, and oblique — are common.

Uses in *Goode's Atlas of Political Geography*: Thematic and reference maps of North and South America, Asia, Africa, Australia, and polar regions.

Lambert Azimuthal Equal-Area Projection

Miller Cylindrical

Type: Cylindrical   Conformal: No   Equal-area: No

Notes: This projection is useful for showing the entire earth in a simple rectangular form. However, polar areas exhibit significant exaggeration of area, a problem common to many cylindrical projections. The projection is named after Osborn Miller, Director of the American Geographical Society, who developed it in 1942 as a compromise projection that is neither conformal nor equal-area.

Uses in *Goode's Atlas of Political Geography*: World climate and time zone maps.

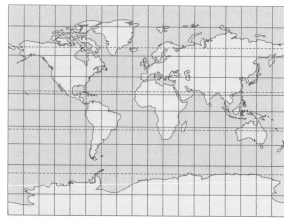

Miller Cylindrical Projection

Sinusoidal

Type: Pseudocylindrical   Conformal: No   Equal-area: Yes

Notes: The straight, evenly spaced parallels on this projection resemble the parallels on cylindrical projections. Unlike cylindrical projections, however, meridians are curved and converge at the poles. This causes significant shape distortion in polar regions. The Sinusoidal is the oldest-known pseudocylindrical projection, dating to the 16th century.

Uses in *Goode's Atlas of Political Geography*: Reference maps of equatorial regions.

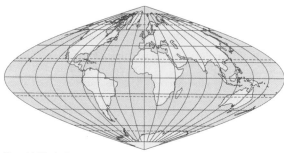

Sinusoidal Projection

Mollweide

Type: Pseudocylindrical   Conformal: No   Equal-area: Yes

Notes: The Mollweide (or Homolographic) projection resembles the Sinusoidal but has less shape distortion in polar areas due to its elliptical (or oval) form. One of several pseudocylindrical projections developed in the 19th century, it is named after Karl Mollweide, an astronomer and mathematician.

Uses in *Goode's Atlas of Political Geography*: Oceanic reference maps.

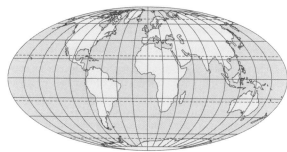

Mollweide Projection

Goode's Interrupted Homolosine

Type: Pseudocylindrical   Conformal: No   Equal-area: Yes

Notes: This projection is a fusion of the Sinusoidal between 40°44'N and S, and the Mollweide between these parallels and the poles. The unique appearance of the projection is due to the introduction of discontinuities in oceanic regions, the goal of which is to reduce distortion for continental landmasses. A condensed version of the projection also exists in which the Atlantic Ocean is compressed in an east-west direction. This modification helps maximize the scale of the map on the page. The Interrupted Homolosine projection is named after J. Paul Goode of the University of Chicago, who developed it in 1923. Goode was an advocate of interrupted projections and, as editor of *Goode's School Atlas*, promoted their use in education.

Uses in *Goode's Atlas of Political Geography*: Small-scale world thematic and reference maps. Both condensed and non-condensed forms are used. An uninterrupted example is used for the Pacific Ocean map.

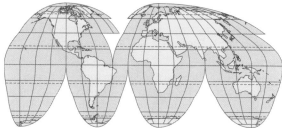

Goode's Interrupted Homolosine Projection

Robinson

Type: Pseudocylindrical   Conformal: No   Equal-area: No

Notes: This projection resembles the Mollweide except that polar regions are flattened and stretched out. While it is neither conformal nor equal-area, both shape and area distortion are relatively low. The projection was developed in 1963 by Arthur Robinson of the University of Wisconsin, at the request of Rand McNally.

Uses in *Goode's Atlas of Political Geography*: World maps where the interrupted nature of Goode's Homolosine would be inappropriate, such as the World Oceanic Environments map.

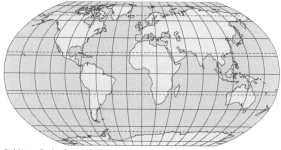

Robinson Projection

Thematic Maps in *Goode's Atlas of Political Geography*

Thematic maps depict a single "theme" such as population density, agricultural productivity, or annual precipitation. The selected theme is presented on a base of locational information, such as coastlines, country boundaries, and major drainage features. The primary purpose of a thematic map is to convey an impression of the overall geographic distribution of the theme. It is usually not the intent of the map to provide exact numerical values. To obtain such information, the graphs and tables accompanying the map should be used.

Goode's Atlas of Political Geography contains many different types of thematic maps. The characteristics of each are summarized below.

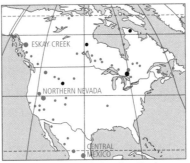
Point symbol map: Detail of Precious Metals

Point Symbol Maps

Point symbol maps are perhaps the simplest type of thematic map. They show features that occur at discrete locations. Examples include earthquakes, nuclear power plants, and minerals-producing areas. The Precious Metals map is an example of a point symbol map showing the locations of areas producing gold, silver, and platinum. A different color is used for each type of metal, while symbol size indicates relative importance.

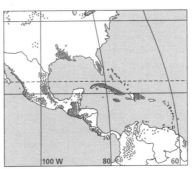
Area symbol map: Detail of Tobacco and Fisheries

Area Symbol Maps

Area symbol maps are useful for delineating regions of interest on the earth's surface. For example, the Tobacco and Fisheries map shows major tobacco-producing regions in one color and important fishing areas in another. On some area symbol maps, different shadings or colors are used to differentiate between major and minor areas.

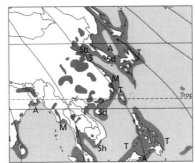
Dot map: Detail of Sugar

Dot Maps

Dot maps show a distribution using a pattern of dots, where each dot represents a certain quantity or amount. For example, on the Sugar map, each dot represents 20,000 metric tons of sugar produced. Different dot colors are used to distinguish cane sugar from beet sugar. Dot maps are an effective way of representing the variable density of geographic phenomena over the earth's surface. This type of map is used extensively in *Goode's Atlas of Political Geography* to show the distribution of agricultural commodities.

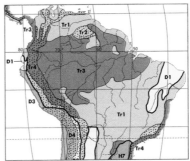
Area class map: Detail of Ecoregions

Area Class Maps

On area class maps, the earth's surface is divided into areas based on different classes or categories of a particular geographic phenomenon. For example, the Ecoregions map differentiates natural landscape categories, such as Tundra, Savanna, and Prairie. Other examples of area class maps in *Goode's Atlas of Political Geography* include Landforms, Climatic Regions, Natural Vegetation, Soils, Agricultural Areas, Languages and Religions.

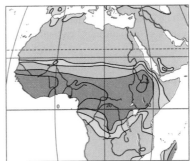
Isoline map: Detail of Precipitation

Isoline Maps

Isoline maps are used to portray quantities that vary smoothly over the surface of the earth. These maps are frequently used for climatic variables such as precipitation and temperature, but a variety of other quantities — from crop yield to population density — can also be treated in this way.

An isoline is a line on the map that joins locations with the same value. For example, the Summer (May to October) Precipitation map contains isolines at 5, 10, 20, and 40 inches. On this map, any 10-inch isoline separates areas that have less than 10 inches of precipitation from areas that have more than 10 inches. Note that the areas between isolines are given different colors to assist in map interpretation.

Proportional Symbol Maps

Proportional symbol maps portray numerical quantities, such as the total population of each state, the total value of agricultural goods produced in different regions, or the amount of hydroelectricity generated in different countries. The symbols on these maps — usually circles — are drawn such that the size of each is proportional to the value at that location. For example the Exports map shows the value of goods exported by each country in the world, in millions of U.S. dollars.

Proportional symbols are frequently subdivided based on the percentage of individual components making up the total. The Exports map uses wedges of different color to show the percentages of various types of exports, such as manufactured articles and raw materials.

Flow Line Maps

Flow line maps show flows between locations. Usually, the thickness of the flow lines is proportional to flow volume. Flows may be physical commodities like petroleum, or less tangible quantities like information. The flow lines on the Mineral Fuels map represent movement of petroleum measured in billions of U.S. dollars. Note that the locations of flow lines may not represent actual physical routes.

Choropleth Maps

Choropleth maps apply distinctive colors to predefined areas, such as counties or states, to represent different quantities in each area. The quantities shown are usually rates, percentages, or densities. For example, the Birth Rate map shows the annual number of births per one thousand people for each country.

Digital Images

Some maps are actually digital images, analogous to the pictures captured by digital cameras. These maps are created from a very fine grid of cells called **pixels**, each of which is assigned a color that corresponds to a specific value or range of values. The population density maps in this atlas are examples of this type. The effect is much like an isoline map, but the isolines themselves are not shown and the resulting geographic patterns are more subtle and variable. This approach is increasingly being used to map environmental phenomena observable from remote sensing systems.

Cartograms

Cartograms deliberately distort map shapes to achieve specific effects. On **area cartograms**, the size of each area, such as a country, is made proportional to its population. Countries with large populations are therefore drawn larger than countries with smaller populations, regardless of the actual size of these countries on the earth.

The world cartogram series in this atlas depicts each country as a rectangle. This is a departure from cartograms in earlier editions of the atlas, which attempted to preserve some of the salient shape characteristics for each country. The advantage of the rectangle method is that it is easier to compare the area of countries when their shapes are consistent.

The cartogram series incorporates choropleth shading on top of the rectangular cartogram base. In this way map readers can make inferences about the relationship between population and another thematic variable, such as HIV-infection rates.

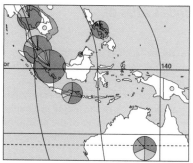
Proportional symbol map: Detail of Exports

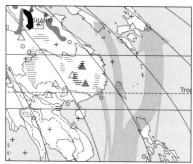
Flow line map: Detail of Mineral Fuels

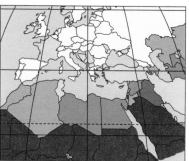
Choropleth map: Detail of Birth Rate

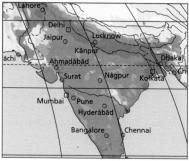
Digital image map: Detail of Population Density

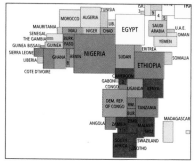
Cartogram: Detail of HIV Infection

GOODE'S

ATLAS OF Political Geography

Map Legend

Political Boundaries

Political maps	Physical maps	
━━ ┄	━━ ┉	International (Demarcated, Undemarcated, and Administrative)
━ ··	━ ··	Disputed de facto
▬ ▬	▬ ▬	Indefinite or Undefined
━·━·	━·━·	Secondary, State, Provincial, etc.

⬜	Parks, Indian Reservations
🏙	City Limits
🗺	Urbanized Areas

Transportation

Political maps	Physical maps	
━━	━━	Railroads
┄┄	┄┄	Railroad Ferries
━━		Major Roads
━━		Minor Roads
┄┄┄		Caravan Routes
✈		Airports

Cultural Features

∿	Dams
┄┄┄	Pipelines
▲	Points of Interest
∴	Ruins

Populated Places

⊙	1,000,000 and over
◎	250,000 to 1,000,000
⊙	100,000 to 250,000
•	25,000 to 100,000
○	Under 25,000
□	Neighborhoods, Sections of Cities
TŌKYŌ	National Capitals
Boise	Secondary Capitals

Note: On maps at 1:20,000,000 and smaller, symbols do not follow the population classification shown above. Some other maps use a slightly different classification, which is shown in a separate legend in the map margin. On all maps, type size indicates the relative importance of the city.

Land Features

△	Peaks, Spot Heights
≈	Passes
▨	Sand
⬭	Contours

Elevation

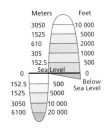

Lakes and Reservoirs

⬭	Fresh Water
⬭	Fresh Water: Intermittent
⬭	Salt Water
⬭	Salt Water: Intermittent

Other Water Features

⬭	Salt Basins, Flats
	Swamps
	Ice Caps and Glaciers
	Rivers
	Intermittent Rivers
	Aqueducts and Canals
	Ship Channels
	Falls
	Rapids
	Springs
△	Water Depths
	Sand Bars
	Reefs
→	Warm Ocean Currents
→	Cold Ocean Currents

The legend above shows the symbols used for the political and physical reference maps in *Goode's Atlas of Political Geography*.

To portray relative areas correctly, uniform map scales have been used wherever possible:

Continents – 1:40,000,000
Countries and regions – between 1:4,000,000 and 1:20,000,000
World, polar areas and oceans – between 1:50,000,000 and 1:100,000,000
Urbanized areas – 1:1,000,000

Elevations on the maps are shown using a combination of shaded relief and hypsometric tints. Shaded relief (or hill-shading) gives a three-dimensional impression of the landscape, while hypsometric tints show elevation ranges in different colors.

The choice of names for mapped features is complicated by the fact that a variety of languages and alphabets are used throughout the world. A local-names policy is used in *Goode's Atlas of Political Geography* for populated places and local physical features. For some major features, an English form of the name is used with the local name given below in parentheses. Examples include Moscow (Moskva), Vienna (Wien) and Naples (Napoli). In countries where more than one official language is used, names are given in the dominant local language. For large physical features spanning international borders, the conventional English form of the name is used. In cases where a non-Roman alphabet is used, names have been transliterated according to accepted practice.

Selected features are also listed in the Index (pp. 113-135), which includes a pronunciation guide. A list of foreign geographic terms is provided in the Glossary (p. 110).

2

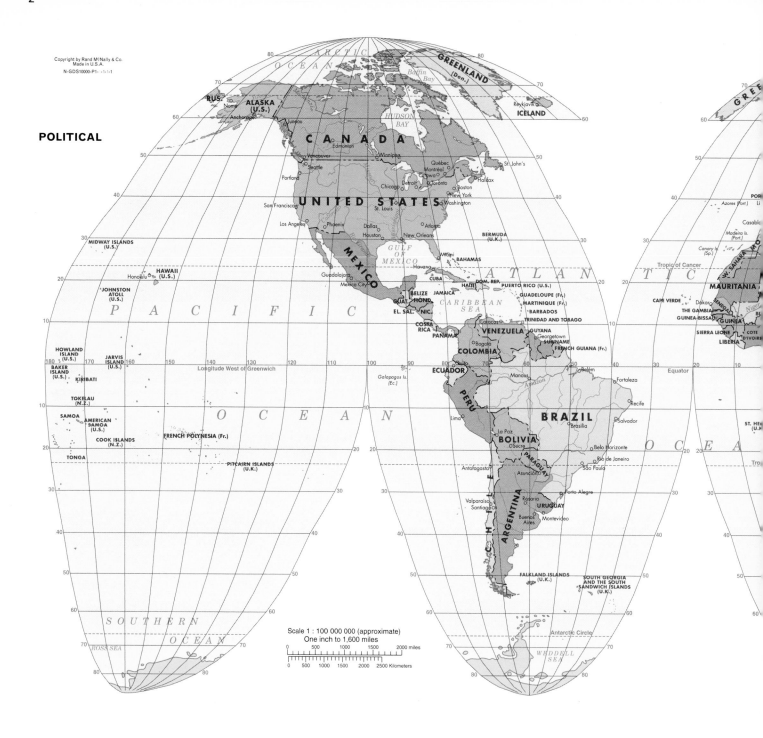

POLITICAL

Copyright by Rand McNally & Co.
Made in U.S.A.
N-GDS10000-P1- -1-1-1

Scale 1 : 100 000 000 (approximate)
One inch to 1,600 miles

0   500   1000   1500   2000 miles

0   500   1000   1500   2000   2500 Kilometers

Comparative Land Areas (Land and inland water. Numbers indicate thousands of square miles.)

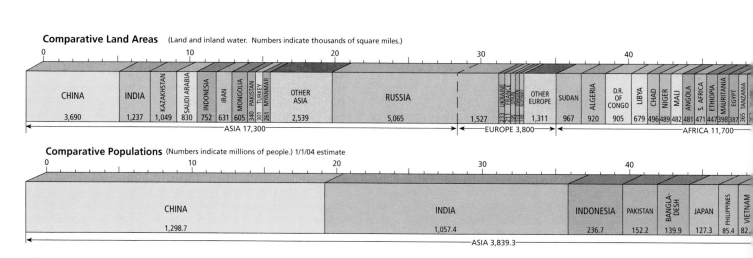

CHINA	INDIA	KAZAKHSTAN	SAUDI ARABIA	INDONESIA	IRAN	MONGOLIA	PAKISTAN	TURKEY	MYANMAR	OTHER ASIA	RUSSIA		UKRAINE	FRANCE	SPAIN	SWEDEN	OTHER EUROPE	SUDAN	ALGERIA	D.R. OF CONGO	LIBYA	CHAD	NIGER	MALI	ANGOLA	S. AFRICA	ETHIOPIA	MAURITANIA	EGYPT	TANZANIA
3,690	1,237	1,049	830	752	631	605	340	301	261	2,539	5,065	1,527	233	211	195	174	1,311	967	920	905	679	496	489	482	481	471	447	398	387	365

← ASIA 17,300 →   ← EUROPE 3,800 →   ← AFRICA 11,700 →

Comparative Populations (Numbers indicate millions of people.) 1/1/04 estimate

CHINA	INDIA	INDONESIA	PAKISTAN	BANGLA-DESH	JAPAN	PHILIPPINES	VIETNAM
1,298.7	1,057.4	236.7	152.2	139.9	127.3	85.4	82.

← ASIA 3,839.3 →

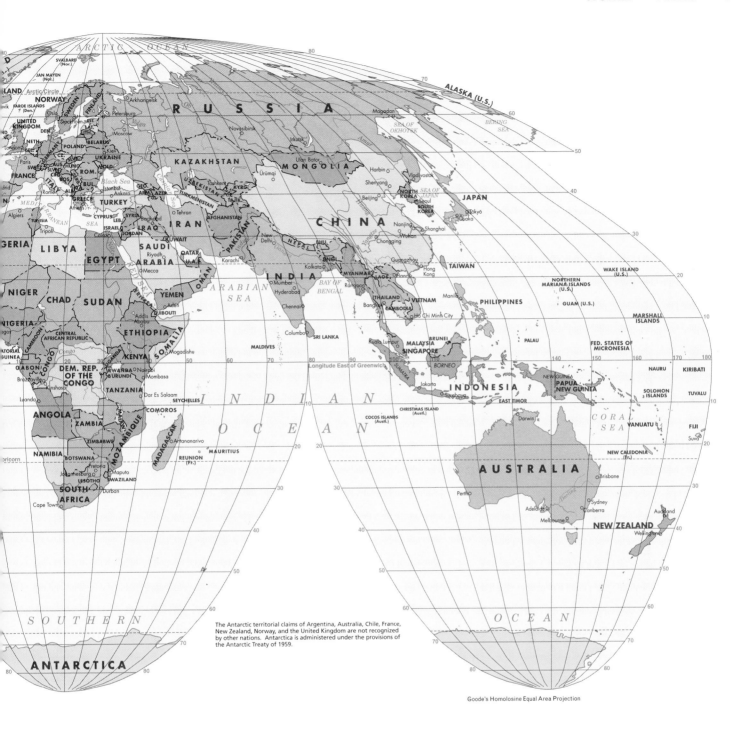

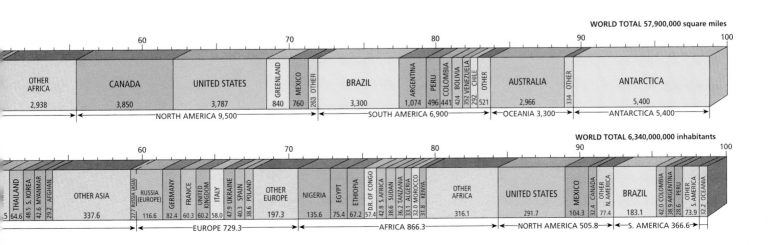

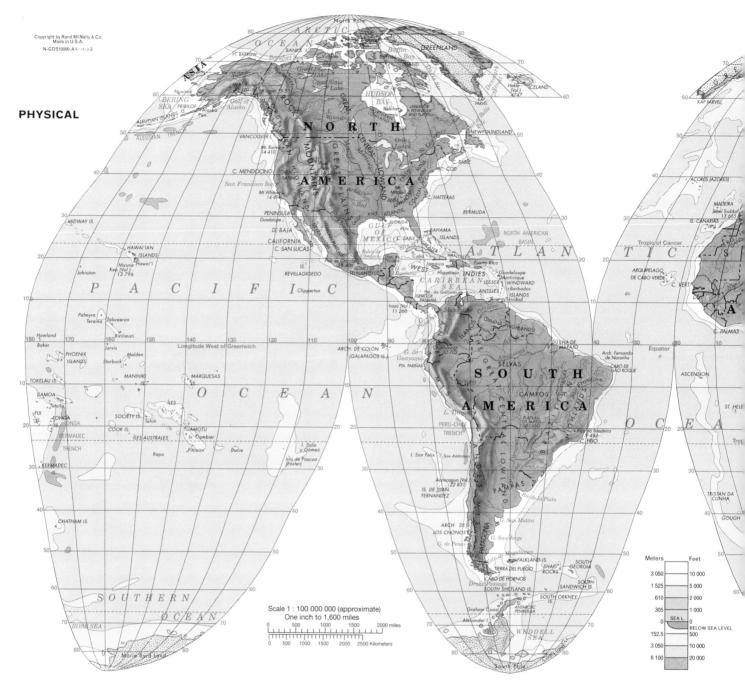

PHYSICAL

Scale 1 : 100 000 000 (approximate)
One inch to 1,600 miles

0   500   1000   1500   2000 miles

0   500   1000   1500   2000   2500 Kilometers

Meters		Feet
3 050		10 000
1 525		5 000
610		2 000
305		1 000
0	SEA L.	0
		BELOW SEA LEVEL
152.5		500
3 050		10 000
6 100		20 000

Land Elevations in Profile

Ocean Depths in Profile

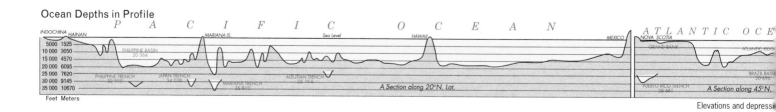

Elevations and depress

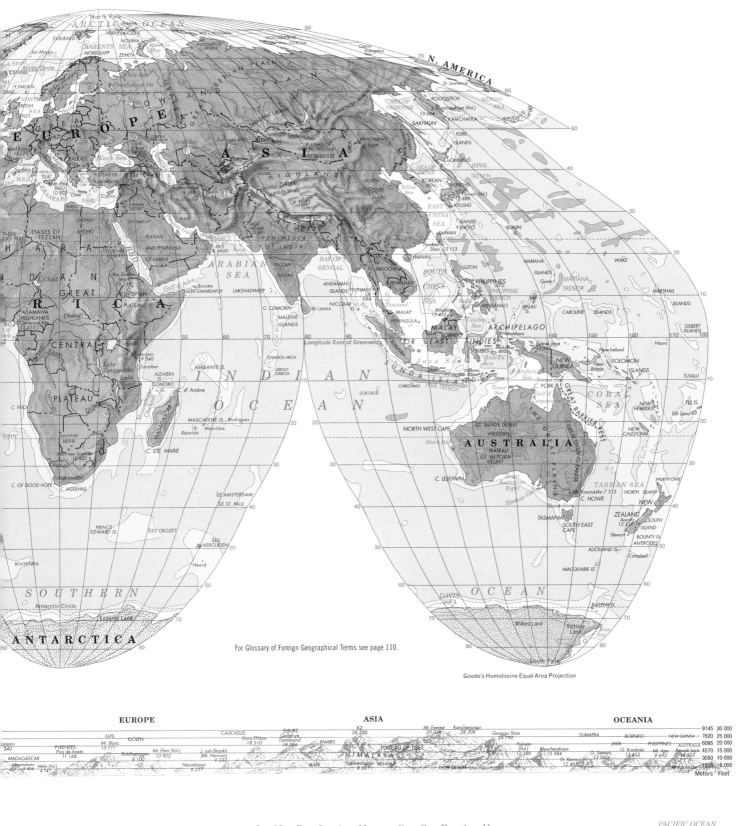

For Glossary of Foreign Geographical Terms see page 110.

Goode's Homolosine Equal Area Projection

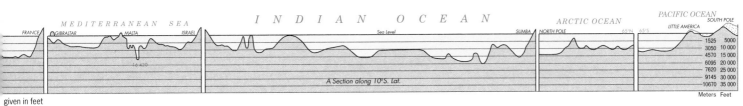

given in feet

Scale 1:72 000 000 at 40° latitude.

120° 105° 90° 75° 60° 45° 30° 15° 0° 15° 30° 45°

Baffin Bay
GREENLAND (Denmark)
Labrador Sea
Labrador Basin
Hudson Bay
CANADA
NORTH AMERICA
Arctic Circle
Irminger Basin
Reykjanes Ridge
ICELAND
Iceland Basin
Norwegian Basin
Norwegian Sea
Lofoten Basin
Barents Sea
NORWAY
FINLAND
North Sea
SWEDEN
EST.
LATVIA
LITH.
RUSSIA
Moscow
NEWFOUNDLAND
IRELAND
UNITED KINGDOM
London
DEN.
Baltic Sea
BELARUS
POLAND
EUROPE
GERMANY
Berlin
Paris
FRANCE
CZ. REP.
UKRAINE
AUS. HUNG.
ROMANIA
Montréal
Chicago
Washington
New York
UNITED STATES
St. Lawrence
West European Basin
Newfoundland Basin
Azores Plateau
AZORES (Port.)
PORT.
Lisbon
SPAIN
Rome
ITALY
BOS. SERB. BUL.
ALB.
GREECE
Athens
TURK.
Mediterranean Sea
North American Basin
BERMUDA (Br.)
Casablanca
MOROCCO
Algiers
TUNISIA
New Orleans
Mississippi
Gulf of Mexico
Mexico Basin
Havana
BAHAMAS
CUBA
Miami
CANARY ISLANDS (Sp.)
Tropic of Cancer
Canary Basin
WESTERN SAHARA
ALGERIA
LIBYA
MEXICO
GUAT.
HOND.
HAITI
DOM. REP.
Santo Domingo
Puerto Rico Trench
8605
Caribbean Sea
A T L A N T I C
MID-ATLANTIC RIDGE
MAURITANIA
MALI
NIGER
CHAD
SUDAN
NIC.
COSTA RICA
PANAMA
Caracas
VENEZUELA
GUYANA
SUR.
FRENCH GUIANA
Orinoco
COLOMBIA
Guiana Basin
CAPE VERDE
Dakar
SENEGAL
GUINEA
Niger
BURKINA FASO
NIGERIA
Lagos
AFRICA
CENTRAL AFRICAN REPUBLIC
Cape Verde Basin
Freetown
LIBERIA
GHANA
COTE D'IVOIRE
Abidjan
CAMEROON
EQUADOR
Manaus
Amazon
O C E A N
Equator
Romanche Gap
Libreville
GABON
CONGO
DEM. REP. OF THE CONGO
Kinshasa
Congo
PERU
Lima
BRAZIL
SOUTH AMERICA
Recife
ASCENSION (St. Hel.)
Guinea Basin
Luanda
Angola
ANGOLA
PERU
Peru Basin
Nazca Ridge
BOLIVIA
Brasília
Brazil Basin
ST. HELENA (Br.)
Angola Basin
ZAMBIA
PARAGUAY
São Paulo
Rio de Janeiro
Tropic of Capricorn
NAMIBIA
BOTSWANA
PACIFIC
CHILE
Santiago
URUGUAY
Buenos Aires
ARGENTINA
Paraná
Bromley Plateau
SOUTH AFRICA
Johannesburg
MID-ATLANTIC RIDGE
Walvis Ridge
Cape Basin
Cape Town
Argentine Basin
TRISTAN DA CUNHA GROUP (St. Hel.)
GOUGH ISLAND (St. Hel.)
Agulhas Basin
O C E A N
Chile Rise
FALKLAND ISLANDS (Br.)
Scotia Ridge
SOUTH GEORGIA (Br.)
South Sandwich Trench
BOUVETOYA (N.)
Atlantic-Indian Ridge
SOUTHEAST Pacific Basin
SOUTH SHETLAND ISLANDS (Br.)
SOUTH ORKNEY ISLANDS (Br.)
SOUTHERN OCEAN
Antarctic Circle
Atlantic-Indian Basin
Weddell Sea
ANTARCTICA

© Rand McNally & Co.
N-GDS14000-A2- -1-1-3

Scale 1:72 000 000 at 40° latitude. ROBINSON PROJECTION

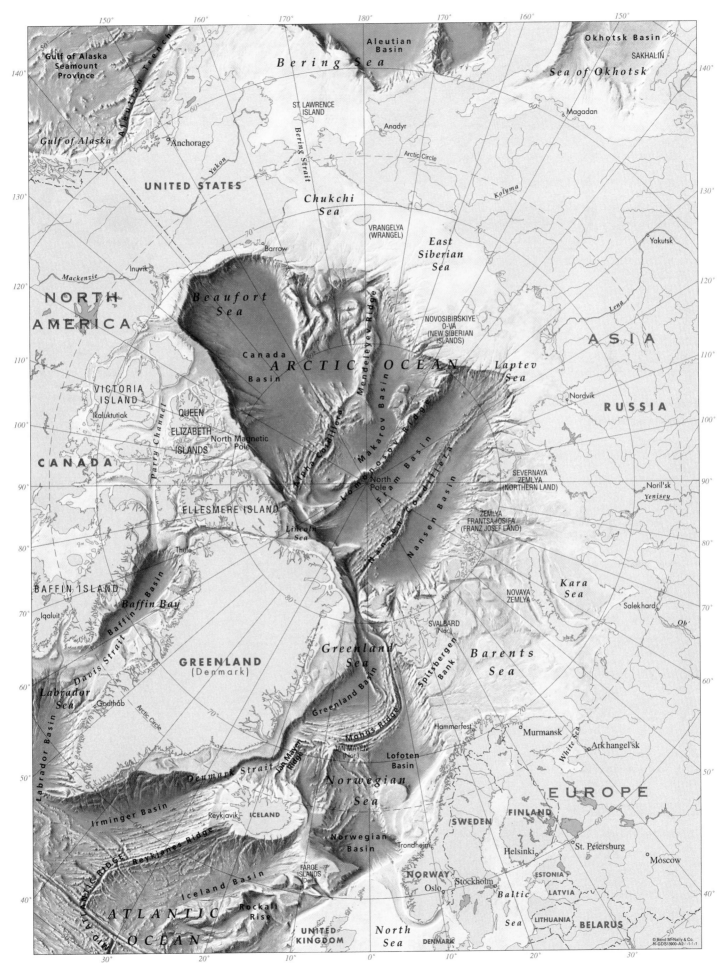

Scale 1:30 000 000. LAMBERT AZIMUTHAL EQUAL AREA PROJECTION

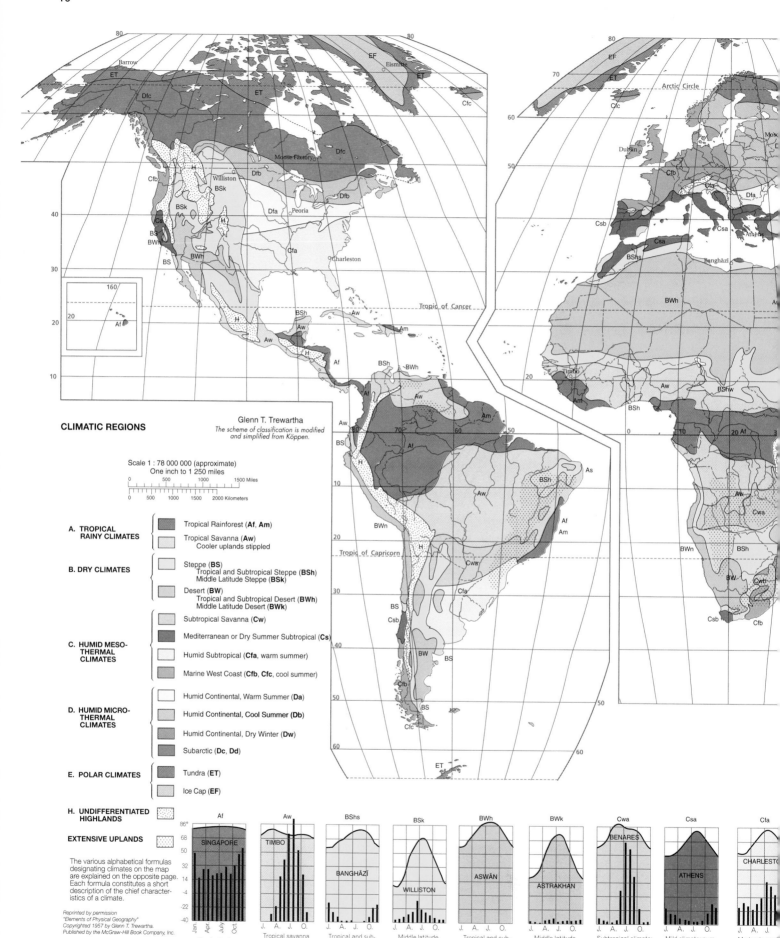

CLIMATIC REGIONS

Glenn T. Trewartha
*The scheme of classification is modified
and simplified from Köppen.*

Scale 1 : 78 000 000 (approximate)
One inch to 1 250 miles

0   500   1000   1500 Miles

0 500 1000 1500 2000 Kilometers

A. TROPICAL RAINY CLIMATES
Tropical Rainforest (**Af, Am**)
Tropical Savanna (**Aw**)
    Cooler uplands stippled

B. DRY CLIMATES
Steppe (**BS**)
    Tropical and Subtropical Steppe (**BSh**)
    Middle Latitude Steppe (**BSk**)
Desert (**BW**)
    Tropical and Subtropical Desert (**BWh**)
    Middle Latitude Desert (**BWk**)

C. HUMID MESO-THERMAL CLIMATES
Subtropical Savanna (**Cw**)
Mediterranean or Dry Summer Subtropical (**Cs**)
Humid Subtropical (**Cfa**, warm summer)
Marine West Coast (**Cfb, Cfc**, cool summer)

D. HUMID MICRO-THERMAL CLIMATES
Humid Continental, Warm Summer (**Da**)
Humid Continental, Cool Summer (**Db**)
Humid Continental, Dry Winter (**Dw**)
Subarctic (**Dc, Dd**)

E. POLAR CLIMATES
Tundra (**ET**)
Ice Cap (**EF**)

H. UNDIFFERENTIATED HIGHLANDS

EXTENSIVE UPLANDS

The various alphabetical formulas
designating climates on the map
are explained on the opposite page.
Each formula constitutes a short
description of the chief characteristics of a climate.

*Reprinted by permission
"Elements of Physical Geography"
Copyrighted 1957 by Glenn T. Trewartha.
Published by the McGraw-Hill Book Company, Inc.*

Copyright by Rand McNally & Co.
Made in U.S.A.
N-GDS10000-C1- 2-3

Af	Aw	BShs	BSk	BWh	BWk	Cwa	Csa	Cfa
SINGAPORE	TIMBO	BANGHĀZĪ	WILLISTON	ASWĀN	ASTRAKHAN	BENARES	ATHENS	CHARLESTON
Tropical rain-forest climate	Tropical savanna climate; with wet and dry seasons	Tropical and sub-tropical steppe climate	Middle latitude steppe climate.	Tropical and sub-tropical desert climate	Middle latitude desert climate	Subtropical climate; winter drought and summer rain	Mild climate; sum-wer drought and winter rain	Moderate conti-tal forest clim mild winters

86°
68
50
32
14
-4
-22
-40

J. A. J. O.

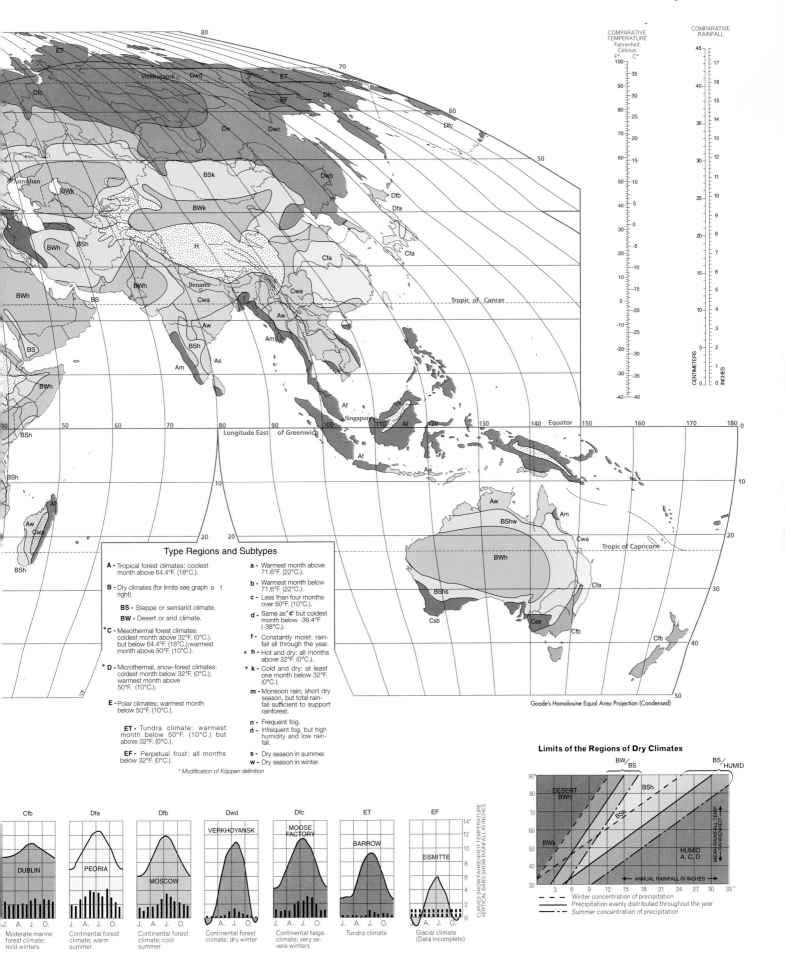

COMPARATIVE
TEMPERATURE
Fahrenheit
Celcius
F° C°

COMPARATIVE
RAINFALL

ET

Verkhoyansk Dwd

ET

Dfc

Dw   Dwc

Dfc

BSk

Dwb

Astrakhan

BWk

BWk

Dfb

Dfa

BWh BSh

BWh

Cfa

Cfa

BWh

Benares

Cwa

BS   Cwa

Aw

Aw   Am

BSh

Am As

BS

Am

Af

BWh

Singapore Af

Tropic of Cancer

Af

BSh

Longitude East of Greenwich

BSh

Af

Af

Aw

Af

Aw

Aw

Am

BShw

Cwa

Tropic of Capricorn

BWh

Cfa

BShs

BSh

Csb

Csa

Cfb

Cfb

Cfb

Goode's Homolosine Equal Area Projection (Condensed)

Type Regions and Subtypes

A - Tropical forest climates: coolest month above 64.4°F. (18°C.).

B - Dry climates (for limits see graph a t right)

 BS - Steppe or semiarid climate.

 BW - Desert or arid climate.

*__C__ - Mesothermal forest climates: coldest month above 32°F. (0°C.). but below 64.4°F. (18°C.);warmest month above 50°F. (10°C.).

*__D__ - Microthermal, snow-forest climates: coldest month below 32°F. (0°C.); warmest month above 50°F. (10°C.).

E - Polar climates; warmest month below 50°F. (10°C.).

 ET - Tundra climate: warmest month below 50°F. (10°C.) but above 32°F. (0°C.).

 EF - Perpetual frost: all months below 32°F. (0°C.).

Modification of Köppen definition

a - Warmest month above 71.6°F. (22°C.).

b - Warmest month below 71.6°F. (22°C.).

c - Less than four months over 50°F. (10°C.).

d - Same as" **c'** but coldest month below -36.4°F (-38°C.).

f - Constantly moist: rainfall all through the year.

* **h** - Hot and dry: all months above 32°F. (0°C.).

* **k** - Cold and dry: at least one month below 32°F. (0°C.).

m - Monsoon rain; short dry season, but total rainfall sufficient to support rainforest.

n - Frequent fog.

ń - Infrequent fog, but high humidity and low rainfall.

s - Dry season in summer.

w - Dry season in winter.

Limits of the Regions of Dry Climates

BW/BS   BS/HUMID

DESERT
BWh   BSh

BS

BWk

HUMID
A, C, D

ANNUAL RAINFALL IN INCHES

MEAN RAINFALL TEMP.
FAHRENHEIT

Winter concentration of precipitation
Precipitation evenly distributed throughout the year
Summer concentration of precipitation

Cfb	Dfa	Dfb	Dwd	Dfc	ET	EF

DUBLIN

PEORIA

MOSCOW

VERKHOYANSK

MOOSE FACTORY

BARROW

EISMITTE

CURVES SHOW FAHRENHEIT TEMPERATURE
VERTICAL BARS SHOW RAINFALL IN INCHES

J. A. J. O.

Moderate marine forest climate; mild winters

Continental forest climate; warm summer

Continental forest climate; cool summer

Continental forest climate; dry winter

Continental taiga climate; very severe winters

Tundra climate

Glacial climate (Data Incomplete)

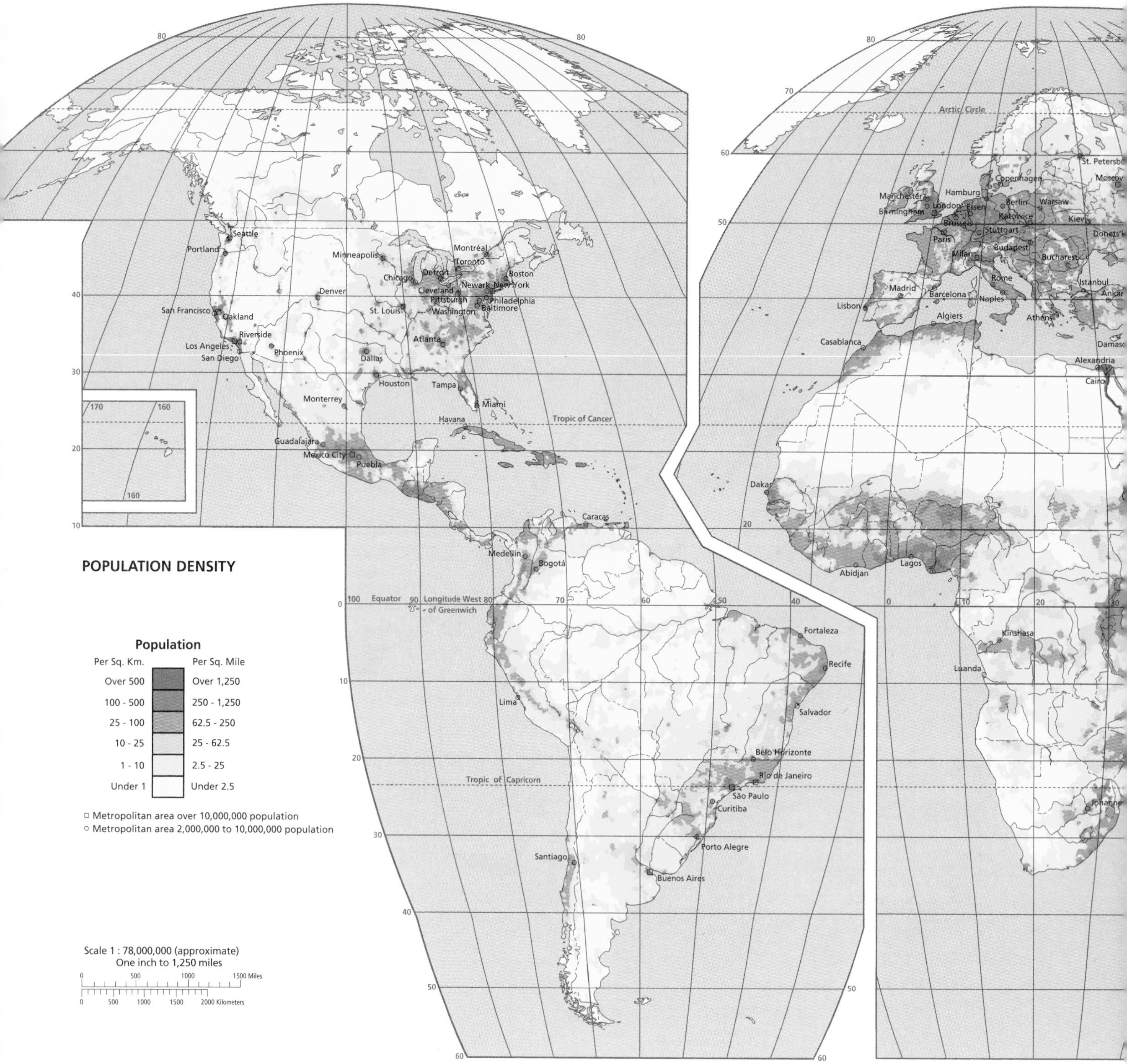

POPULATION DENSITY

Population

Per Sq. Km.	Per Sq. Mile
Over 500	Over 1,250
100 - 500	250 - 1,250
25 - 100	62.5 - 250
10 - 25	25 - 62.5
1 - 10	2.5 - 25
Under 1	Under 2.5

□ Metropolitan area over 10,000,000 population
○ Metropolitan area 2,000,000 to 10,000,000 population

Scale 1 : 78,000,000 (approximate)
One inch to 1,250 miles

0   500   1000   1500 Miles

0   500   1000   1500   2000 Kilometers

Map labels

Seattle, Portland, Minneapolis, Montréal, Toronto, Chicago, Detroit, Cleveland, Pittsburgh, Boston, Newark, New York, Philadelphia, Baltimore, Denver, St. Louis, Washington, San Francisco, Oakland, Riverside, Los Angeles, San Diego, Phoenix, Atlanta, Dallas, Houston, Tampa, Monterrey, Miami, Havana, Guadalajara, Mexico City, Puebla, Caracas, Medellín, Bogotá, Fortaleza, Recife, Lima, Salvador, Belo Horizonte, Rio de Janeiro, São Paulo, Curitiba, Porto Alegre, Santiago, Buenos Aires

St. Petersbo, Moscow, Copenhagen, Hamburg, Manchester, Berlin, Warsaw, Birmingham, London, Essen, Katowice, Kiev, Brussels, Stuttgart, Donets'k, Paris, Milan, Budapest, Madrid, Rome, Bucharest, Barcelona, Naples, Istanbul, Ankar, Lisbon, Athens, Casablanca, Algiers, Damasc, Alexandria, Cairo, Dakar, Lagos, Abidjan, Kinshasa, Luanda, Johanne

Arctic Circle, Tropic of Cancer, Equator, Longitude West of Greenwich, Tropic of Capricorn

Largest Countries of the World 1950, 2000, 2050

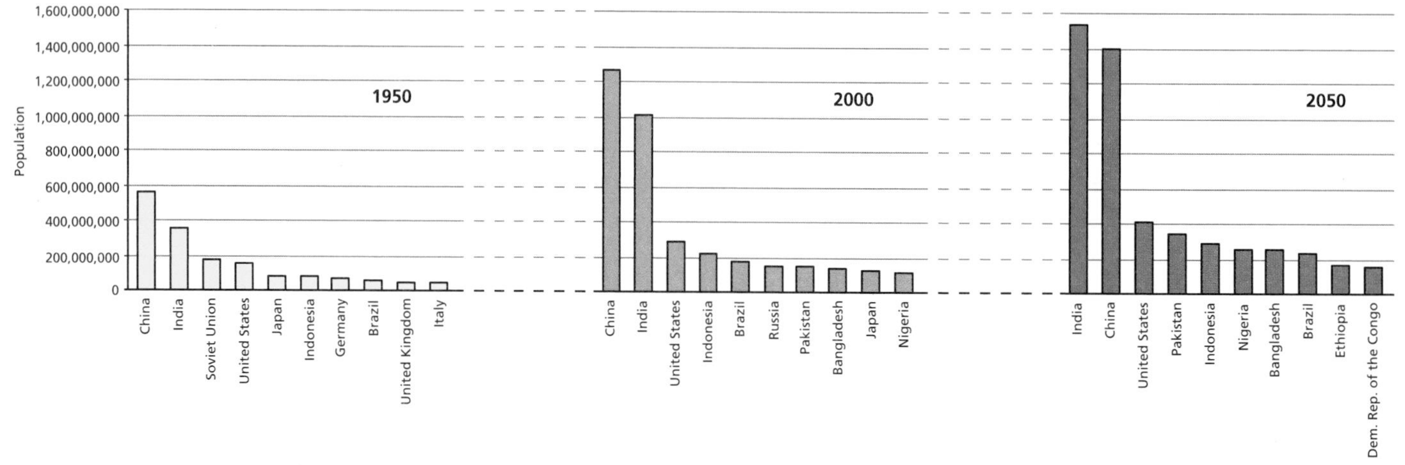

1950
- China
- India
- Soviet Union
- United States
- Japan
- Indonesia
- Germany
- Brazil
- United Kingdom
- Italy

2000
- China
- India
- United States
- Indonesia
- Brazil
- Russia
- Pakistan
- Bangladesh
- Japan
- Nigeria

2050
- India
- China
- United States
- Pakistan
- Indonesia
- Nigeria
- Bangladesh
- Brazil
- Ethiopia
- Dem. Rep. of the Congo

Population axis: 0 to 1,600,000,000

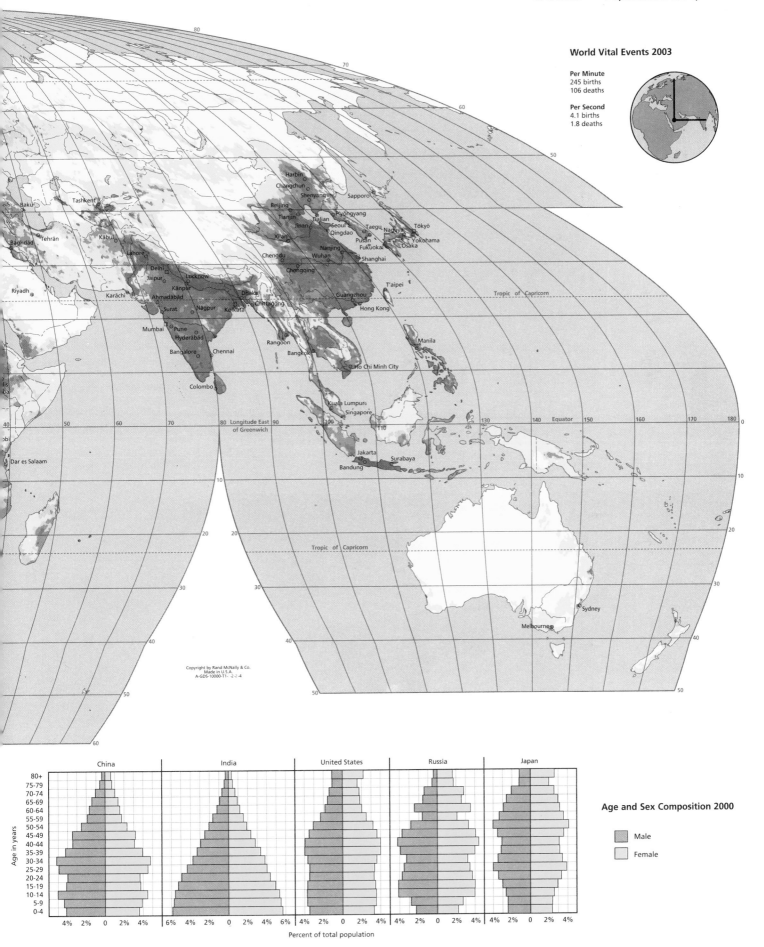

World Vital Events 2003

Per Minute
245 births
106 deaths

Per Second
4.1 births
1.8 deaths

Age and Sex Composition 2000

Male
Female

China   India   United States   Russia   Japan

Age in years

Percent of total population

Copyright by Rand McNally & Co.
Made in U.S.A.
A-GDS-10000-T1- -2-2-4

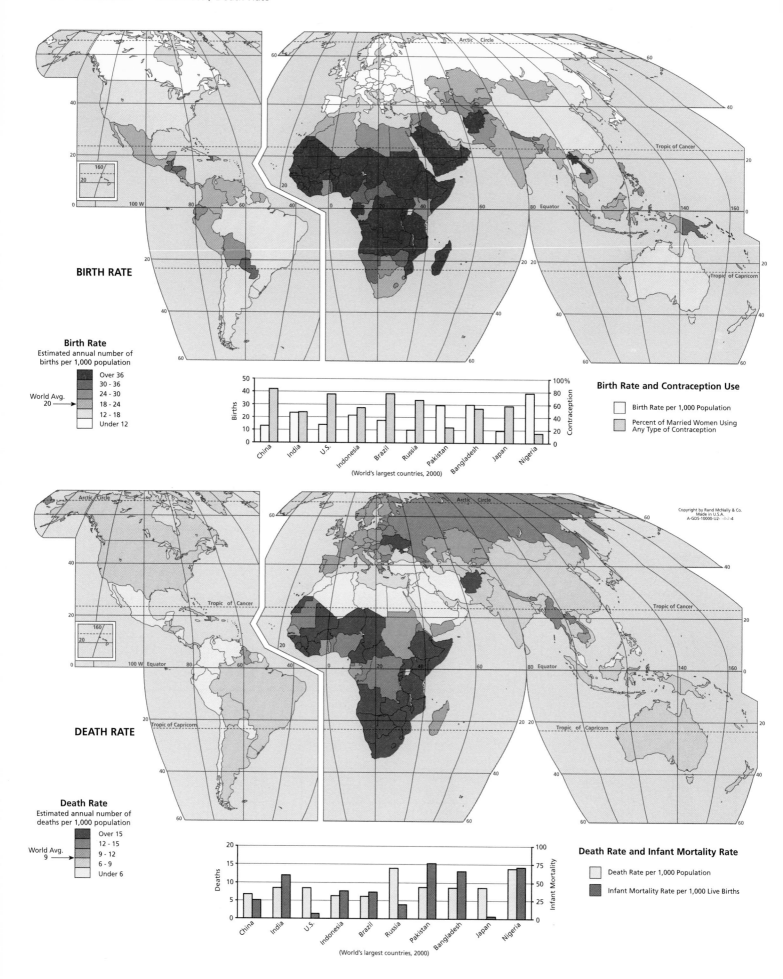

BIRTH RATE

Birth Rate
Estimated annual number of
births per 1,000 population

Over 36
30 - 36
24 - 30
World Avg. → 18 - 24
20   12 - 18
Under 12

Birth Rate and Contraception Use

☐ Birth Rate per 1,000 Population

☐ Percent of Married Women Using
Any Type of Contraception

Births

China India U.S. Indonesia Brazil Russia Pakistan Bangladesh Japan Nigeria

(World's largest countries, 2000)

DEATH RATE

Death Rate
Estimated annual number of
deaths per 1,000 population

Over 15
12 - 15
World Avg. → 9 - 12
9   6 - 9
Under 6

Copyright by Rand McNally & Co.
Made in U.S.A.
A-GD5-10000-U2- -3-2-4

Death Rate and Infant Mortality Rate

☐ Death Rate per 1,000 Population

☐ Infant Mortality Rate per 1,000 Live Births

Deaths

China India U.S. Indonesia Brazil Russia Pakistan Bangladesh Japan Nigeria

(World's largest countries, 2000)

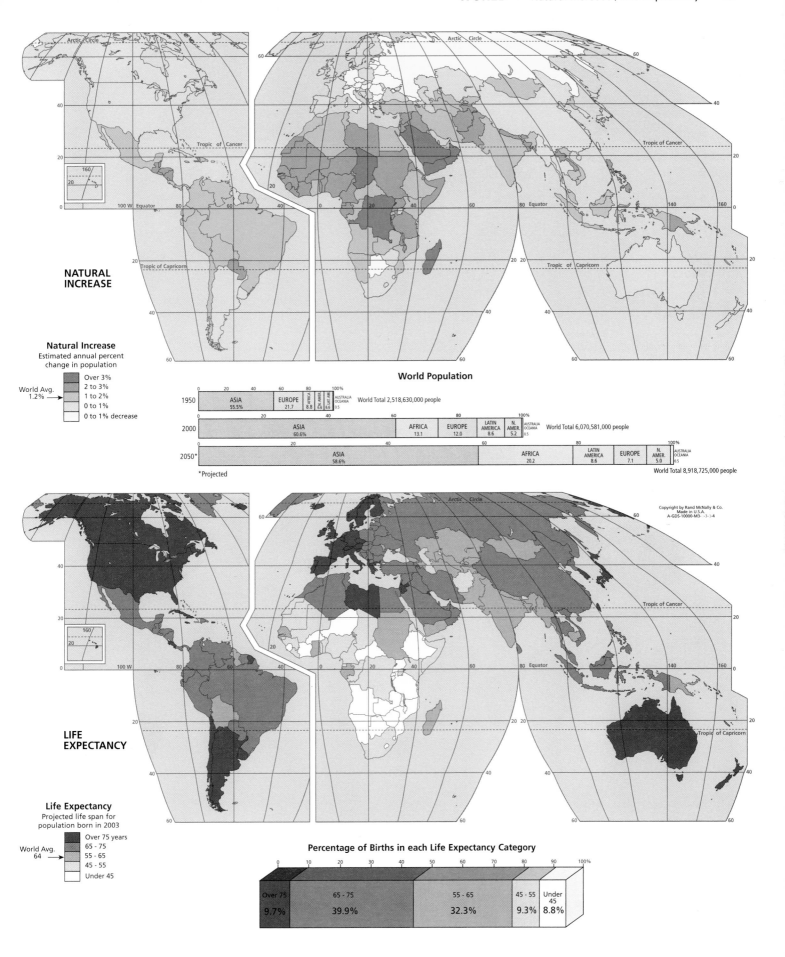

NATURAL
INCREASE

Natural Increase
Estimated annual percent
change in population

World Avg.
1.2%

- Over 3%
- 2 to 3%
- 1 to 2%
- 0 to 1%
- 0 to 1% decrease

World Population

1950   ASIA 55.5%   EUROPE 21.7   AFRICA 8.8   N. AMER.   LAT. AM. 6.6   AUSTRALIA OCEANIA 0.5   World Total 2,518,630,000 people

2000   ASIA 60.6%   AFRICA 13.1   EUROPE 12.0   LATIN AMERICA 8.6   N. AMER. 5.2   AUSTRALIA OCEANIA 0.5   World Total 6,070,581,000 people

2050*   ASIA 58.6%   AFRICA 20.2   LATIN AMERICA 8.6   EUROPE 7.1   N. AMER. 5.0   AUSTRALIA OCEANIA 0.5   World Total 8,918,725,000 people

*Projected

Copyright by Rand McNally & Co.
Made in U.S.A.
A-GDS-10000-M3- -3- 3-4

LIFE
EXPECTANCY

Life Expectancy
Projected life span for
population born in 2003

World Avg.
64

- Over 75 years
- 65 - 75
- 55 - 65
- 45 - 55
- Under 45

Percentage of Births in each Life Expectancy Category

Over 75	65 - 75	55 - 65	45 - 55	Under 45
9.7%	39.9%	32.3%	9.3%	8.8%

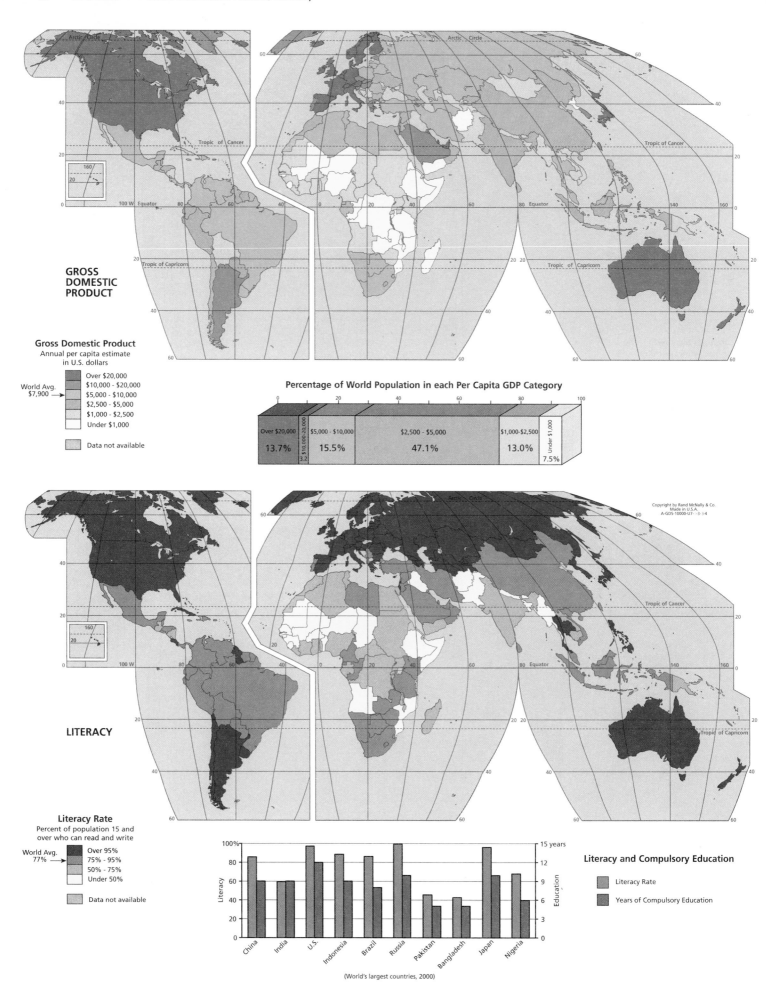

**GROSS
DOMESTIC
PRODUCT**

Gross Domestic Product
Annual per capita estimate
in U.S. dollars

World Avg.
$7,900 →

- Over $20,000
- $10,000 - $20,000
- $5,000 - $10,000
- $2,500 - $5,000
- $1,000 - $2,500
- Under $1,000

Data not available

Percentage of World Population in each Per Capita GDP Category

Over $20,000	$10,000-20,000	$5,000 - $10,000	$2,500 - $5,000	$1,000-$2,500	Under $1,000
13.7%	3.2	15.5%	47.1%	13.0%	7.5%

LITERACY

Literacy Rate
Percent of population 15 and
over who can read and write

World Avg.
77% →

- Over 95%
- 75% - 95%
- 50% - 75%
- Under 50%

Data not available

Copyright by Rand McNally & Co.
Made in U.S.A.
A-GDS-10000-U7- --3- :-4

Literacy and Compulsory Education

- Literacy Rate
- Years of Compulsory Education

(World's largest countries, 2000)

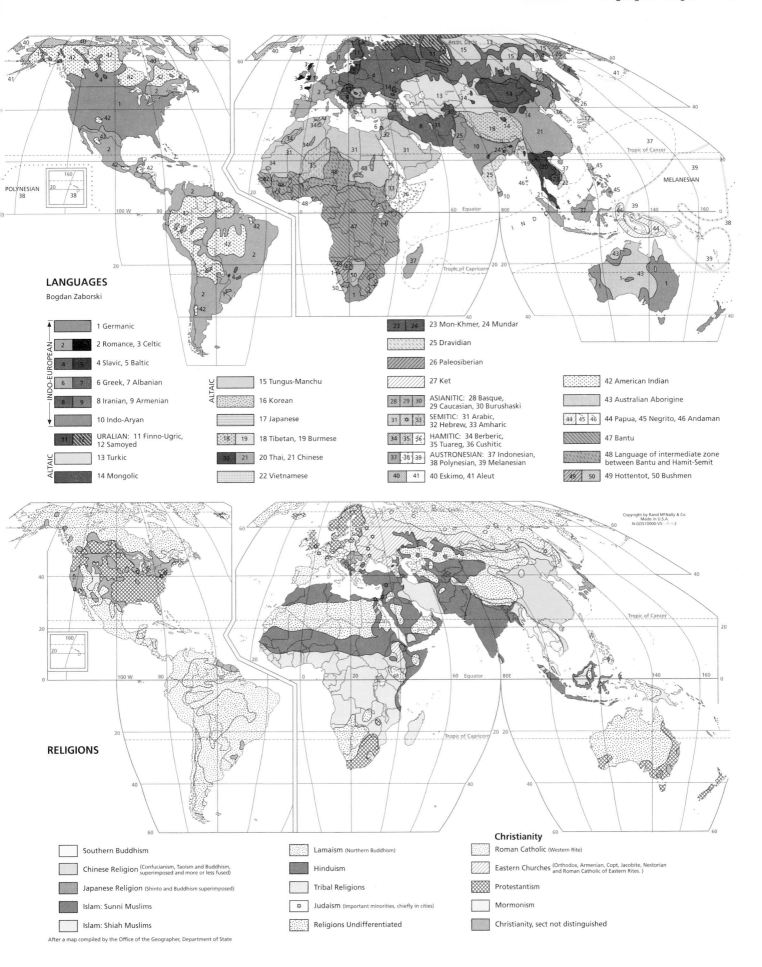

LANGUAGES

Bogdan Zaborski

POLYNESIAN
38

MELANESIAN

INDO-EUROPEAN
- 1 Germanic
- 2 Romance, 3 Celtic
- 4 Slavic, 5 Baltic
- 6 Greek, 7 Albanian
- 8 Iranian, 9 Armenian
- 10 Indo-Aryan

URALIAN: 11 Finno-Ugric, 12 Samoyed

ALTAIC
- 13 Turkic
- 14 Mongolic
- 15 Tungus-Manchu
- 16 Korean
- 17 Japanese
- 18 Tibetan, 19 Burmese
- 20 Thai, 21 Chinese
- 22 Vietnamese

- 23 Mon-Khmer, 24 Mundar
- 25 Dravidian
- 26 Paleosiberian
- 27 Ket
- ASIANITIC: 28 Basque, 29 Caucasian, 30 Burushaski
- SEMITIC: 31 Arabic, 32 Hebrew, 33 Amharic
- HAMITIC: 34 Berberic, 35 Tuareg, 36 Cushitic
- AUSTRONESIAN: 37 Indonesian, 38 Polynesian, 39 Melanesian
- 40 Eskimo, 41 Aleut

- 42 American Indian
- 43 Australian Aborigine
- 44 Papua, 45 Negrito, 46 Andaman
- 47 Bantu
- 48 Language of intermediate zone between Bantu and Hamit-Semit
- 49 Hottentot, 50 Bushmen

Copyright by Rand McNally & Co.
Made in U.S.A.
N-GDS10000-VS- -1-1-2

RELIGIONS

- Southern Buddhism
- Chinese Religion (Confucianism, Taoism and Buddhism, superimposed and more or less fused)
- Japanese Religion (Shinto and Buddhism superimposed)
- Islam: Sunni Muslims
- Islam: Shiah Muslims

- Lamaism (Northern Buddhism)
- Hinduism
- Tribal Religions
- ✡ Judaism (Important minorities, chiefly in cities)
- Religions Undifferentiated

Christianity
- Roman Catholic (Western Rite)
- Eastern Churches (Orthodox, Armenian, Copt, Jacobite, Nestorian and Roman Catholic of Eastern Rites.)
- Protestantism
- Mormonism
- Christianity, sect not distinguished

After a map compiled by the Office of the Geographer, Department of State

URBANIZED POPULATION

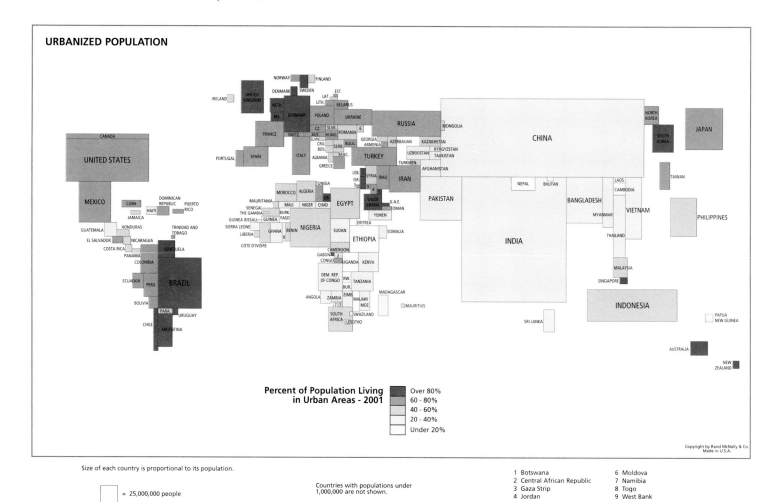

Percent of Population Living
in Urban Areas - 2001

- Over 80%
- 60 - 80%
- 40 - 60%
- 20 - 40%
- Under 20%

Size of each country is proportional to its population.

☐ = 25,000,000 people

Countries with populations under
1,000,000 are not shown.

1 Botswana
2 Central African Republic
3 Gaza Strip
4 Jordan
5 Kuwait

6 Moldova
7 Namibia
8 Togo
9 West Bank

NUTRITION

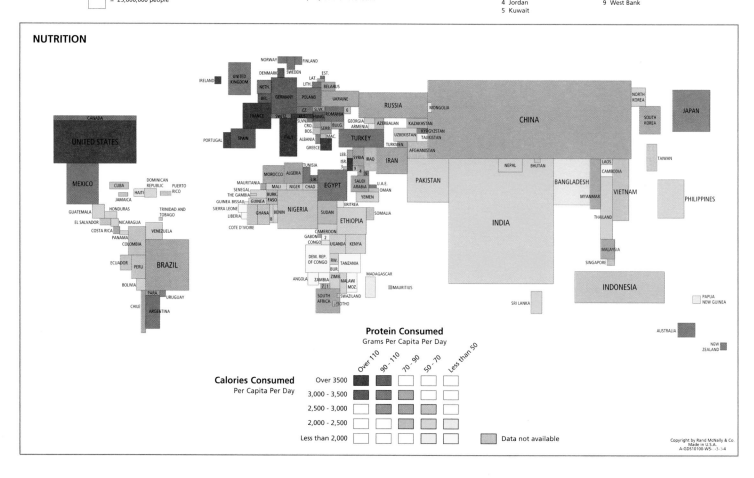

Protein Consumed
Grams Per Capita Per Day

	Over 110	90 - 110	70 - 90	50 - 70	Less than 50
Calories Consumed Per Capita Per Day					
Over 3500					
3,000 - 3,500					
2,500 - 3,000					
2,000 - 2,500					
Less than 2,000					

☐ Data not available

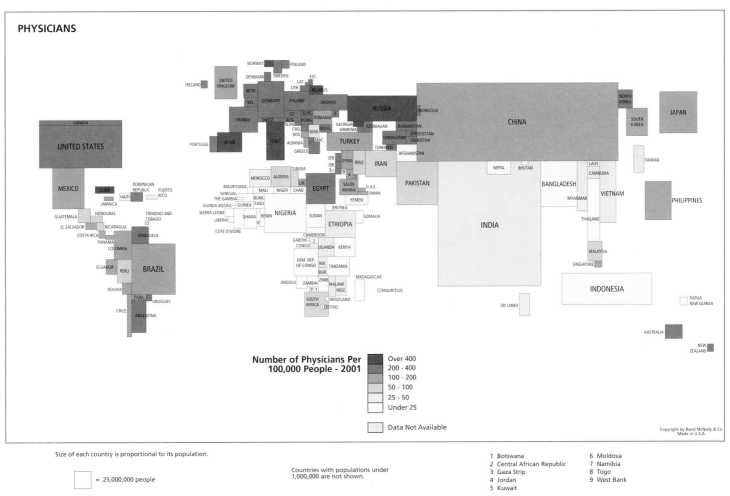

PHYSICIANS

Number of Physicians Per
100,000 People - 2001

- Over 400
- 200 - 400
- 100 - 200
- 50 - 100
- 25 - 50
- Under 25

Data Not Available

Copyright by Rand McNally & Co.
Made in U.S.A.

Size of each country is proportional to its population.

= 25,000,000 people

Countries with populations under
1,000,000 are not shown.

1 Botswana	6 Moldova
2 Central African Republic	7 Namibia
3 Gaza Strip	8 Togo
4 Jordan	9 West Bank
5 Kuwait	

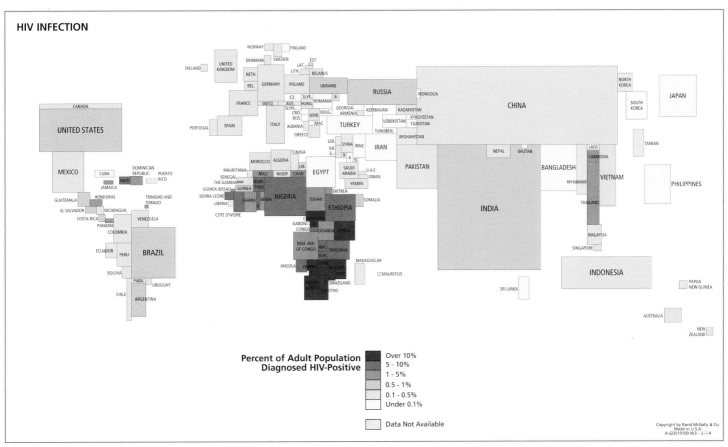

HIV INFECTION

Percent of Adult Population
Diagnosed HIV-Positive

- Over 10%
- 5 - 10%
- 1 - 5%
- 0.5 - 1%
- 0.1 - 0.5%
- Under 0.1%

Data Not Available

Copyright by Rand McNally & Co.
Made in U.S.A.
A-GDS10100-W3- -3-3-4

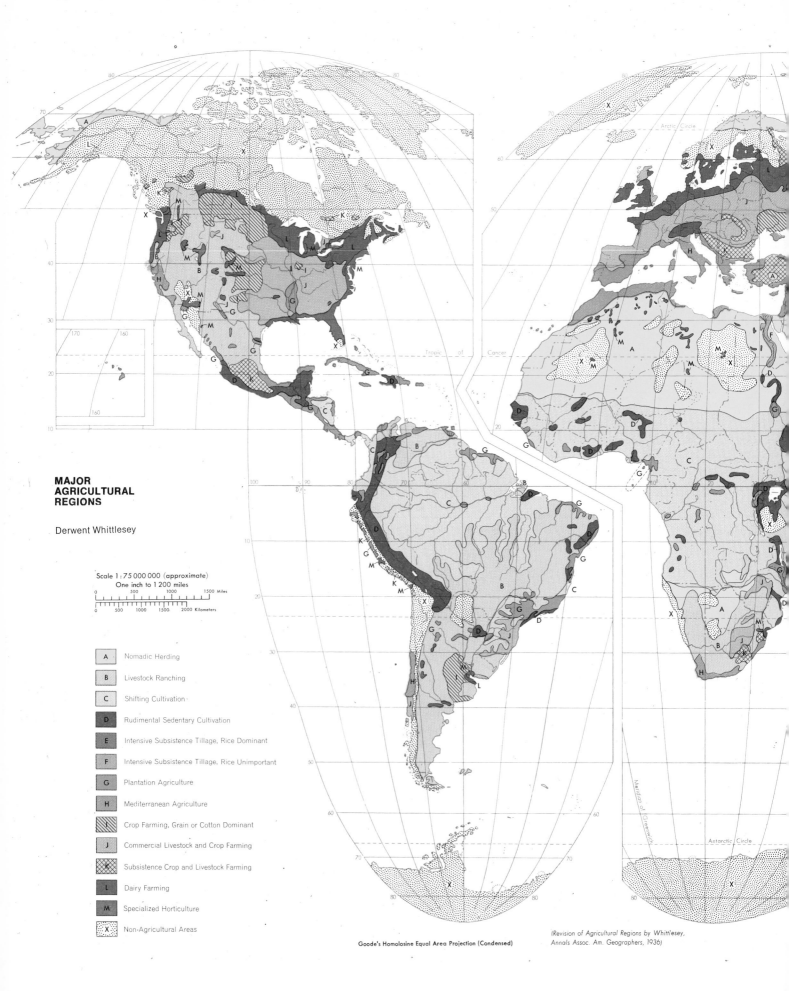

**MAJOR
AGRICULTURAL
REGIONS**

Derwent Whittlesey

Scale 1:75 000 000 (approximate)
One inch to 1 200 miles

0   500   1000   1500 Miles

0   500   1000   1500   2000 Kilometers

A	Nomadic Herding
B	Livestock Ranching
C	Shifting Cultivation
D	Rudimental Sedentary Cultivation
E	Intensive Subsistence Tillage, Rice Dominant
F	Intensive Subsistence Tillage, Rice Unimportant
G	Plantation Agriculture
H	Mediterranean Agriculture
I	Crop Farming, Grain or Cotton Dominant
J	Commercial Livestock and Crop Farming
K	Subsistence Crop and Livestock Farming
L	Dairy Farming
M	Specialized Horticulture
X	Non-Agricultural Areas

Goode's Homolosine Equal Area Projection (Condensed)

(Revision of Agricultural Regions by Whittlesey,
Annals Assoc. Am. Geographers, 1936)

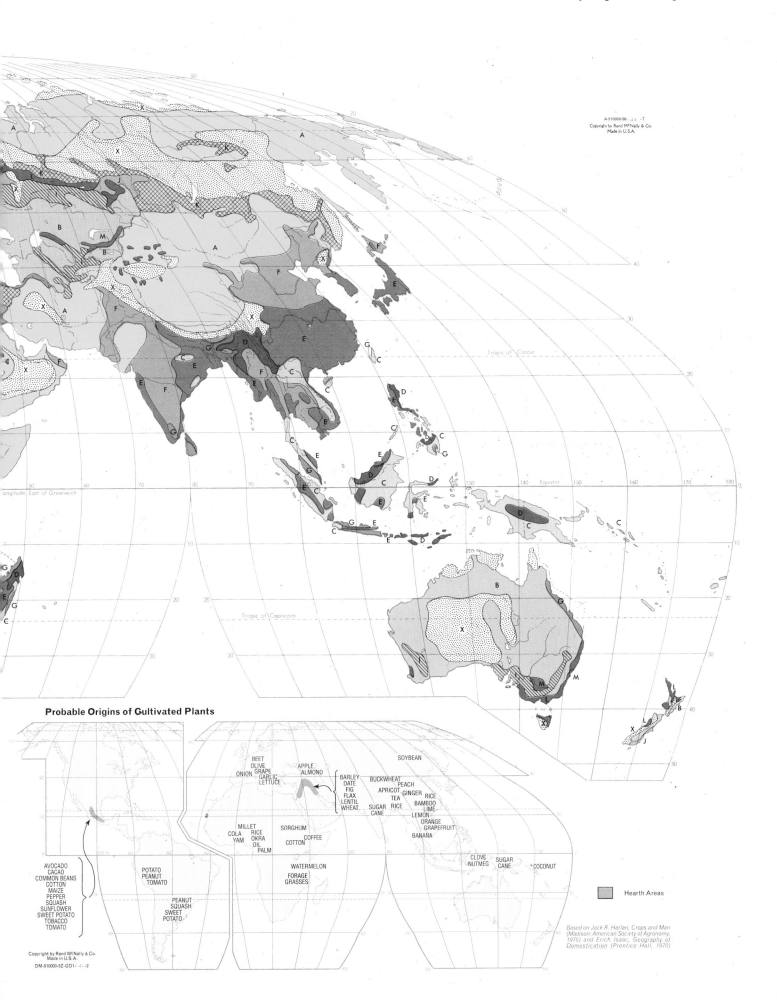

A-510000-56-⸱2 4 -7
Copyright by Rand M⁰Nally & Co.
Made in U.S.A.

Tropic of Cancer

Equator

Tropic of Capricorn

Longitude East of Greenwich

Probable Origins of Cultivated Plants

BEET
OLIVE
GRAPE
ONION GARLIC
LETTUCE

APPLE
ALMOND

SOYBEAN

BARLEY
DATE
FIG
FLAX
LENTIL
WHEAT

BUCKWHEAT
PEACH
APRICOT
GINGER
TEA
RICE

SUGAR
CANE RICE

PEACH

RICE

BAMBOO
LIME
LEMON
ORANGE
GRAPEFRUIT
BANANA

MILLET
RICE
COLA OKRA
YAM OIL
PALM

SORGHUM

COFFEE
COTTON

WATERMELON

FORAGE
GRASSES

CLOVE
NUTMEG
SUGAR
CANE
COCONUT

AVOCADO
CACAO
COMMON BEANS
COTTON
MAIZE
PEPPER
SQUASH
SUNFLOWER
SWEET POTATO
TOBACCO
TOMATO

POTATO
PEANUT
TOMATO

PEANUT
SQUASH
SWEET
POTATO

Hearth Areas

*Based on Jack R. Harlan, Crops and Man
(Madison: American Society of Agronomy,
1975) and Erich Isaac, Geography of
Domestication (Prentice Hall, 1970)*

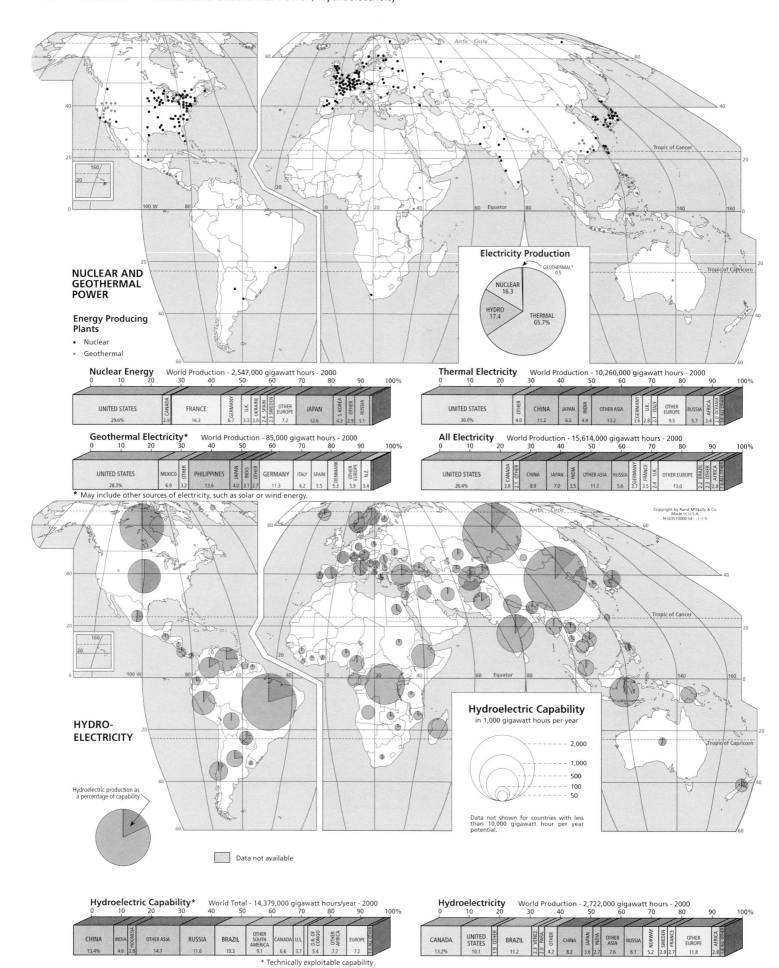

NUCLEAR AND GEOTHERMAL POWER

Energy Producing Plants

- Nuclear
- Geothermal

Electricity Production

- GEOTHERMAL* 0.5
- NUCLEAR 16.3
- HYDRO 17.4
- THERMAL 65.7%

Nuclear Energy
World Production - 2,547,000 gigawatt hours - 2000

0	10	20	30	40	50	60	70	80	90	100%

UNITED STATES 29.6% | CANADA 2.9 | FRANCE 16.3 | GERMANY 6.7 | U.K. 3.3 | UKRAINE 3.0 | SPAIN 2.4 | SWEDEN 2.3 | OTHER EUROPE 7.2 | JAPAN 12.6 | S. KOREA 4.3 | OTHER 2.9 | RUSSIA 5.1

Geothermal Electricity*
World Production - 85,000 gigwatt hours - 2000

0	10	20	30	40	50	60	70	80	90	100%

UNITED STATES 28.3% | MEXICO 6.9 | OTHER 3.2 | PHILIPPINES 13.6 | JAPAN 4.0 | INDO. 3.1 | ITALY 2.7 | GERMANY 11.3 | ITALY 6.2 | SPAIN 5.5 | DENMARK 5.3 | OTHER EUROPE 5.9 | N.Z. 3.4

* May include other sources of electricity, such as solar or wind energy.

Thermal Electricity
World Production - 10,260,000 gigawatt hours - 2000

0	10	20	30	40	50	60	70	80	90	100%

UNITED STATES 30.0% | OTHER 4.0 | CHINA 11.2 | JAPAN 6.5 | INDIA 4.4 | OTHER ASIA 13.2 | GERMANY 3.6 | U.K. 2.8 | ITALY 2.1 | OTHER EUROPE 9.5 | RUSSIA 5.7 | AFRICA 3.4 | OCEANIA 2.0 | ALL OTHER 1.6

All Electricity
World Production - 15,614,000 gigawatt hours - 2000

0	10	20	30	40	50	60	70	80	90	100%

UNITED STATES 26.4% | CANADA 3.8 | OTHER 2.1 | CHINA 8.9 | JAPAN 7.0 | INDIA 3.5 | OTHER ASIA 11.1 | RUSSIA 5.6 | GERMANY 3.7 | FRANCE 3.5 | U.K. 2.4 | OTHER EUROPE 13.0 | BRAZIL 2.2 | OTHER 2.3 | AFRICA 2.8 | ALL OTHER 1.6

Copyright by Rand McNally & Co.
Made in U.S.A.
N-GDS10000-54- -3-4-5

HYDRO-ELECTRICITY

Hydroelectric Capability
in 1,000 gigawatt hours per year

- 2,000
- 1,000
- 500
- 100
- 50

Data not shown for countries with less than 10,000 gigawatt hour per year potential.

Hydroelectric production as a percentage of capability

Data not available

Hydroelectric Capability*
World Total - 14,379,000 gigawatt hours/year - 2000

0	10	20	30	40	50	60	70	80	90	100%

CHINA 13.4% | INDIA 4.6 | INDONESIA 2.8 | OTHER ASIA 14.7 | RUSSIA 11.6 | BRAZIL 10.3 | OTHER SOUTH AMERICA 9.1 | CANADA 6.6 | U.S. 3.7 | D.R. OF CONGO 5.4 | OTHER AFRICA 7.7 | EUROPE 7.2 | ALL OTHER 1.6

* Technically exploitable capability

Hydroelectricity
World Production - 2,722,000 gigawatt hours - 2000

0	10	20	30	40	50	60	70	80	90	100%

CANADA 13.2% | UNITED STATES 10.1 | OTHER 1.9 | BRAZIL 11.2 | VENEZ. 2.3 | PARA. 2.0 | OTHER 4.2 | CHINA 8.2 | JAPAN 3.6 | INDIA 2.7 | OTHER ASIA 7.6 | RUSSIA 6.1 | NORWAY 5.2 | SWEDEN 2.9 | FRANCE 2.7 | OTHER EUROPE 11.8 | AFRICA 2.8 | ALL OTHER 1.6

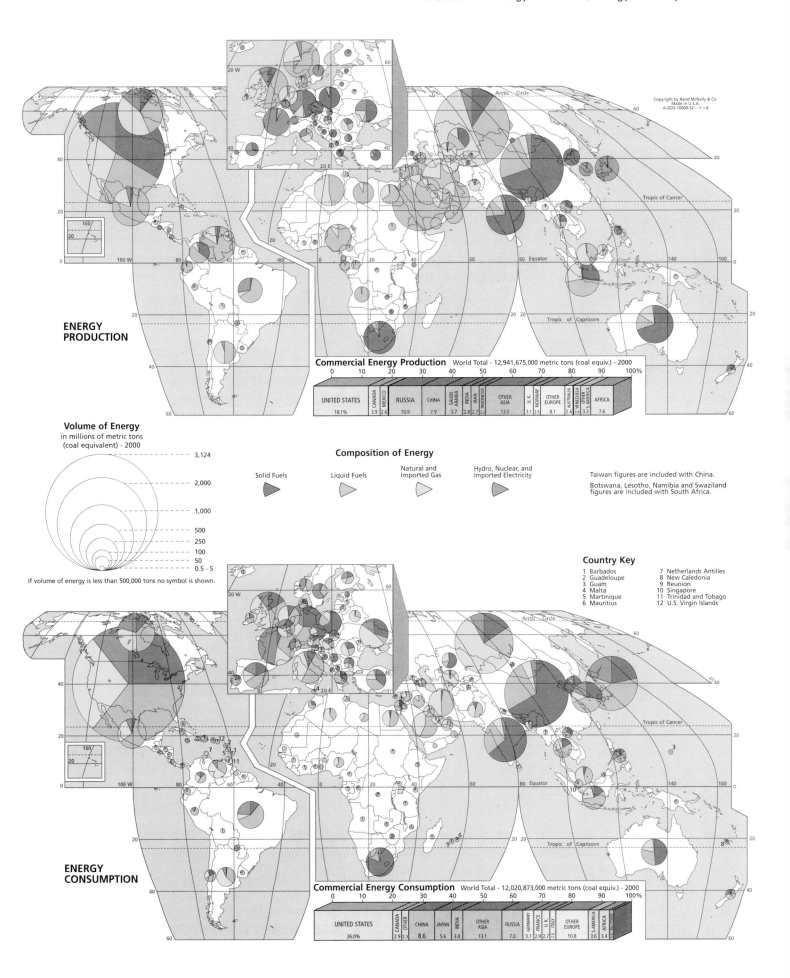

ENERGY PRODUCTION

Commercial Energy Production   World Total - 12,941,675,000 metric tons (coal equiv.) - 2000

	UNITED STATES	CANADA	MEXICO	RUSSIA	CHINA	SAUDI ARABIA	INDIA	IRAN	INDONESIA	OTHER ASIA	U.K.	NORWAY	OTHER EUROPE	AUSTRALIA	VENEZUELA	OTHER S. AMERICA	AFRICA
	18.1%	3.9	2.6	10.9	7.9	5.7	2.8	2.7	2.2	13.0	3.1	2.5	8.1	2.6	2.4	3.3	7.6

Volume of Energy
in millions of metric tons
(coal equivalent) - 2000

- 3,124
- 2,000
- 1,000
- 500
- 250
- 100
- 50
- 0.5 - 5

If volume of energy is less than 500,000 tons no symbol is shown.

Composition of Energy

Solid Fuels   Liquid Fuels   Natural and Imported Gas   Hydro, Nuclear, and Imported Electricity

Taiwan figures are included with China.

Botswana, Lesotho, Namibia and Swaziland figures are included with South Africa.

Country Key

1 Barbados	7 Netherlands Antilles
2 Guadeloupe	8 New Caledonia
3 Guam	9 Reunion
4 Malta	10 Singapore
5 Martinique	11 Trinidad and Tobago
6 Mauritius	12 U.S. Virgin Islands

ENERGY CONSUMPTION

Commercial Energy Consumption   World Total - 12,020,873,000 metric tons (coal equiv.) - 2000

	UNITED STATES	CANADA	OTHER	CHINA	JAPAN	INDIA	OTHER ASIA	RUSSIA	GERMANY	FRANCE	U.K.	ITALY	OTHER EUROPE	S. AMERICA	AFRICA	ALL OTHER
	26.0%	2.9	2.3	8.6	5.6	3.8	13.1	7.0	3.7	2.9	2.7	2.3	10.8	3.6	3.4	

24

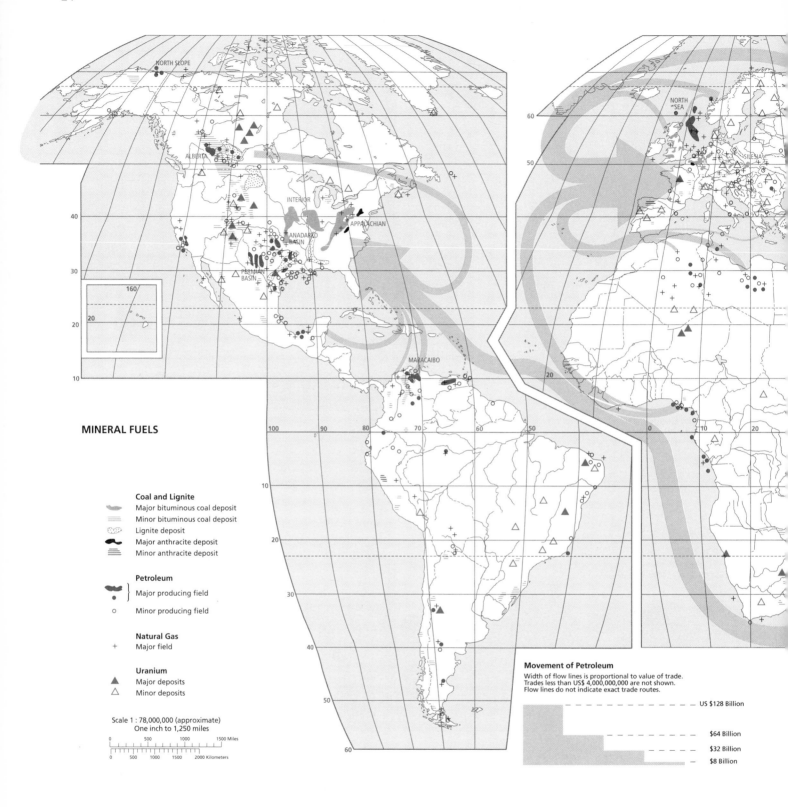

MINERAL FUELS

Coal and Lignite
- Major bituminous coal deposit
- Minor bituminous coal deposit
- Lignite deposit
- Major anthracite deposit
- Minor anthracite deposit

Petroleum
} Major producing field

○ Minor producing field

Natural Gas
+ Major field

Uranium
▲ Major deposits
△ Minor deposits

Scale 1 : 78,000,000 (approximate)
One inch to 1,250 miles

0   500   1000   1500 Miles

0   500   1000   1500   2000 Kilometers

Movement of Petroleum
Width of flow lines is proportional to value of trade.
Trades less than US$ 4,000,000,000 are not shown.
Flow lines do not indicate exact trade routes.

— — — — US $128 Billion

– – – – $64 Billion

– – – – $32 Billion

— — $8 Billion

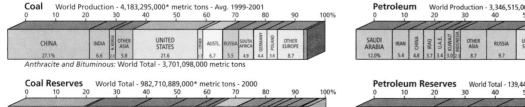

Coal   World Production - 4,183,295,000* metric tons - Avg. 1999-2001

| 0 | 10 | 20 | 30 | 40 | 50 | 60 | 70 | 80 | 90 | 100% |

CHINA	INDIA	N. KOREA	OTHER ASIA	UNITED STATES	OTHER	AUSTL.	RUSSIA	SOUTH AFRICA	POLAND	OTHER EUROPE
27.1%	6.6	2.0	5.8	21.6	3.7	6.7	5.5	4.9	3.6	8.7

Anthracite and Bituminous: World Total - 3,701,098,000 metric tons

Petroleum   World Production - 3,346,515,000** metric tons (24,606,731,000 barrels) - Avg. 1999-200

| 0 | 10 | 20 | 30 | 40 | 50 | 60 | 70 | 80 | 90 | 100% |

SAUDI ARABIA	IRAN	CHINA	IRAQ	U.A.E.	KUWAIT	INDONESIA	OTHER ASIA	RUSSIA	UNITED STATES	MEXICO	CANADA	NORWAY	U.K.	VENEZ.	OTHER S. AMERICA	NIGERIA	LIBYA	OTHER AFRICA
12.0%	5.4	4.8	3.7	3.4	3.0	2.1	8.7	9.7	8.7	4.5	2.9	4.6	3.6	4.4	4.8	3.2	2.0	5.8

Coal Reserves   World Total - 982,710,889,000* metric tons - 2000

| 0 | 10 | 20 | 30 | 40 | 50 | 60 | 70 | 80 | 90 | 100% |

UNITED STATES	RUSSIA	CHINA	INDIA	KAZAKH.	OTHER	AUSTL.	GERMANY	UKRAINE	POLAND	OTHER	SOUTH AFRICA	S. AMER.
25.3%	16.0	11.7	8.6	3.5	2.1	8.4	6.7	3.5	3.5	5.0		2.1

Anthracite and Bituminous: World Total - 518,203,342,000 metric tons
*Includes anthracite, bituminous, and lignite coal

Petroleum Reserves   World Total - 139,445,735,000** metric tons (1,025,336,289,000 barrels) - 2002

| 0 | 10 | 20 | 30 | 40 | 50 | 60 | 70 | 80 | 90 | 100% |

SAUDI ARABIA	IRAQ	KUWAIT	IRAN	U.A.E.	CHINA	OTHER ASIA	VENEZUELA	OTHER	RUSSIA	LIBYA	NIGERIA	OTHER	MEXICO	U.S.	EUROPE
25.5%	11.1	9.5	9.2	7.8	2.6	4.8	6.2	1.7	5.0	2.6	2.6	2.8	2.4	2.2	2.4

**Crude Petroleum

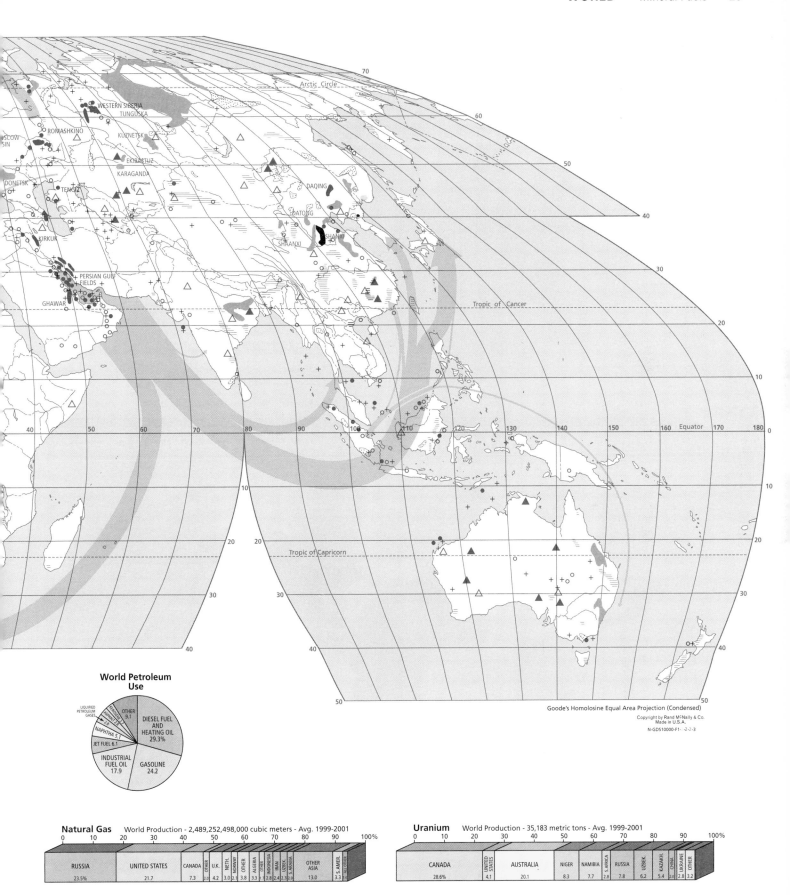

70

Arctic Circle

60

WESTERN SIBERIA
TUNGUSKA

ROMASHKINO

KUZNETSK

SCOW
SIN

50

EKIBASTUZ
KARAGANDA

DAQING

DONETSK

40

TENGIZ

DATONG

SHAANXI   SHANXI

30

KIRKUK

PERSIAN GULF
FIELDS

Tropic of Cancer   20

GHAWAR

10

40   50   60   70   80   90   100   110   120   130   140   150   160   Equator   170   180   0

10

10   20

20   10   Tropic of Capricorn

30

30   40

**World Petroleum
Use**

LIQUIFIED
PETROLEUM
GASES   OTHER
9.1

KEROSENE

ASPHALT 2.8

DIESEL FUEL
AND
HEATING OIL
29.3%

NAPHTHA 5.1

JET FUEL 6.1

INDUSTRIAL
FUEL OIL
17.9

GASOLINE
24.2

40

50

Goode's Homolosine Equal Area Projection (Condensed)

Copyright by Rand McNally & Co.
Made in U.S.A.
N-GDS10000-F1- -2-2-3

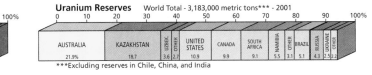

Natural Gas World Production - 2,489,252,498,000 cubic meters - Avg. 1999-2001

0	10	20	30	40	50	60	70	80	90	100%

RUSSIA	UNITED STATES	CANADA	OTHER	U.K.	NETH.	NORWAY	OTHER	ALGERIA	OTHER	INDONESIA	IRAN	UZBEK.	S. ARABIA	OTHER ASIA	S. AMER.	ALL OTHER
23.5%	21.7	7.3	2.0	4.2	3.0	2.1	3.8	3.3	1.1	2.8	2.4	2.3	2.0	13.0	3.3	1.0

Natural Gas Reserves World Total - 161,226,133,894,000 cubic meters - 2002

0	10	20	30	40	50	60	70	80	90	100%

RUSSIA	IRAN	QATAR	S. ARABIA	U.A.E.	IRAQ	OTHER ASIA	U.S.	OTHER	ALGERIA	NIGERIA	VENEZ.	EUROPE	ALL OTHER
29.7%	15.4	11.1	3.9	3.7	2.0	12.0	3.2	2.1	2.9	2.2	2.2	3.6	3.2

Uranium World Production - 35,183 metric tons - Avg. 1999-2001

0	10	20	30	40	50	60	70	80	90	100%

CANADA	UNITED STATES	AUSTRALIA	NIGER	NAMIBIA	S. AFRICA	RUSSIA	UZBEK.	KAZAKH.	CHINA	UKRAINE	OTHER
28.6%	4.1	20.1	8.3	7.7	2.8	7.8	6.2	5.4	2.0	2.8	3.2

Uranium Reserves World Total - 3,183,000 metric tons*** - 2001

0	10	20	30	40	50	60	70	80	90	100%

AUSTRALIA	KAZAKHSTAN	UZBEK.	OTHER	UNITED STATES	CANADA	SOUTH AFRICA	NAMIBIA	OTHER	BRAZIL	RUSSIA	UKRAINE	OTHER
21.9%	18.7	3.6	2.7	10.9	9.9	9.1	5.5	3.1	5.1	4.3	2.5	2.2

***Excluding reserves in Chile, China, and India

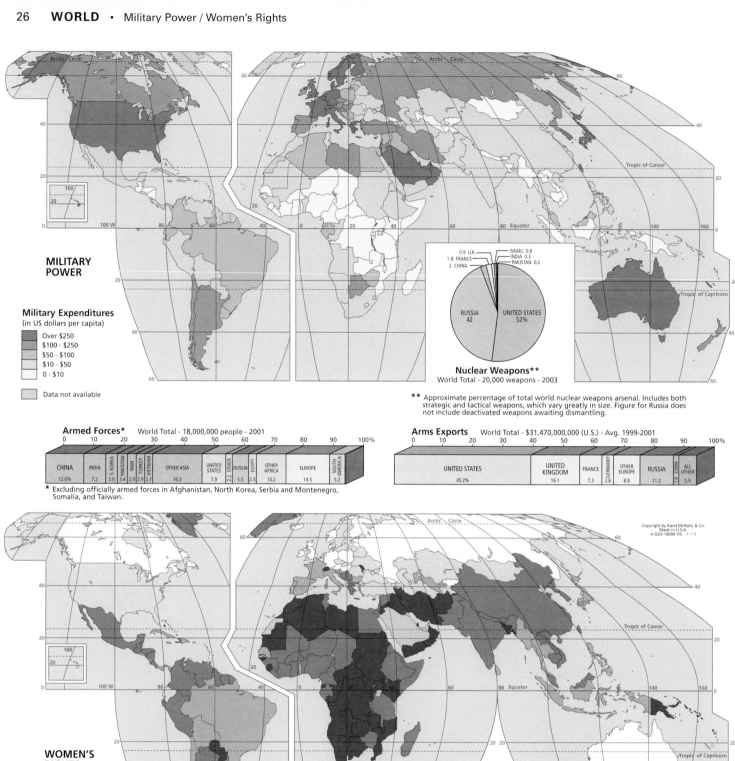

MILITARY POWER

Military Expenditures
(in US dollars per capita)

- Over $250
- $100 - $250
- $50 - $100
- $10 - $50
- 0 - $10

- Data not available

Nuclear Weapons**
World Total - 20,000 weapons - 2003

- 0.9 U.K.
- 1.8 FRANCE
- 2 CHINA
- ISRAEL 0.8
- INDIA 0.3
- PAKISTAN 0.2
- RUSSIA 42
- UNITED STATES 52%

** Approximate percentage of total world nuclear weapons arsenal. Includes both strategic and tactical weapons, which vary greatly in size. Figure for Russia does not include deactivated weapons awaiting dismantling.

Armed Forces*
World Total - 18,000,000 people - 2001

0	10	20	30	40	50	60	70	80	90	100%

| CHINA 12.6% | INDIA 7.2 | S. KOREA 3.8 | PAKISTAN 3.4 | IRAN 2.9 | TURKEY 2.9 | VIETNAM 2.7 | OTHER ASIA 16.3 | UNITED STATES 7.9 | OTHER 2.2 | RUSSIA 5.5 | EGYPT 2.5 | OTHER AFRICA 10.2 | EUROPE 14.5 | SOUTH AMERICA 5.2 |

* Excluding officially armed forces in Afghanistan, North Korea, Serbia and Montenegro, Somalia, and Taiwan.

Arms Exports
World Total - $31,470,000,000 (U.S.) - Avg. 1999-2001

0	10	20	30	40	50	60	70	80	90	100%

| UNITED STATES 45.2% | UNITED KINGDOM 16.1 | FRANCE 7.3 | GERMANY 3.8 | OTHER EUROPE 8.8 | RUSSIA 11.2 | CHINA 1.8 | ALL OTHER 5.9 |

Copyright by Rand McNally & Co.
Made in U.S.A.
A-GDS-10000-Y6- -1-1-1

WOMEN'S RIGHTS

Voting Rights
Year women received the right to vote

- After 1960
- 1946 - 1960
- 1931 - 1945
- 1919 - 1930
- Before 1919

- Not Applicable*

*Women are not allowed to vote in Kuwait. Neither women nor men are allowed to vote in Brunei, Saudi Arabia, United Arab Emirates, or Western Sahara.

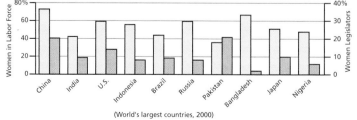

Women's Economic Activity and Legislative Participation Rates

- Percentage of women aged 15 and above in the economically active labor force
- Percentage of seats in national legislature held by women

(World's largest countries, 2000)

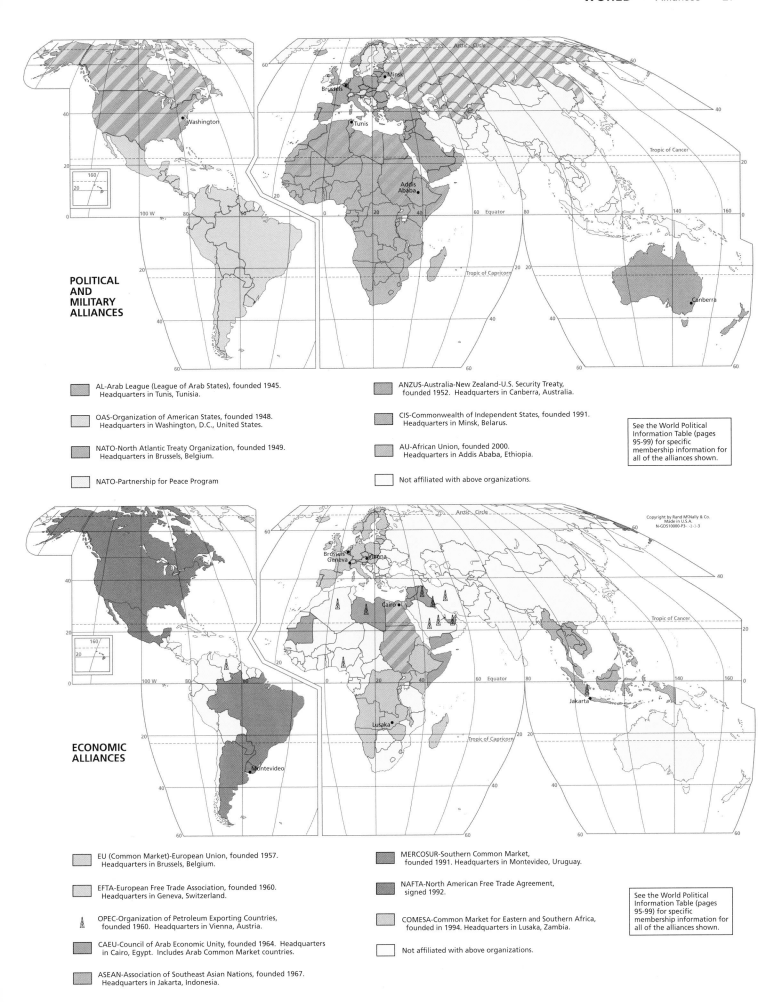

**POLITICAL
AND
MILITARY
ALLIANCES**

AL-Arab League (League of Arab States), founded 1945.
Headquarters in Tunis, Tunisia.

OAS-Organization of American States, founded 1948.
Headquarters in Washington, D.C., United States.

NATO-North Atlantic Treaty Organization, founded 1949.
Headquarters in Brussels, Belgium.

NATO-Partnership for Peace Program

ANZUS-Australia-New Zealand-U.S. Security Treaty,
founded 1952. Headquarters in Canberra, Australia.

CIS-Commonwealth of Independent States, founded 1991.
Headquarters in Minsk, Belarus.

AU-African Union, founded 2000.
Headquarters in Addis Ababa, Ethiopia.

Not affiliated with above organizations.

See the World Political
Information Table (pages
95-99) for specific
membership information for
all of the alliances shown.

Copyright by Rand McNally & Co.
Made in U.S.A.
N-GDS10000-P3- -3-3-3

**ECONOMIC
ALLIANCES**

EU (Common Market)-European Union, founded 1957.
Headquarters in Brussels, Belgium.

EFTA-European Free Trade Association, founded 1960.
Headquarters in Geneva, Switzerland.

OPEC-Organization of Petroleum Exporting Countries,
founded 1960. Headquarters in Vienna, Austria.

CAEU-Council of Arab Economic Unity, founded 1964. Headquarters
in Cairo, Egypt. Includes Arab Common Market countries.

ASEAN-Association of Southeast Asian Nations, founded 1967.
Headquarters in Jakarta, Indonesia.

MERCOSUR-Southern Common Market,
founded 1991. Headquarters in Montevideo, Uruguay.

NAFTA-North American Free Trade Agreement,
signed 1992.

COMESA-Common Market for Eastern and Southern Africa,
founded in 1994. Headquarters in Lusaka, Zambia.

Not affiliated with above organizations.

See the World Political
Information Table (pages
95-99) for specific
membership information for
all of the alliances shown.

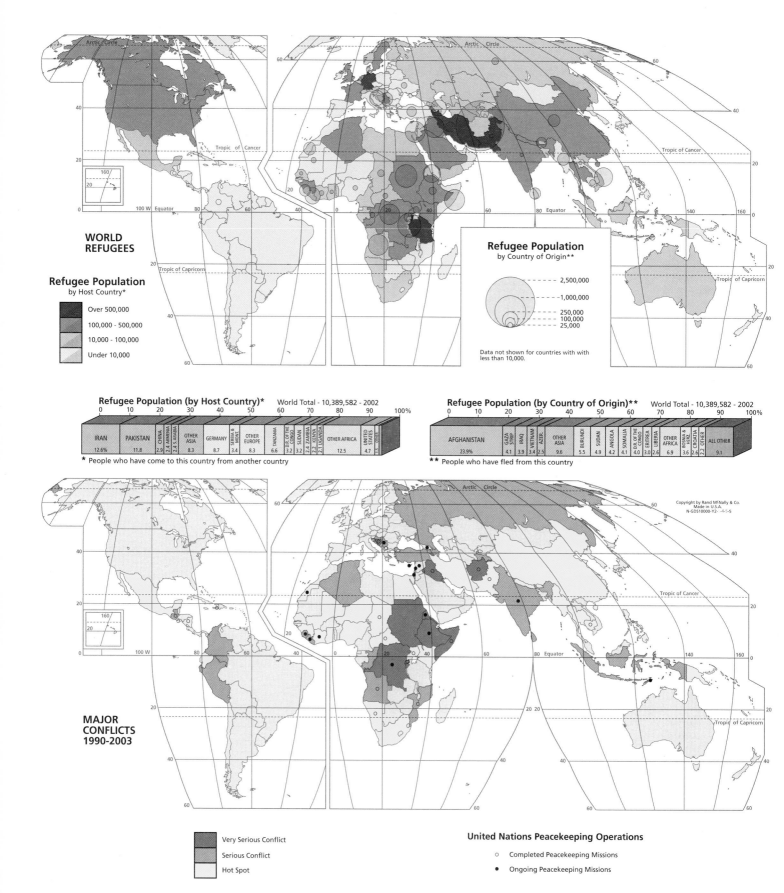

WORLD REFUGEES

Refugee Population
by Host Country*

- Over 500,000
- 100,000 - 500,000
- 10,000 - 100,000
- Under 10,000

Refugee Population
by Country of Origin**

- 2,500,000
- 1,000,000
- 250,000
- 100,000
- 25,000

Data not shown for countries with with less than 10,000.

Refugee Population (by Host Country)*   World Total - 10,389,582 - 2002

IRAN	PAKISTAN	CHINA	ARMENIA	S. ARABIA	OTHER ASIA	GERMANY	SERBIA & MONT.	OTHER EUROPE	TANZANIA	D.R. OF THE CONGO	SUDAN	ZAMBIA	KENYA	UGANDA	OTHER AFRICA	UNITED STATES	OTHER
12.6%	11.8	2.9	2.4	2.4	8.3	8.7	3.4	8.3	6.6	3.2	3.2	2.4	2.2	2.1	12.5	4.7	1.5

* People who have come to this country from another country

Refugee Population (by Country of Origin)**   World Total - 10,389,582 - 2002

AFGHANISTAN	GAZA STRIP	IRAQ	VIETNAM	AZER.	OTHER ASIA	BURUNDI	SUDAN	ANGOLA	SOMALIA	D.R. OF THE CONGO	ERITREA	LIBERIA	OTHER AFRICA	BOSNIA & HERZ.	CROATIA	OTHER	ALL OTHER
23.9%	4.1	3.9	3.4	2.5	9.6	5.5	4.9	4.2	4.1	4.0	3.0	2.6	6.9	3.6	2.6	2.2	9.1

** People who have fled from this country

MAJOR CONFLICTS 1990-2003

Copyright by Rand McNally & Co.
Made in U.S.A.
N-GDS10000-Y2- -4-5-5

- Very Serious Conflict
- Serious Conflict
- Hot Spot

United Nations Peacekeeping Operations

- ○ Completed Peacekeeping Missions
- ● Ongoing Peacekeeping Missions

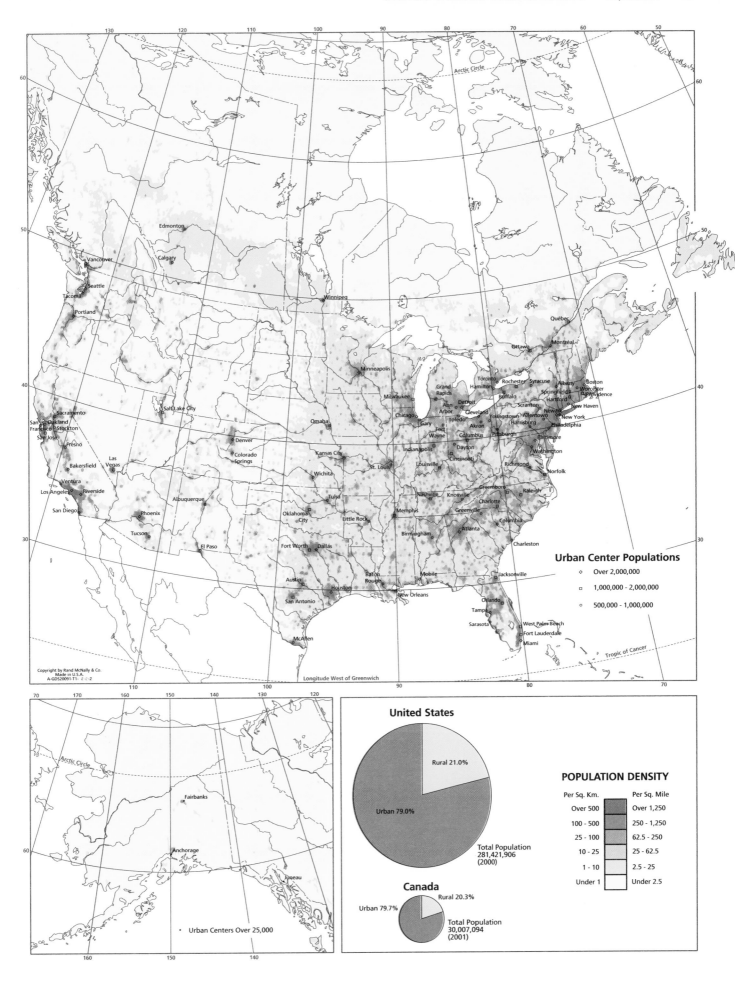

Urban Center Populations

◇ Over 2,000,000

□ 1,000,000 - 2,000,000

○ 500,000 - 1,000,000

• Urban Centers Over 25,000

United States

Rural 21.0%

Urban 79.0%

Total Population
281,421,906
(2000)

Canada

Rural 20.3%

Urban 79.7%

Total Population
30,007,094
(2001)

POPULATION DENSITY

Per Sq. Km.	Per Sq. Mile
Over 500	Over 1,250
100 - 500	250 - 1,250
25 - 100	62.5 - 250
10 - 25	25 - 62.5
1 - 10	2.5 - 25
Under 1	Under 2.5

Copyright by Rand McNally & Co.
Made in U.S.A.
A-GDS20091-T1-

Longitude West of Greenwich

WHITE POPULATION

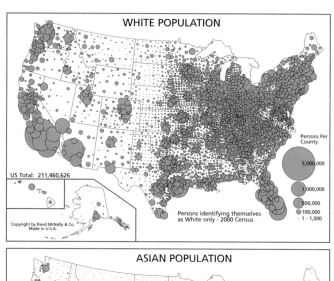

US Total: 211,460,626

Persons Per County
5,000,000
1,000,000
500,000
100,000
· 1 - 1,000

Persons identifying themselves as White only - 2000 Census

Copyright by Rand McNally & Co.
Made in U.S.A.

AFRICAN AMERICAN POPULATION

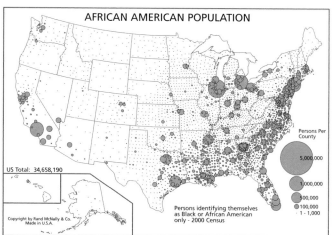

US Total: 34,658,190

Persons Per County
5,000,000
1,000,000
500,000
100,000
1 - 1,000

Persons identifying themselves as Black or African American only - 2000 Census

Copyright by Rand McNally & Co.
Made in U.S.A.

ASIAN POPULATION

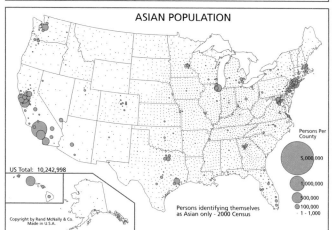

US Total: 10,242,998

Persons Per County
5,000,000
1,000,000
500,000
100,000
· 1 - 1,000

Persons identifying themselves as Asian only - 2000 Census

Copyright by Rand McNally & Co.
Made in U.S.A.

AMERICAN INDIAN AND ALASKA NATIVE POPULATION

US Total: 2,475,956

Persons Per County
5,000,000
1,000,000
500,000
100,000
· 1 - 1,000

Persons identifying themselves as American Indian or Alaska Native only - 2000 Census

Copyright by Rand McNally & Co.
Made in U.S.A.

NATIVE HAWAIIAN AND PACIFIC ISLANDER POPULATION

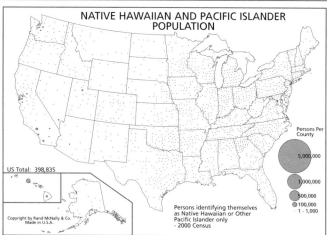

US Total: 398,835

Persons Per County
5,000,000
1,000,000
500,000
100,000
· 1 - 1,000

Persons identifying themselves as Native Hawaiian or Other Pacific Islander only - 2000 Census

Copyright by Rand McNally & Co.
Made in U.S.A.

SOME OTHER RACE

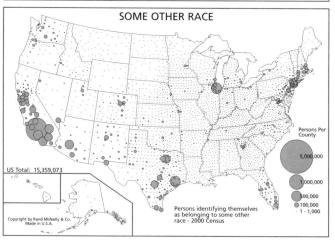

US Total: 15,359,073

Persons Per County
5,000,000
1,000,000
500,000
100,000
· 1 - 1,000

Persons identifying themselves as belonging to some other race - 2000 Census

Copyright by Rand McNally & Co.
Made in U.S.A.

TWO OR MORE RACES

US Total: 6,826,228

Persons Per County
5,000,000
1,000,000
500,000
100,000
· 1 - 1,000

Persons identifying themselves as belonging to two or more races - 2000 Census

Copyright by Rand McNally & Co.
Made in U.S.A.

HISPANIC POPULATION (ANY RACE)

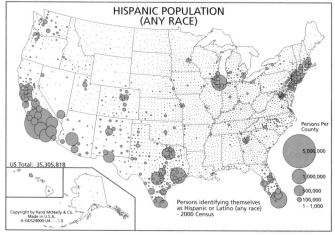

US Total: 35,305,818

Persons Per County
5,000,000
1,000,000
500,000
100,000
· 1 - 1,000

Persons identifying themselves as Hispanic or Latino (any race) - 2000 Census

Copyright by Rand McNally & Co.
Made in U.S.A.
A-GDS24000-U4---3-3

POPULATION CHANGE

Entire United States: 13.2%

Over 50%
25 - 50
10 - 25
0 - 10
-10 to 0
Under -10

Copyright by Rand McNally & Co.
Made in U.S.A.

Percentage increase or decrease in county population from 1990 to 2000

INTER-STATE POPULATION SHIFTS

Entire United States: 40.0%

Over 50%
40 - 50
30 - 40
20 - 30
Under 20%

Copyright by Rand McNally & Co.
Made in U.S.A.

Percentage of county native-born population not residing in state of birth - 2000 Census

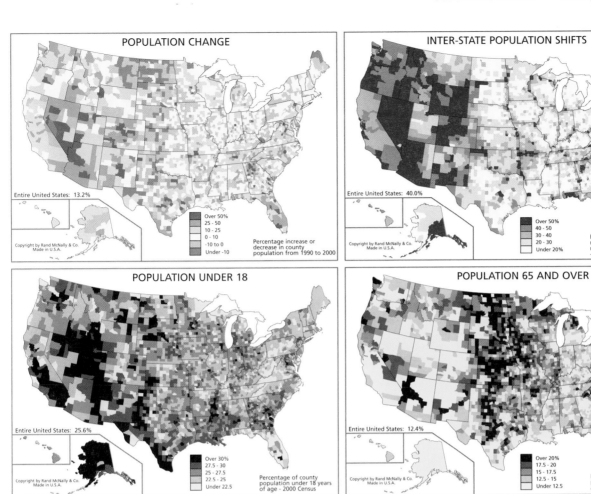

POPULATION UNDER 18

Entire United States: 25.6%

Over 30%
27.5 - 30
25 - 27.5
22.5 - 25
Under 22.5

Copyright by Rand McNally & Co.
Made in U.S.A.

Percentage of county population under 18 years of age - 2000 Census

POPULATION 65 AND OVER

Entire United States: 12.4%

Over 20%
17.5 - 20
15 - 17.5
12.5 - 15
Under 12.5

Copyright by Rand McNally & Co.
Made in U.S.A.

Percentage of county population aged 65 and over - 2000 Census

EDUCATIONAL ATTAINMENT RATE

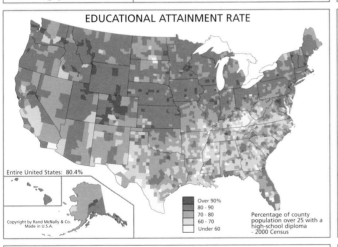

Entire United States: 80.4%

Over 90%
80 - 90
70 - 80
60 - 70
Under 60

Copyright by Rand McNally & Co.
Made in U.S.A.

Percentage of county population over 25 with a high-school diploma - 2000 Census

COLLEGE ENROLLMENT RATE

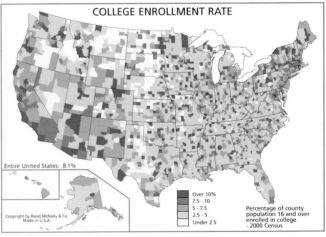

Entire United States: 8.1%

Over 10%
7.5 - 10
5 - 7.5
2.5 - 5
Under 2.5

Copyright by Rand McNally & Co.
Made in U.S.A.

Percentage of county population 16 and over enrolled in college - 2000 Census

COMMUTING TIME

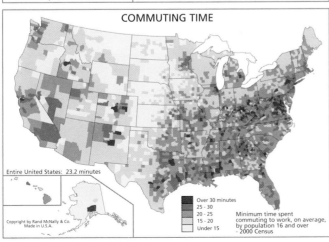

Entire United States: 23.2 minutes

Over 30 minutes
25 - 30
20 - 25
15 - 20
Under 15

Copyright by Rand McNally & Co.
Made in U.S.A.

Minimum time spent commuting to work, on average, by population 16 and over - 2000 Census

MEDIAN DECADE OF HOUSE CONSTRUCTION

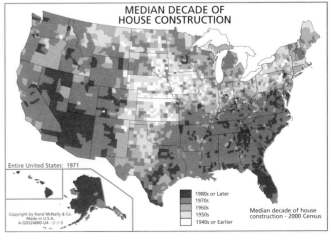

Entire United States: 1971

1980s or Later
1970s
1960s
1950s
1940s or Earlier

Copyright by Rand McNally & Co.
Made in U.S.A.
A-GDS24000-U4- -3-1-3

Median decade of house construction - 2000 Census

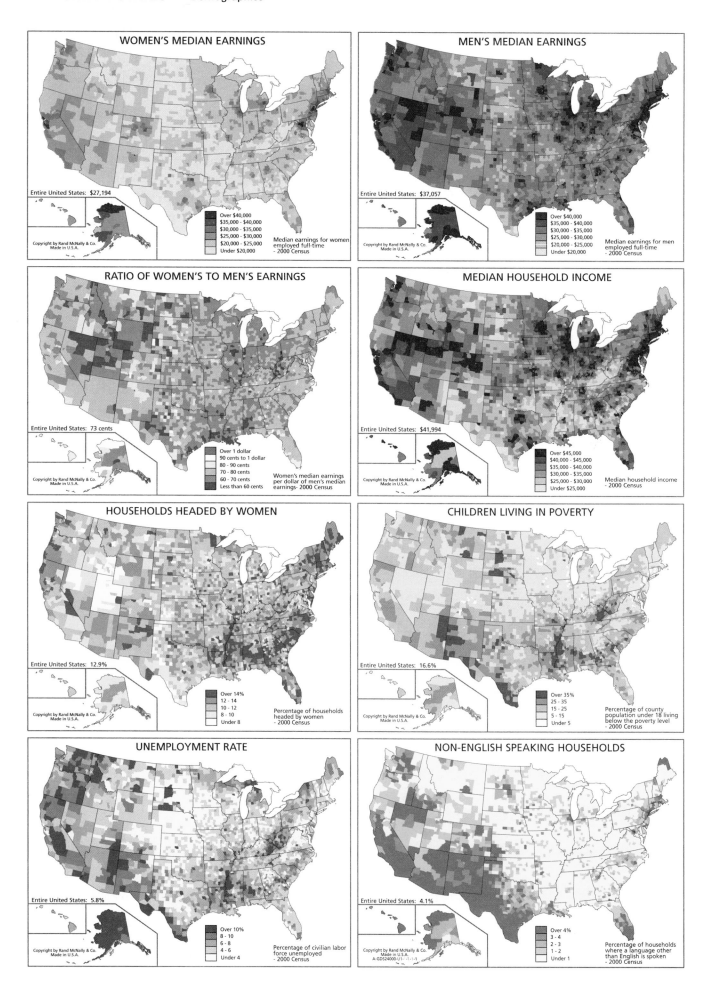

WOMEN'S MEDIAN EARNINGS

Entire United States: $27,194

Copyright by Rand McNally & Co.
Made in U.S.A.

Over $40,000
$35,000 - $40,000
$30,000 - $35,000
$25,000 - $30,000
$20,000 - $25,000
Under $20,000

Median earnings for women
employed full-time
- 2000 Census

MEN'S MEDIAN EARNINGS

Entire United States: $37,057

Copyright by Rand McNally & Co.
Made in U.S.A.

Over $40,000
$35,000 - $40,000
$30,000 - $35,000
$25,000 - $30,000
$20,000 - $25,000
Under $20,000

Median earnings for men
employed full-time
- 2000 Census

RATIO OF WOMEN'S TO MEN'S EARNINGS

Entire United States: 73 cents

Copyright by Rand McNally & Co.
Made in U.S.A.

Over 1 dollar
90 cents to 1 dollar
80 - 90 cents
70 - 80 cents
60 - 70 cents
Less than 60 cents

Women's median earnings
per dollar of men's median
earnings- 2000 Census

MEDIAN HOUSEHOLD INCOME

Entire United States: $41,994

Copyright by Rand McNally & Co.
Made in U.S.A.

Over $45,000
$40,000 - $45,000
$35,000 - $40,000
$30,000 - $35,000
$25,000 - $30,000
Under $25,000

Median household income
- 2000 Census

HOUSEHOLDS HEADED BY WOMEN

Entire United States: 12.9%

Copyright by Rand McNally & Co.
Made in U.S.A.

Over 14%
12 - 14
10 - 12
8 - 10
Under 8

Percentage of households
headed by women
- 2000 Census

CHILDREN LIVING IN POVERTY

Entire United States: 16.6%

Copyright by Rand McNally & Co.
Made in U.S.A.

Over 35%
25 - 35
15 - 25
5 - 15
Under 5

Percentage of county
population under 18 living
below the poverty level
- 2000 Census

UNEMPLOYMENT RATE

Entire United States: 5.8%

Copyright by Rand McNally & Co.
Made in U.S.A.

Over 10%
8 - 10
6 - 8
4 - 6
Under 4

Percentage of civilian labor
force unemployed
- 2000 Census

NON-ENGLISH SPEAKING HOUSEHOLDS

Entire United States: 4.1%

Copyright by Rand McNally & Co.
Made in U.S.A.
A-GDS24000-U1- -1 -1 -1

Over 4%
3 - 4
2 - 3
1 - 2
Under 1

Percentage of households
where a language other
than English is spoken
- 2000 Census

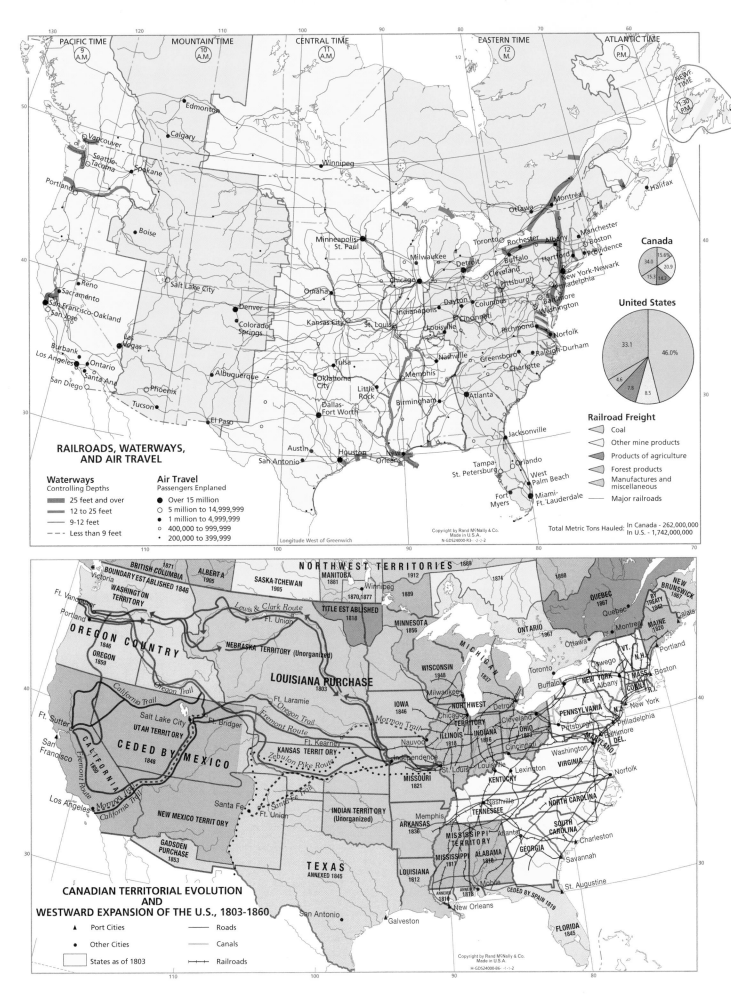

PACIFIC TIME
9 A.M.

MOUNTAIN TIME
10 A.M.

CENTRAL TIME
11 A.M.

EASTERN TIME
12 M.

ATLANTIC TIME
1 P.M.

NEWF. TIME
1:30 P.M.

Edmonton

Vancouver
Calgary

Seattle-Tacoma
Spokane

Portland
Winnipeg

Boise

Reno
Salt Lake City
Minneapolis St. Paul
Milwaukee
Detroit
Ottawa
Montréal
Halifax

Sacramento
San Francisco-Oakland
San Jose

Denver
Omaha
Chicago
Cleveland
Toronto
Rochester
Albany
Buffalo
Pittsburgh
Manchester
Boston
Providence
Hartford
New York-Newark
Philadelphia

Las Vegas
Colorado Springs
Kansas City
St. Louis
Indianapolis
Dayton
Columbus
Cincinnati
Louisville
Baltimore
Washington
Richmond
Norfolk

Burbank
Ontario
Los Angeles
Santa Ana
San Diego

Phoenix
Albuquerque
Tulsa
Oklahoma City
Memphis
Nashville
Greensboro
Charlotte
Raleigh-Durham

Tucson
El Paso
Little Rock
Birmingham
Atlanta

Austin
Dallas-Fort Worth
Houston
San Antonio
New Orleans
Jacksonville
Tampa-St. Petersburg
Orlando
West Palm Beach
Fort Myers
Miami-Ft. Lauderdale

RAILROADS, WATERWAYS, AND AIR TRAVEL

Waterways
Controlling Depths
▬ 25 feet and over
▬ 12 to 25 feet
— 9-12 feet
- - - Less than 9 feet

Air Travel
Passengers Enplaned
● Over 15 million
○ 5 million to 14,999,999
• 1 million to 4,999,999
○ 400,000 to 999,999
· 200,000 to 399,999

Canada
34.0   15.6%
15.3   20.9
      14.2

United States
33.1   46.0%
4.6   7.8   8.5

Railroad Freight
◁ Coal
◁ Other mine products
◀ Products of agriculture
◁ Forest products
◁ Manufactures and miscellaneous
— Major railroads

Longitude West of Greenwich

Total Metric Tons Hauled: In Canada - 262,000,000
In U.S. - 1,742,000,000

BRITISH COLUMBIA 1871
BOUNDARY ESTABLISHED 1846
Victoria
ALBERTA 1905
SASKATCHEWAN 1905
MANITOBA 1881
NORTHWEST TERRITORIES 1889
1912
1874
1898
QUEBEC 1867
NEW BRUNSWICK 1867
BY TREATY 1842

WASHINGTON TERRITORY
Ft. Vancouver
Portland
Lewis & Clark Route
Ft. Union
TITLE ESTABLISHED 1818
Winnipeg
1870,1877
1889
MINNESOTA 1856
ONTARIO 1867
Quebec
Montreal
Ottawa
MAINE 1820
Portland
Calais

OREGON COUNTRY
OREGON 1846
OREGON 1859
Oregon Trail
NEBRASKA TERRITORY (Unorganized)
LOUISIANA PURCHASE 1803
MICHIGAN
Toronto
Buffalo
NEW YORK
Albany
Oswego
VT.
N.H.
MASS.
CONN.
R.I.
Boston

California Trail
Ft. Laramie
Oregon Trail
Fremont Route
Mormon Trail
WISCONSIN 1848
1831
Milwaukee
Detroit
Cleveland
NORTHWEST TERRITORY
PENNSYLVANIA
Pittsburgh
Philadelphia
N.J.
New York

Ft. Sutter
Salt Lake City
UTAH TERRITORY
Ft. Bridger
Fremont Route
IOWA 1846
Chicago
ILLINOIS 1818
Nauvoo
INDIANA 1816
OHIO 1803
Cincinnati
Washington
MD.
DEL.
Baltimore

San Francisco
CALIFORNIA 1850
CEDED BY MEXICO 1848
Ft. Kearney
KANSAS TERRITORY
Zebulon Pike Route
Independence
St. Louis
Louisville
Lexington
KENTUCKY
VIRGINIA
Norfolk

Los Angeles
Mormon Trail
California Trail
NEW MEXICO TERRITORY
Santa Fe
Ft. Union
Santa Fe Trail
MISSOURI 1821
TENNESSEE
Nashville
NORTH CAROLINA
SOUTH CAROLINA
Charleston

GADSDEN PURCHASE 1853
INDIAN TERRITORY (Unorganized)
ARKANSAS 1836
Memphis
MISSISSIPPI TERRITORY
MISSISSIPPI 1817
ALABAMA 1818
Atlanta
GEORGIA
Savannah

TEXAS ANNEXED 1845
LOUISIANA 1812
ANNEXED 1818
ANNEXED 1819
New Orleans
CEDED BY SPAIN 1819
St. Augustine
FLORIDA 1845

CANADIAN TERRITORIAL EVOLUTION AND WESTWARD EXPANSION OF THE U.S., 1803-1860

San Antonio
Galveston

▲ Port Cities
● Other Cities
▭ States as of 1803
— Roads
═ Canals
—┼┼— Railroads

Scale 1:12,000,000. One inch to 190 miles.
One centimeter to 120 kilometers. Albers Conic Projection

| 0 | 50 | 100 | 200 | 300 | 400 Miles |

| 0 | 50 | 100 | 150 | 200 | 300 | 400 | 500 | 600 Kilometers |

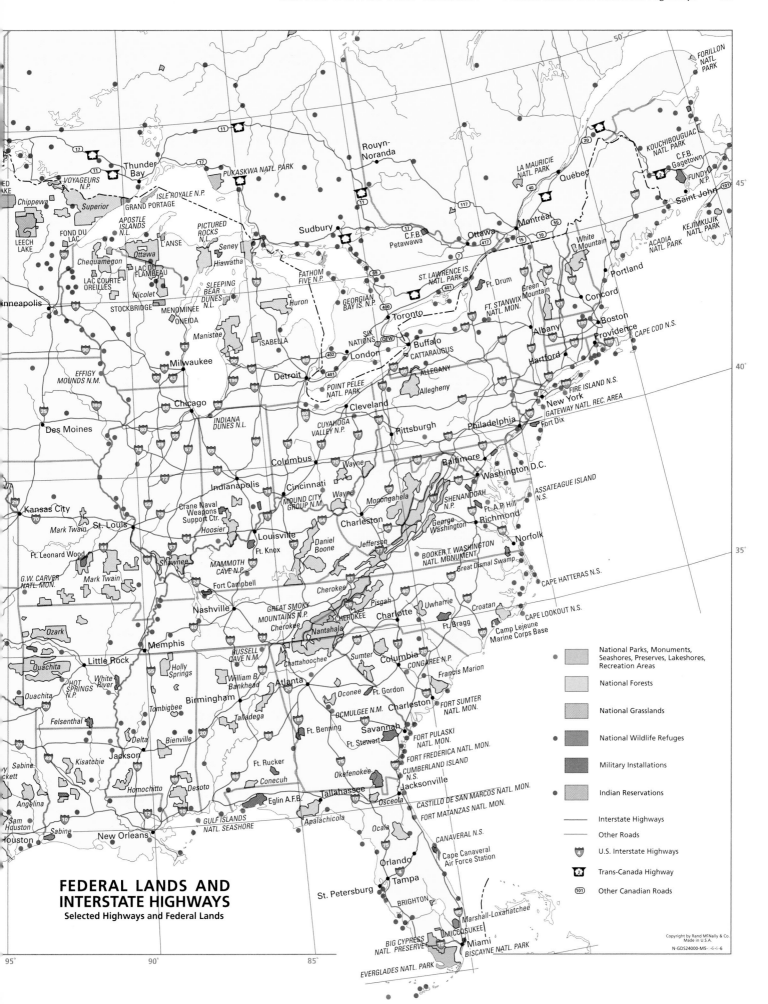

FEDERAL LANDS AND INTERSTATE HIGHWAYS
Selected Highways and Federal Lands

National Parks, Monuments, Seashores, Preserves, Lakeshores, Recreation Areas	
National Forests	
National Grasslands	
National Wildlife Refuges	
Military Installations	
Indian Reservations	
Interstate Highways	
Other Roads	
U.S. Interstate Highways	
Trans-Canada Highway	
Other Canadian Roads	

Copyright by Rand McNally & Co.
Made in U.S.A.

N-GDS24000-M5- -6-ä-6

Scale 1:40 000 000; one inch to 630 miles. Lambert's Azimuthal Equal Area Projection
Elevations and depressions are given in feet

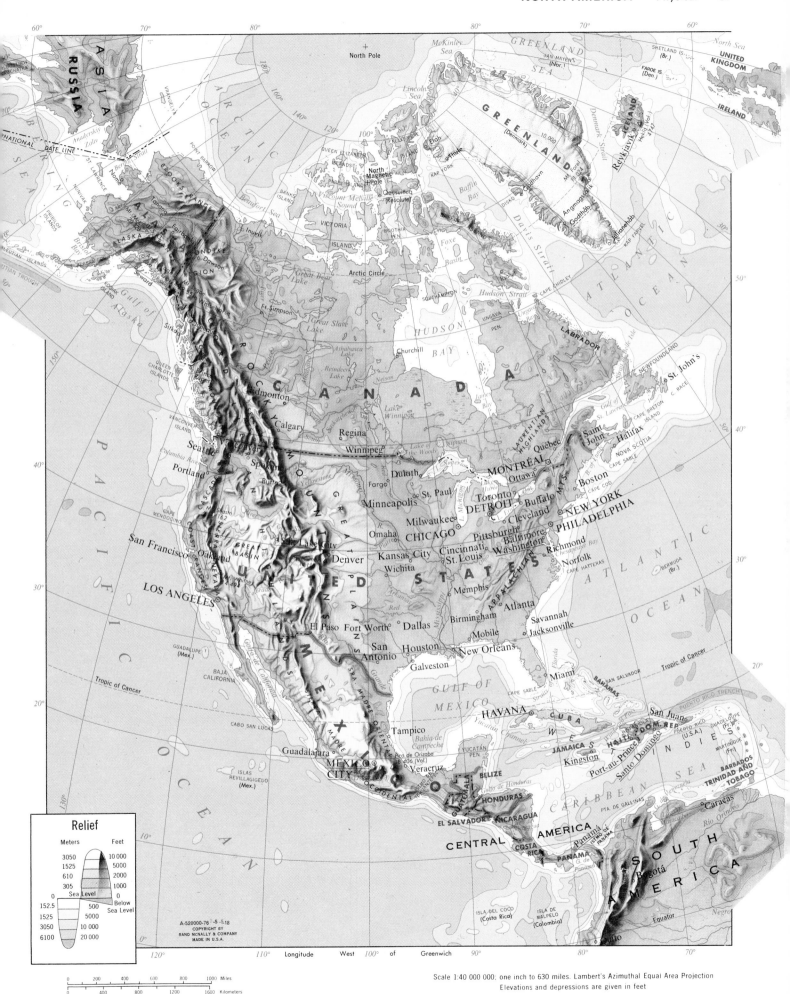

ASIA
RUSSIA
North Pole
GREENLAND SEA
JAN MAYEN (Nor.)
SHETLAND IS. (Br.)
North Sea
UNITED KINGDOM
IRELAND
FAROE IS. (Den.)

McKinley Sea
INTERNATIONAL DATE LINE
Anadyrskiy Zaliv
WRANGEL I.
BERING STRAIT
Nome
POINT BARROW
BROOKS RANGE
Fairbanks
ALASKA
Anchorage
Seward
KODIAK ISLAND
Gulf of Alaska
Sitka
ALEUTIAN ISLANDS
ALEUTIAN TRENCH
PRIBILOF ISLANDS
Bristol Bay
NUNIVAK
ST. LAWRENCE
BERING SEA

ARCTIC OCEAN
Lincoln Sea
ELLESMERE ISLAND
Etah
Thule
KAP YORK
GREENLAND (Denmark)
10,000
Godhavn
Angmagssalik
Godthåb
Julianehåb
KAP FARVEL
ICELAND
Reykjavík
Hekla (Vol.) 4747

BANKS ISLAND
QUEEN ELIZABETH ISLANDS
VISCOUNT MELVILLE SOUND
PARRY ISLANDS
North Magnetic Pole
Qeqertarsuaq (Resolute)
BOOTHIA PEN.
Baffin Bay

VICTORIA ISLAND
Inuvik
Great Bear Lake
Ft. Simpson
Great Slave Lake
Athabasca Lake
Reindeer Lake
Churchill
Arctic Circle

CANADA

HUDSON BAY
SOUTHAMPTON
Foxe Basin
Hudson Strait
CAPE CHIDLEY
UNGAVA BAY
UNGAVA PEN.
LABRADOR
James Bay

Davis Strait
ATLANTIC OCEAN
NEWFOUNDLAND
St. John's
C. RACE
CAPE BRETON ISLAND
Gulf of St. Lawrence
CAPE SABLE
LAURENTIAN HIGHLANDS

ROCKY MOUNTAINS
Edmonton
Calgary
Regina
Lake Winnipeg
Winnipeg
Lake of the Woods
Nelson
Saskatchewan

QUEEN CHARLOTTE ISLANDS
VANCOUVER ISLAND
Vancouver
Seattle
Spokane
Columbia River
Portland
CAPE MENDOCINO
CASCADE RANGE
COAST RANGES
San Francisco
Oakland
SIERRA NEVADA
Mt. Whitney 14,491
LOS ANGELES

Duluth
Fargo
St. Paul
Minneapolis
Milwaukee
Omaha
CHICAGO
LAKE SUPERIOR
Toronto
Ottawa
MONTRÉAL
Québec
Saint John
Halifax
NOVA SCOTIA
LAKE HURON
LAKE MICHIGAN
DETROIT
Cleveland
Buffalo
Boston
CAPE COD
NEW YORK
PHILADELPHIA
LAKE ONTARIO
LAKE ERIE
Pittsburgh
Cincinnati
St. Louis
Baltimore
Washington
Richmond
Norfolk
Chesapeake Bay
CAPE HATTERAS
BERMUDA (Br.)

UNITED STATES
GREAT BASIN
Salt Lake City
Denver
Kansas City
Wichita
Memphis
GREAT PLAINS
Platte
Arkansas
Red
APPALACHIANS

San Francisco
Whitney 14,491
Colorado
El Paso
Fort Worth
Dallas
San Antonio
Houston
Galveston
New Orleans
Mobile
Birmingham
Atlanta
Savannah
Jacksonville
Miami
Florida
CAPE SABLE

SIERRA MADRE OCCIDENTAL
SIERRA MADRE ORIENTAL
MEXICO
Tampico
Veracruz
Pico de Orizaba 18,406 (Vol.)
MEXICO CITY 17,887
Guadalajara
CABO SAN LUCAS
BAJA CALIFORNIA
Golfo de California
GUADALUPE (Mex.)
ISLAS REVILLAGIGEDO (Mex.)
Tropic of Cancer

GULF OF MEXICO
Bahía de Campeche
YUCATÁN PEN.
Yucatán Channel
HAVANA
CUBA
BAHAMAS
SAN SALVADOR
Straits of Florida
San Juan
PUERTO RICO (U.S.A.)
PUERTO RICO TRENCH
DOM. REP.
HAITI
JAMAICA
Kingston
Port-au-Prince
Santo Domingo
HISPANIOLA
GUADELOUPE (Fr.)
MARTINIQUE (Fr.)
WEST INDIES
BARBADOS
TRINIDAD AND TOBAGO
CARIBBEAN SEA

BELIZE
Golfo de Honduras
GUATEMALA
HONDURAS
EL SALVADOR
NICARAGUA
COSTA RICA
PANAMA
ISMO. DE PANAMÁ
G. de Panamá
CENTRAL AMERICA
PTA. DE GALLINAS
Caracas
Río Orinoco
Bogotá
SOUTH AMERICA
Equator
Negro
ISLA DEL COCO (Costa Rica)
ISLA DE MALPELO (Colombia)

PACIFIC OCEAN
Tropic of Cancer

Relief

Meters		Feet
3050		10 000
1525		5000
610		2000
305		1000
0	Sea Level	0
152.5		500
1525	Below Sea Level	5000
3050		10 000
6100		20 000

A-520000-76 -5 -5-18
COPYRIGHT BY
RAND McNALLY & COMPANY
MADE IN U.S.A.

Longitude West of Greenwich

Scale 1:40 000 000; one inch to 630 miles. Lambert's Azimuthal Equal Area Projection
Elevations and depressions are given in feet

0 200 400 600 800 1000 Miles
0 400 800 1200 1600 Kilometers

38

Scale 1:12 000 000; one inch to 190 miles. Polyconic Projection
Elevations and depressions are given in feet

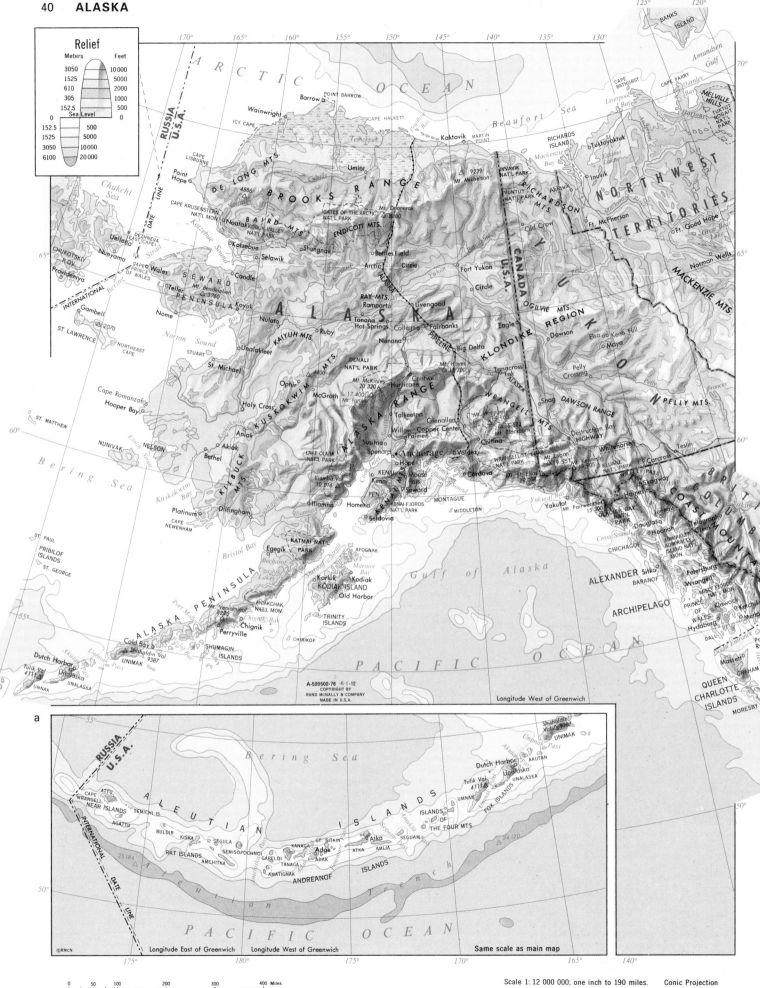

Relief

Meters	Feet
3050	10 000
1525	5000
610	2000
305	1000
152.5	500
0	Sea Level

152.5	500
1525	5000
3050	10 000
6100	20 000

A-520502-76 -6-6-12
COPYRIGHT BY
RAND McNALLY & COMPANY
MADE IN U.S.A.

a

Scale 1: 12 000 000; one inch to 190 miles.   Conic Projection

Elevations and depressions are given in feet

Same scale as main map

0   50   100   200   300   400 Miles
0   100   200   300   400   500   600 Kilometers

®RMCN.   Longitude East of Greenwich   Longitude West of Greenwich

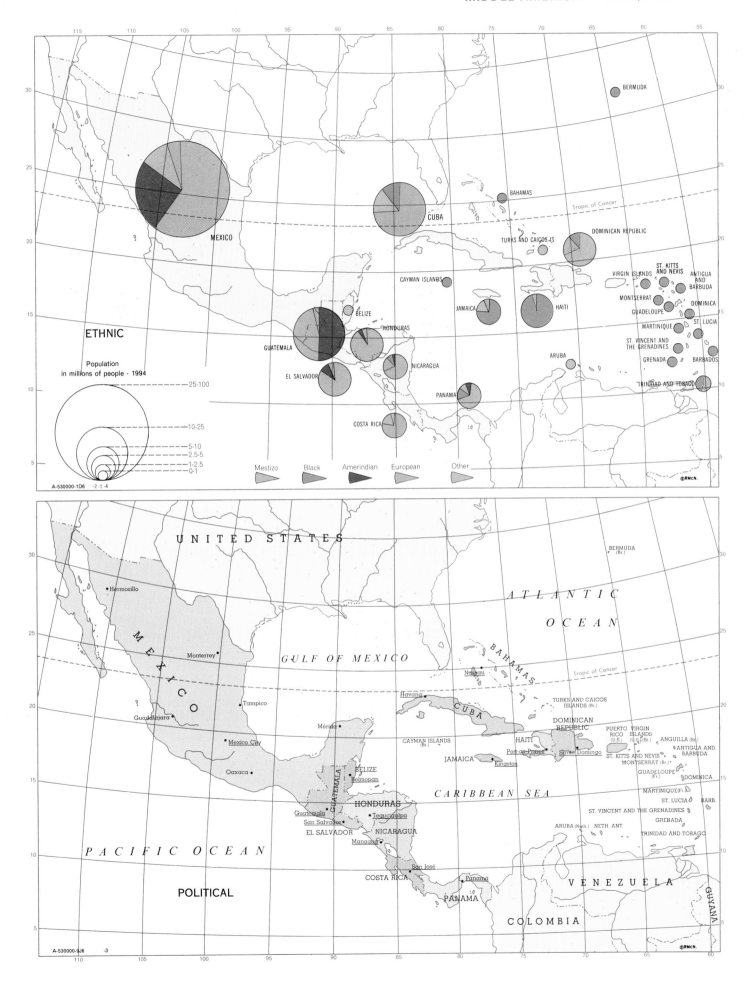

ETHNIC

Population
in millions of people - 1994

25-100

10-25

5-10
2.5-5
1-2.5
0-1

A-530000-1D6   -2 -2 -4

Mestizo   Black   Amerindian   European   Other

©RMCN.

MEXICO

BAHAMAS

CUBA

TURKS AND CAICOS IS.

DOMINICAN REPUBLIC

CAYMAN ISLANDS

VIRGIN ISLANDS   ST. KITTS
AND NEVIS   ANTIGUA
AND
BARBUDA

MONTSERRAT   DOMINICA

BELIZE   GUADELOUPE

HONDURAS   JAMAICA   HAITI   MARTINIQUE   ST. LUCIA

GUATEMALA   ST. VINCENT AND
THE GRENADINES

EL SALVADOR   NICARAGUA   ARUBA   GRENADA   BARBADOS

PANAMA   TRINIDAD AND TOBAGO

COSTA RICA

BERMUDA

Tropic of Cancer

POLITICAL

UNITED STATES

ATLANTIC

OCEAN

BERMUDA
(Br.)

Hermosillo

MEXICO

Monterrey

GULF OF MEXICO

BAHAMAS

Nassau

Tropic of Cancer

Tampico

TURKS AND CAICOS
ISLANDS (Br.)

Guadalajara

Havana

CUBA

DOMINICAN
REPUBLIC

PUERTO   VIRGIN
RICO   ISLANDS
(U.S.)   (U.S.)(Br.)   ANGUILLA (Br.)

Mérida

CAYMAN ISLANDS
(Br.)

HAITI   Santo Domingo

ANTIGUA AND
BARBUDA

Mexico City

Port-au-Prince

JAMAICA   Kingston

ST. KITTS AND NEVIS
MONTSERRAT (Br.)

Oaxaca

BELIZE   GUADELOUPE
(Fr.)

Belmopan

CARIBBEAN SEA   MARTINIQUE(Fr.)

GUATEMALA

HONDURAS   DOMINICA

Guatemala   Tegucigalpa   ST. LUCIA   BARB.

San Salvador   ST. VINCENT AND THE GRENADINES

EL SALVADOR   NICARAGUA   GRENADA

Managua   ARUBA (Neth.)   NETH. ANT.

PACIFIC OCEAN   TRINIDAD AND TOBAGO

San José   VENEZUELA

COSTA RICA   Panamá

POLITICAL   PANAMA   GUYANA

COLOMBIA

A-530000-9J6   -3   ©RMCN.

a

PANAMA

Scale 1:1 000 000
0   10 Miles
0 4 8 12 16 Kilometers

Scale 1:16 000 000; one inch to 250 miles. Polyconic Projection
Elevations and depressions are given in feet

A-530000-76g 9-27
COPYRIGHT BY
RAND McNALLY & COMPANY
MADE IN U.S.A.

b

ATLANTIC OCEAN

Arecibo   San Juan
Aguadilla   Bayamón   CABEZAS DE   ST. THOMAS   TORTOLA
PTA. HIGUERO   San Juan   SAN JUAN   (U.S.A.)   (Br.)
Utuado   Fajardo   Charlotte   CULEBRA   Amalie   ST. JOHN
PUERTO RICO   Caguas   Vieques   (U.S.A.)
Mayagüez   (U.S.A.)   Cayey   Humacao   VIEQUES
Coamo
CABO ROJO   Ponce   Salinas   Guayama

CARIBBEAN   SEA   Christiansted
SAINT CROIX   (U.S.A.)

Scale 1:4 000 000
0   10   20   30   40 Miles
0   10 20 30 40 50 60 Kilometers
©RMcN

c

65°   LITTLE   64°50'
OUTER BRASS   HANS LOLLICK
INNER BRASS   HANS LOLLICK
STORMY PT.   PICARA PT   THATCH CAY   GRASS   CAY
ST. THOMAS   (U.S.A.)   Crown Mt.   1558   18°
Charlotte Amalie   20'
WATER   (St. Thomas)   Nadir
FLAMINGO PT   St. Thomas   Harbor
©RMcN   Scale 1:500 000

Relief

Meters	Feet
3050	10 000
1525	5000
610	2000
305	1000
152.5	500
0	Sea Level
152.5	500
1525	5000
3050	10 000
6100	20 000

0   50   100   200   300   400   500 Miles
0   100   200   400   600   800 Kilometers

Cities and Towns
0 to 50,000   ○
50,000 to 500,000   ⊙
500,000 to 1,000,000   ◉
1,000,000 and over

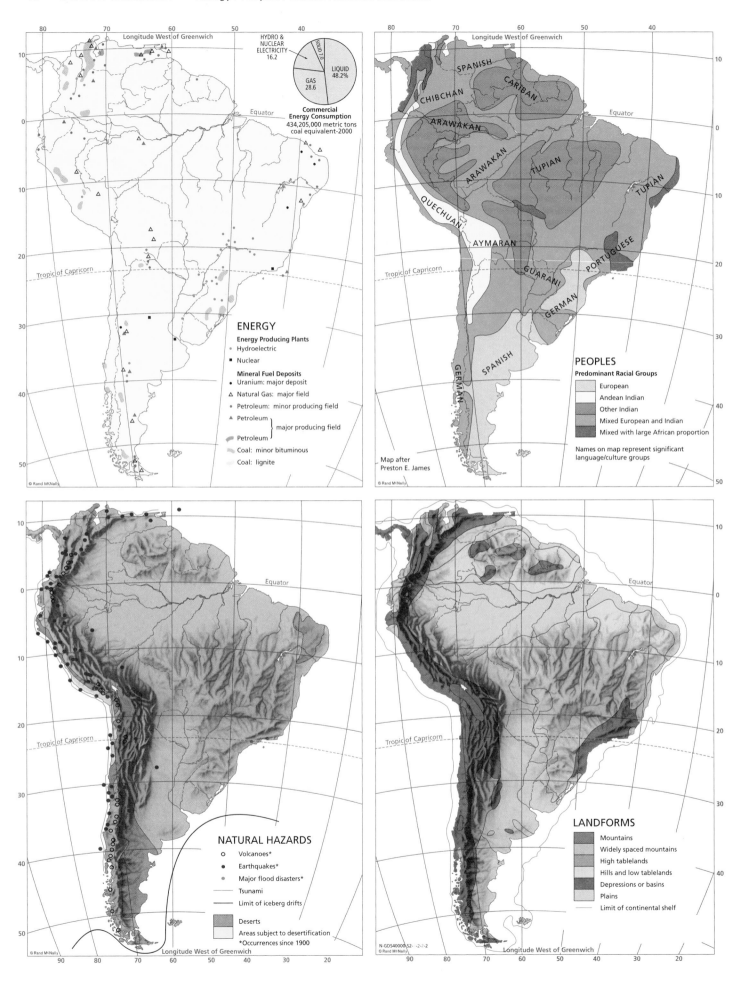

Longitude West of Greenwich

HYDRO &
NUCLEAR
ELECTRICITY
16.2

SOLID 7.0

GAS
28.6

LIQUID
48.2%

**Commercial
Energy Consumption**
434,205,000 metric tons
coal equivalent-2000

Equator

Tropic of Capricorn

© Rand McNally

ENERGY

Energy Producing Plants
- • Hydroelectric
- ■ Nuclear

Mineral Fuel Deposits
- • Uranium: major deposit
- △ Natural Gas: major field
- • Petroleum: minor producing field
- ▲ Petroleum: } major producing field
- ▬ Petroleum: }
- ▬ Coal: minor bituminous
- ▬ Coal: lignite

Longitude West of Greenwich

SPANISH

CHIBCHAN

CARIBAN

ARAWAKAN

ARAWAKAN

TUPIAN

TUPIAN

QUECHUAN

AYMARAN

PORTUGUESE

GUARANI

GERMAN

SPANISH

GERMAN

Equator

Tropic of Capricorn

Map after
Preston E. James

© Rand McNally

PEOPLES

Predominant Racial Groups
- ▢ European
- ▢ Andean Indian
- ▢ Other Indian
- ▢ Mixed European and Indian
- ▢ Mixed with large African proportion

Names on map represent significant
language/culture groups

Equator

Tropic of Capricorn

© Rand McNally

NATURAL HAZARDS

- ○ Volcanoes*
- • Earthquakes*
- • Major flood disasters*
- ⎯ Tsunami
- ▬ Limit of iceberg drifts
- ▢ Deserts
- ▢ Areas subject to desertification
- *Occurrences since 1900

Longitude West of Greenwich

Equator

Tropic of Capricorn

LANDFORMS

- ▢ Mountains
- ▢ Widely spaced mountains
- ▢ High tablelands
- ▢ Hills and low tablelands
- ▢ Depressions or basins
- ▢ Plains
- ⎯ Limit of continental shelf

N-GDS40000-S2- -2-/-2

© Rand McNally

Longitude West of Greenwich

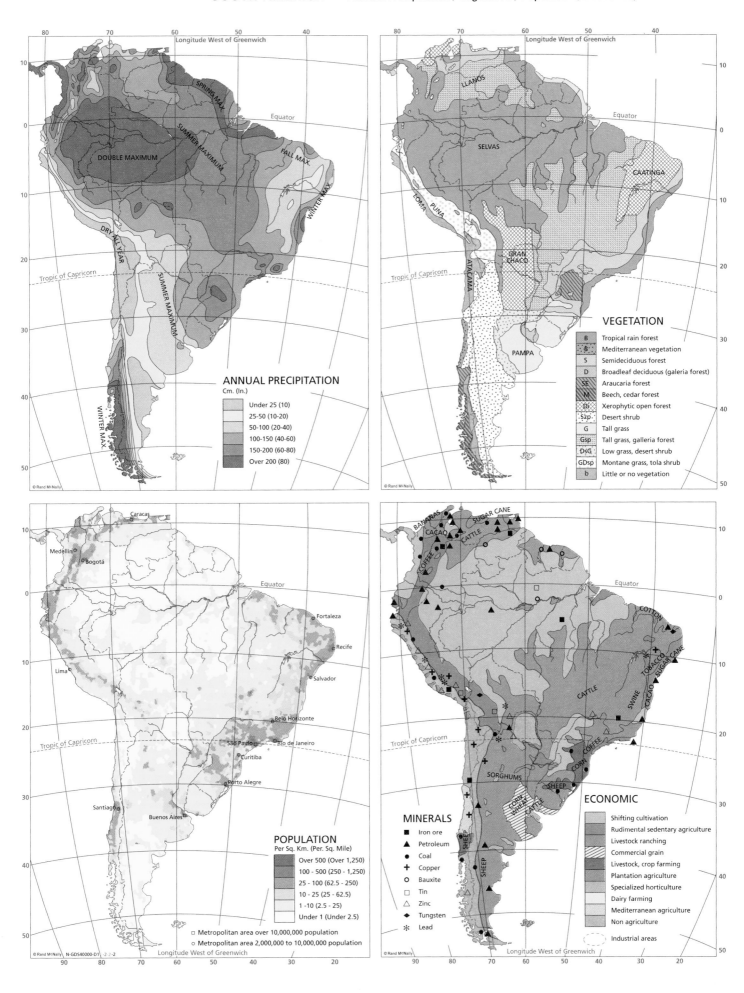

ANNUAL PRECIPITATION
Cm. (In.)

- Under 25 (10)
- 25-50 (10-20)
- 50-100 (20-40)
- 100-150 (40-60)
- 150-200 (60-80)
- Over 200 (80)

VEGETATION

B	Tropical rain forest
B	Mediterranean vegetation
S	Semideciduous forest
D	Broadleaf deciduous (galeria forest)
SE	Araucaria forest
M	Beech, cedar forest
Di	Xerophytic open forest
Szp	Desert shrub
G	Tall grass
Gsp	Tall grass, galleria forest
DsG	Low grass, desert shrub
GDsp	Montane grass, tola shrub
b	Little or no vegetation

POPULATION
Per Sq. Km. (Per. Sq. Mile)

- Over 500 (Over 1,250)
- 100 - 500 (250 - 1,250)
- 25 - 100 (62.5 - 250)
- 10 - 25 (25 - 62.5)
- 1 -10 (2.5 - 25)
- Under 1 (Under 2.5)

□ Metropolitan area over 10,000,000 population
o Metropolitan area 2,000,000 to 10,000,000 population

MINERALS

- ■ Iron ore
- ▲ Petroleum
- ● Coal
- + Copper
- ○ Bauxite
- □ Tin
- △ Zinc
- ◆ Tungsten
- ✳ Lead

ECONOMIC

- Shifting cultivation
- Rudimental sedentary agriculture
- Livestock ranching
- Commercial grain
- Livestock, crop farming
- Plantation agriculture
- Specialized horticulture
- Dairy farming
- Mediterranean agriculture
- Non agriculture

- Industrial areas

Tropic of Cancer

HAVANA

Bahía de Campeche

PEN. DE YUCATÁN

Yucatán Channel

CUBA

KEY WEST

HISPANIOLA

JAMAICA

Gulf of Honduras

PUERTO RICO TRENCH

PUERTO RICO (U.S.A.)

San Juan

GUADELOUPE (Fr.)

MARTINIQUE (Fr.)

NORTH AMERICAN BASIN

ATLANTIC OCEAN

CARIBBEAN SEA

WEST INDIES

BARBADOS

CENTRAL

Lago de Nicaragua

AMERICA

Panamá

PUNTA DE GALLINAS

Golfo del Darién

IST. DEO. PAN.

Golfo de Panamá

Barranquilla
Cartagena

Maracaibo

Valencia

La Guaira

CARACAS

Mérida

Ciudad Bolívar

TRINIDAD AND TOBAGO

Port of Spain

ISLA DEL COCO (Costa Rica)

ISLA DE MALPELO (Colombia)

Medellín

LLANOS

VENEZUELA

Cerro Icutú 7800

Georgetown

Paramaribo

Cayenne

BOGOTÁ

GUYANA

SURINAME FR. GUIANA

Nevado del Tolima 17 110

Boa Vista do Rio Branco

GUIANA HIGHLANDS

COLOMBIA

Guaviare

Quito

Cotopaxi 19 347

Guaviare

ILHA DE MARAJÓ

Equator

ROCEDOS SÃO PEDRO E SÃO PAULO (Brazil)

ECUADOR

Guayaquil

Chimborazo 20 702

Japurá

Negro

Manaus (Manáos)

Amazon (Amazonas)

Belém (Pará)

São Luís (Maranhão)

ARCHIPIÉLAGO DE COLÓN (GALÁPAGOS ISLANDS) (Ec.)

Putumayo

Iquitos

Leticia

Solimões

Fortaleza (Ceará)

ARQUIPÉLAGO FERNANDO DE NORONHA (Brazil)

Golfo de Guayaquil

Juruá

Purus

Madeira

Tapajós

Xingu

Teresina

Natal

Chiclayo

Trujillo

Porto Velho

Rio

João Pessoa (Paraíba)

CABO DE SÃO ROQUE

RECIFE (Pernambuco)

Nevs. Huascarán 22 133

Rio Branco

B R A Z I L

SERRADO

Maceió

PERU

LIMA

ANDES

Cusco

Volcán Misti

CHAPADA DE MATO GROSSO

Cuiabá

Brasília

Salvador (Bahia)

Callao

Arequipa

La Paz

Nev. Illimani 20 741

BOLIVIA

Sucre

Potosí

Diamantina

Belo Horizonte

Pico da Bandeira 9482

BRAZILIAN HIGHLANDS

Mollendo

Lago Titicaca

Iquique

CHACO

PARAGUAY

Vitória

Antofagasta

Salta

Bermejo

Asunción

SÃO PAULO

CABO FRIO

Tropic of Capricorn

ISLA DE SAN FÉLIX (Chile)

ISLA DE SAN AMBROSIO (Chile)

Cerro Azul Copiapó

Tucumán

GRAN

Iguassú Falls

Santos

RIO DE JANEIRO

Copiapó

Corrientes

ARGENTINA

Florianópolis

Coquimbo

ISLAS DE JUAN FERNÁNDEZ (Chile)

Córdoba

Santa Fe

Salto

Porto Alegre

Valparaíso

SANTIAGO

Mendoza

Rosario

URUGUAY

Rio Grande

Concepción

BUENOS AIRES

La Plata

MONTEVIDEO

PAMPAS

Río de la Plata

ATLANTIC OCEAN

Bahía Blanca

Valdivia

Colorado

Viedma

Puerto Montt

Golfo San Matias

ISLA DE CHILOÉ

ARCHIPIÉLAGO DE LOS CHONOS

Comodoro Rivadavia

Golfo San Jorge

Monte Valentín 9314

PACIFIC OCEAN

WELLINGTON

FALKLAND IS. (ISLAS MALVINAS) (Br.)

HANOVER

Río Gallegos

Stanley

Punta Arenas

Estrecho de Magallanes

DESOLACIÓN

TIERRA DEL FUEGO

Mt. Sarmiento 8100

ISLA DE LOS ESTADOS

CABO DE HORNOS (CAPE HORN)

Drake Passage

SOUTH GEORGIA (Br.)

SOUTH SANDWICH TRENCH

SOUTH ORKNEY IS. (Br.)

SOUTH SHETLAND ISLANDS (Br.)

JOINVILLE

SOUTH SANDWICH ISLANDS (Br.)

ANTARCTIC PENINSULA

JAMES ROSS

Longitude West of Greenwich

Antarctic Circle

40,000 SQ MI AREA

0   300   600
Miles

0   200   400   600   800   1000 Miles
0   400   800   1200   1600 Kilometers

Scale 1:40,000,000; one inch to 630 miles. Lambert's Azimuthal, Equal Area Projection
Elevations and depressions are given in feet

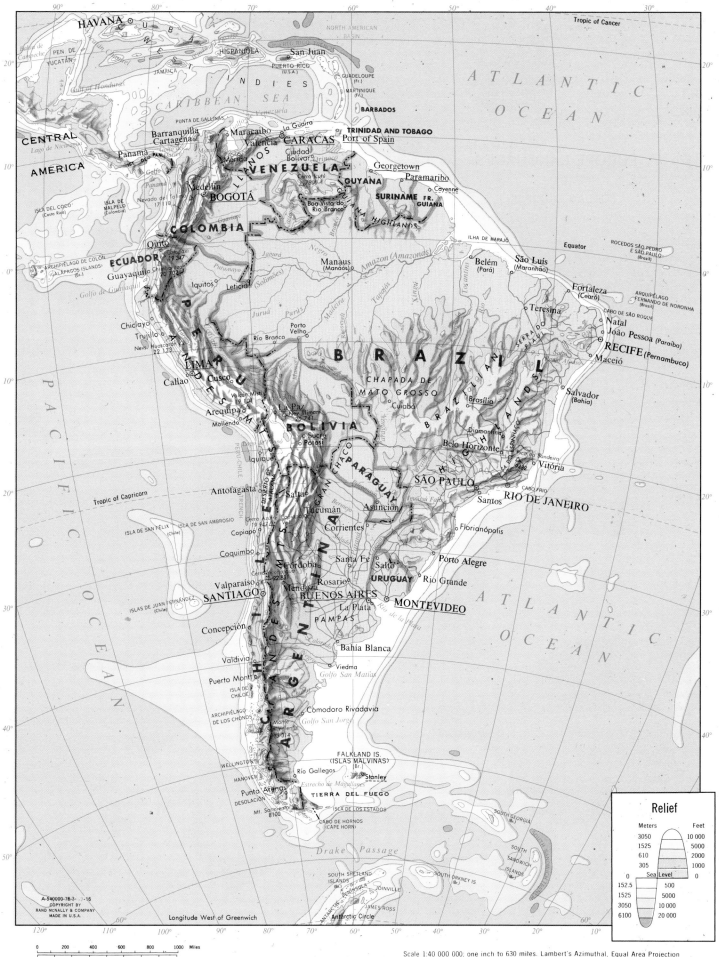

HAVANA · CUBA

PEN. DE YUCATÁN

Bahía de Campeche

HISPANIOLA
San Juan
PUERTO RICO (U.S.A.)
JAMAICA
GUADELOUPE (Fr.)
MARTINIQUE (Fr.)

WEST INDIES

CARIBBEAN SEA

NORTH AMERICAN BASIN

ATLANTIC OCEAN

Tropic of Cancer

Gulf of Honduras

CENTRAL

AMERICA

Lago de Nicaragua

ISLA DEL COCO (Costa Rica)

Panamá

Golfo de Panamá

ISLA DE MALPELO (Colombia)

ARCHIPIÉLAGO DE COLÓN (GALÁPAGOS ISLANDS) (Ec.)

PACIFIC OCEAN

PUNTA DE GALLINAS

Barranquilla
Cartagena
Maracaibo
Valencia   CARACAS
Mérida
BOGOTÁ
Medellín

VENEZUELA

Ciudad Bolívar
Cerro Icutú 7900

BARBADOS
TRINIDAD AND TOBAGO
Port of Spain

La Guaira

Georgetown
Paramaribo
Cayenne

GUYANA
SURINAME
FR. GUIANA

GUIANA HIGHLANDS

COLOMBIA

Nevado del Tolima 17,110

Guaviare

Quito
Cotopaxi 19,347
ECUADOR
Chimborazo 20,702
Guayaquil
Golfo de Guayaquil

Putumayo
Iquitos
Leticia

Manaus (Manáos)
Amazon (Amazonas)
Negro

Belém (Pará)
São Luís (Maranhão)

ILHA DE MARAJÓ

Equator
ROCEDOS SÃO PEDRO E SÃO PAULO (Brazil)

Japurá
(Solimões)

Chiclayo
Trujillo
Nevs. Huascarán 22,132

Jurúa   Purús
Porto Velho
Río Branco

Madeira
Tapajós
Xingu
Tocantins

Fortaleza (Ceará)
ARQUIPÉLAGO FERNANDO DE NORONHA (Brazil)

Teresina

CABO DE SÃO ROQUE
Natal
João Pessoa (Paraíba)
RECIFE (Pernambuco)
Maceió

PERU

LIMA
Callao
Cusco
Volcán Misti 19,101
Arequipa
Mollendo

BRAZIL

CHAPADA DE MATO GROSSO

Cuiabá

SERRA DO ESPINHAÇO

Salvador (Bahia)

Diamantina

BOLIVIA

La Paz
Nev. Illimani
Sucre
Potosí

Brasília

Belo Horizonte
Pico da Bandeira 9492
Vitória

BRAZILIAN HIGHLANDS

PERU-CHILE TRENCH

Iquique

GRAN CHACO

PARAGUAY

SÃO PAULO
CABO FRIO
Santos
RIO DE JANEIRO

Tropic of Capricorn
Antofagasta

ISLA DE SAN FÉLIX (Chile)
ISLA DE SAN AMBROSIO (Chile)

Cerro Azul 19,947
Copiapó
Copiapó

CHACO

Salta
Tucumán
Asunción

Corrientes

Iguassú Falls

Florianópolis

Coquimbo

CÓRDOBA

Santa Fe
Salto
Rosario
Porto Alegre
Río Grande

Valparaíso
SANTIAGO
Mendoza
Cerro Aconcagua 22,831
BUENOS AIRES
La Plata

URUGUAY
MONTEVIDEO

ISLAS DE JUAN FERNÁNDEZ (Chile)

Concepción

PAMPAS

Colorado

Valdivia

ARGENTINA

Bahía Blanca
Viedma
Golfo San Matías

Puerto Montt
ISLA DE CHILOÉ

ARCHIPIÉLAGO DE LOS CHONOS

Monte Valentín 13,314

Comodoro Rivadavia
Golfo San Jorge

FALKLAND IS. (ISLAS MALVINAS) (Br.)

WELLINGTON
HANOVER
Río Gallegos
Stanley

Punta Arenas
DESOLACIÓN
Mt. Sarmiento 8100

Estrecho de Magallanes

ISLA DE LOS ESTADOS

TIERRA DEL FUEGO

SOUTH GEORGIA (Br.)

CABO DE HORNOS (CAPE HORN)

Drake Passage

SOUTH SANDWICH ISLANDS

SOUTH ORKNEY IS (Br.)

SOUTH SHETLAND ISLANDS (Br.)

JOINVILLE
JAMES ROSS

ATLANTIC OCEAN

Antarctic Circle

Longitude West of Greenwich

Relief	
Meters	Feet
3050	10 000
1525	5000
610	2000
305	1000
0	Sea Level	0
152.5	500
1525	5000
3050	10 000
6100	20 000

Miles
0   200   400   600   800   1000

Kilometers
0   400   800   1200   1600

Scale 1:40 000 000; one inch to 630 miles. Lambert's Azimuthal, Equal Area Projection
Elevations and depressions are given in feet

EUROPE LANGUAGES
BY
BOGDAN ZABORSKI

Scale 1:16,500,000; one inch to 260 miles Conic Projection

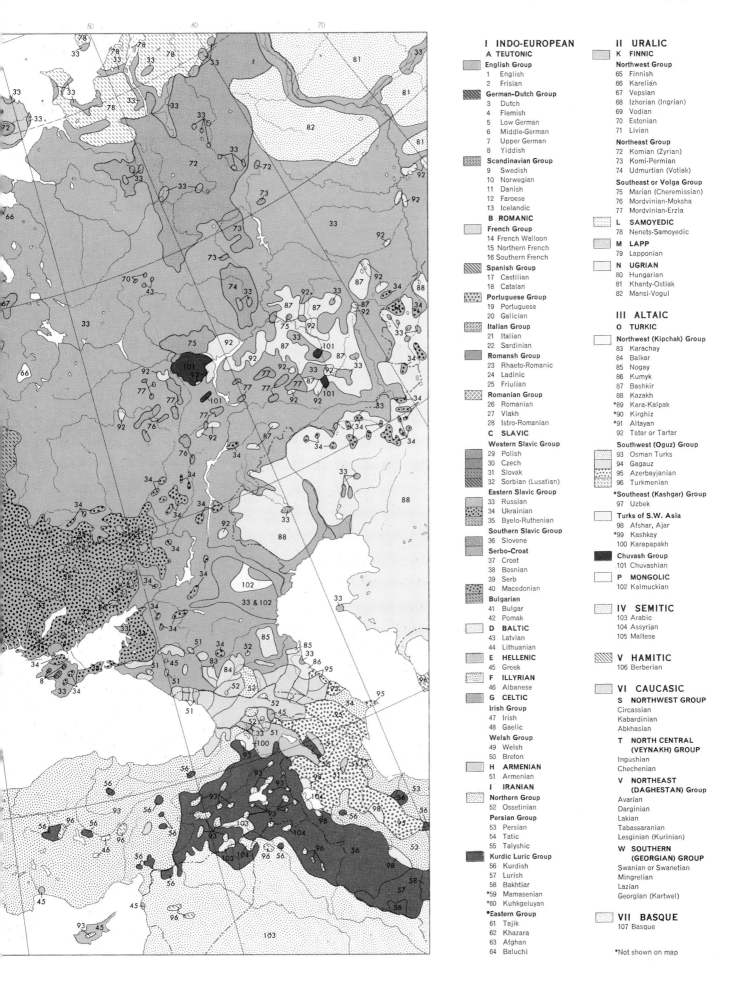

I INDO-EUROPEAN

A TEUTONIC
English Group
1 English
2 Frisian

German-Dutch Group
3 Dutch
4 Flemish
5 Low German
6 Middle-German
7 Upper German
8 Yiddish

Scandinavian Group
9 Swedish
10 Norwegian
11 Danish
12 Faroese
13 Icelandic

B ROMANIC
French Group
14 French Walloon
15 Northern French
16 Southern French

Spanish Group
17 Castilian
18 Catalan

Portuguese Group
19 Portuguese
20 Galician

Italian Group
21 Italian
22 Sardinian

Romansh Group
23 Rhaeto-Romanic
24 Ladinic
25 Friulian

Romanian Group
26 Romanian
27 Vlakh
28 Istro-Romanian

C SLAVIC
Western Slavic Group
29 Polish
30 Czech
31 Slovak
32 Sorbian (Lusatian)

Eastern Slavic Group
33 Russian
34 Ukrainian
35 Byelo-Ruthenian

Southern Slavic Group
36 Slovene

Serbo-Croat
37 Croat
38 Bosnian
39 Serb
40 Macedonian

Bulgarian
41 Bulgar
42 Pomak

D BALTIC
43 Latvian
44 Lithuanian

E HELLENIC
45 Greek

F ILLYRIAN
46 Albanese

G CELTIC
Irish Group
47 Irish
48 Gaelic

Welsh Group
49 Welsh
50 Breton

H ARMENIAN
51 Armenian

I IRANIAN
Northern Group
52 Ossetinian

Persian Group
53 Persian
54 Tatic
55 Talyshic

Kurdic Luric Group
56 Kurdish
57 Lurish
58 Bakhtiar
*59 Mamasenian
*60 Kuhkgeluyan

*Eastern Group
61 Tajik
62 Khazara
63 Afghan
64 Baluchi

II URALIC

K FINNIC
Northwest Group
65 Finnish
66 Karelian
67 Vepsian
68 Izhorian (Ingrian)
69 Vodian
70 Estonian
71 Livian

Northeast Group
72 Komian (Zyrian)
73 Komi-Permian
74 Udmurtian (Votiak)

Southeast or Volga Group
75 Marian (Cheremissian)
76 Mordvinian-Moksha
77 Mordvinian-Erzia

L SAMOYEDIC
78 Nenets-Samoyedic

M LAPP
79 Lapponian

N UGRIAN
80 Hungarian
81 Khanty-Ostiak
82 Mansi-Vogul

III ALTAIC

O TURKIC
Northwest (Kipchak) Group
83 Karachay
84 Balkar
85 Nogay
86 Kumyk
87 Bashkir
88 Kazakh
*89 Kara-Kalpak
*90 Kirghiz
*91 Altayan
92 Tatar or Tartar

Southwest (Oguz) Group
93 Osman Turks
94 Gagauz
95 Azerbayjanian
96 Turkmenian

*Southeast (Kashgar) Group
97 Uzbek

Turks of S.W. Asia
98 Afshar, Ajar
*99 Kashkey
100 Karapapakh

Chuvash Group
101 Chuvashian

P MONGOLIC
102 Kalmuckian

IV SEMITIC
103 Arabic
104 Assyrian
105 Maltese

V HAMITIC
106 Berberian

VI CAUCASIC

S NORTHWEST GROUP
Circassian
Kabardinian
Abkhasian

T NORTH CENTRAL (VEYNAKH) GROUP
Ingushian
Chechenian

V NORTHEAST (DAGHESTAN) Group
Avarian
Darginian
Lakian
Tabassaranian
Lesginian (Kurinian)

W SOUTHERN (GEORGIAN) GROUP
Swanian or Swanetian
Mingrelian
Lazian
Georgian (Kartwel)

VII BASQUE
107 Basque

*Not shown on map

40,000 SQ MI
AREA

0   100   200
Miles

Scale 1: 16 000 000; one inch to 250 miles. Conic Projection
Elevations and depressions are given in feet

Longitude West of Greenwich   0°   Longitude East of Greenwich

0 50 100   200   300   400   500 Miles
0   100   200   400   600   800 Kilometers

52

Scale 1:10 000 000; one inch to 160 miles. Bonne's Projection
Elevations and depressions are given in feet

The Turkish Republic of Northern Cyprus
unilaterally declared its independence
on Nov. 15, 1983.

Areas occupied by Israel since 1967.

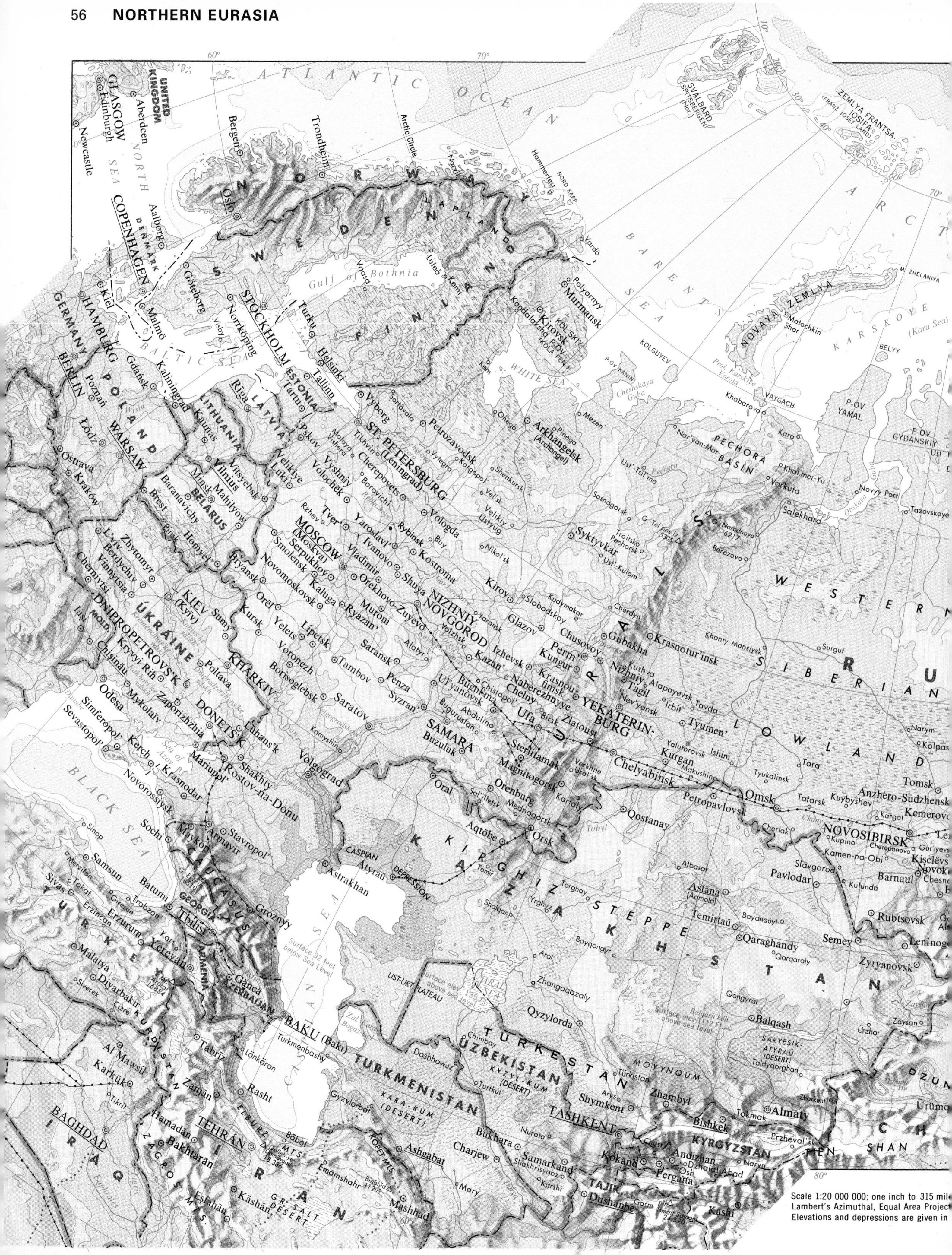

ATLANTIC OCEAN

ARCTIC OCEAN

SVALBARD (SPITSBERGEN) (NOR.)

ZEMLYA FRANTSA IOSIFA (FRANZ JOSEF LAND)

NOVAYA ZEMLYA

KARSKOYE (Kara Sea)

BARENTS SEA

Arctic Circle

NORTH SEA

UNITED KINGDOM

Glasgow
Edinburgh
Aberdeen
Newcastle

GERMANY
HAMBURG
BERLIN
Kiel
DENMARK
COPENHAGEN
Aalborg
Göteborg
Malmö

Bergen
Trondheim
Oslo
Stavanger

N O R W A Y
S W E D E N
STOCKHOLM
Norrköping

FINLAND
Helsinki
Turku
Tallinn
LAPLAND

Hammerfest
NORD KAPP
Vardø
Murmansk
Polyarnyy
Kirovsk
KOLA P-OV (KOLA PEN.)
Kandalaksha

WHITE SEA
Arkhangelsk (Archangel)

POLAND
WARSAW
Poznań
Gdańsk
Łódź
Kraków
Ostrava

BALTIC SEA
Kaliningrad
LITHUANIA
Kaunas
Vilnius
Minsk
BELARUS
Baranavichy
Brest
Pinsk
Homyel'

LATVIA
Rīga
ESTONIA
Tartu
Pskov
Velikiye Luki

St. Petersburg (Leningrad)
Vyborg
Cherepovets
Vologda
Tver'
Rybinsk
Yaroslavl'
Ivanovo
Kostroma
MOSCOW (Moskva)
Serpukhov
Smolensk
Kaluga
Ryazan'
Vladimir
Orekhovo-Zuyevo
Murom

Petrozavodsk
ONEGA

Syktyvkar
Ukhta
Pechora

PECHORA BASIN
Vorkuta
Nar'yan-Mar

YAMAL P-OV
GYDANSKIY P-OV
Salekhard
Berezovo

U R A L

WESTERN SIBERIAN LOWLAND

UKRAINE
KIEV (Kyiv)
L'viv
Zhytomyr
Vinnytsia
Chernivtsi
Chișinău
MOLDOVA
Iași
Odesa
Mykolaiv
DNIPROPETROVSK
Kryvyi Rih
Zaporizhzhia
KHARKIV
Sumy
Poltava
DONETSK
Luhans'k
Mariupol
Shakhty
Rostov-na-Donu

Bryansk
Orël
Kursk
Yelets
Lipetsk
Voronezh
Borisoglebsk
Tambov
Penza
Saransk
Saratov
Syzran'
Ul'yanovsk

NIZHNIY NOVGOROD
Kirov
Glazov
Izhevsk
Kazan'
Perm'
Kungur
Chusovoy
Gubakha
Krasnotur'insk
Khanty Mansiysk
Surgut

YEKATERINBURG
Nizhniy Tagil
Neyva
Zlatoust
Ufa
Chelny
Naberezhnyye
Sterlitamak
Magnitogorsk
Tyumen'
Irbit
Tavda
Ishim
Tobol'sk
Narym
Kolpas

SAMARA
Buzuluk
Oral
Orenburg
Orsk
Qostanay
Petropavlovsk
Chelyabinsk
Kurgan
Omsk
Tatarsk
Kuybyshev
Barabinsk
Anzhero-Sudzhensk
Tomsk
Kemerovo
NOVOSIBIRSK

BLACK SEA
Sevastopol'
Simferopol'
Kerch
Krasnodar
Novorossiysk
Sochi

CAUCASUS
Stavropol'
Armavir
Maykop
Grozny
GEORGIA
Tbilisi
Batumi
Sukhumi
ARMENIA
Yerevan
AZERBAIJAN
BAKU (Baki)
Gəncə

TURKEY
Samsun
Trabzon
Sivas
Tokat
Giresun
Erzurum
Erzincan
Malatya
Diyarbakır
Kars

CASPIAN SEA
Astrakhan
Atyraū
CASPIAN DEPRESSION
Volgograd
Kamyshin

KAZAKHSTAN
KIRGHIZ STEPPE
Aqtöbe
Aral
Shalqar
Torghay
Arys
Astana (Aqmola)
Qaraghandy
Temirtaū
Balqash
Balqash koli
Pavlodar
Semey
Barnaul
Rubtsovsk

UZBEKISTAN
TASHKENT
Nukus
Bukhara
Samarkand
Qyzylorda
Zhangaqazaly
KYZYL KUM (DESERT)
TURKESTAN
Shymkent
Zhambyl
Almaty
Bishkek
KYRGYZSTAN
Osh
Fergana
Andijon
Kokand

TURKMENISTAN
Dashhowuz
Turkmenbashy
KARA-KUM (DESERT)
Ashgabat
Mary
Charjew

UST-URT PLATEAU

ARAL SEA

TAJIKISTAN
Dushanbe

IRAN
TEHRĀN
Tabriz
Rasht
Hamadān
Mashhad
ELBURZ MTS.
ZAGROS MTS.

IRAQ
BAGHDAD
Al Mawṣil
Karkūk

TIEN SHAN

DZUNGARIA
Ürümqi

CHINA

Scale 1:20 000 000; one inch to 315 miles
Lambert's Azimuthal, Equal Area Projection
Elevations and depressions are given in feet

Relief

Meters		Feet
3050		10 000
1525		5000
610		2000
305		1000
152.5		500
0	Sea Level	0
152.5		Below 500
1525		Sea Level 5000
3050		10 000

ARCTIC OCEAN

SEVERNAYA ZEMLYA
(NORTHERN LAND)

M. CHELYUSKIN

P-OV
GORY
TAYMYR
BYRRANGA

Nordvik

Taymyr
Khatangskiy
Zaliv.

BOL'SHOY
BEGICHEV

Ust'-Olenek

Khatanga

Noril'sk

GORY
PUTORANA

garka

Turukhansk

Nizhnyaya Tunguska

Yartsevo

G. Polkan
3543

Yeniseysk

S S I A

ETSK Krasnoyarsk
Bogotol
Kansk Tayshet Bratsk
Balakhta
uznetski Nizhneudinsk Tulun
Piramida
10801
Minusinsk Cheremkhovo
Abakan Munku Angarsk
Sordyk
SAYAN 11457 Irkutsk
KHREBET Kyren
SAYAN
TANNU-OLA Petrovsk-Zabaykal'skiy
Ust' Gorodok Kyakhta
Nuur

Har Us Nuur
Hovd
Uliastay
HANGAYN NURUU
KHANGAIN NURUU
MTS.
Tsast Bogd
13419

MONGOLIA

N A

Hami

LAPTEV SEA

DE-LONGA
FADDEYA
NOVAYA SIBIR'
NOVOSIBIRSKIYE O-VA
(NEW SIBERIAN ISLANDS)
KOTEL'NYY MALYY LYAKHOVSKIYE
LYAKHOVSKIYE
M. SVYATOY
NOS
M. BIOR-
KHAYA

Tiksi

Bulun

Olenek

Zhigansk

VERKHOYANSKIY KHREBET

Verkhoyansk

Abyy

Lena

Vilyuysk

Suntar

Vilyuy

Yakutsk

Olëkminsk

Aldan

Tommot

Muknuyy
Peleduy
Vitim
PATOM
PLATEAU
Bodaybo
Golets-
Purpula
5377
Golets-
Skalistyy
9186

Kirensk

Ilimsk

Nizhne-Angarsk

Zhigalovo
Kachuga
Borguzin

BAYKAL'SKIY KHREBET

Oz. Baykal
(Lake Baikal)
Surface elev 1535 ft
above sea level

Ulan-Ude

YABLONOVYY KHREBET

Aginskoye

Aksha

Ulan Bator
(Ulaanbaatar)

Öndörhaan

Selenge

Orhon

Kerulen

Sayr Usa

GOBI OR SHAMO
(DESERT)

Zhangjiakou

Fengzhen

BEIJING

Baoding

EAST SIBERIAN SEA

M. SHELAGSKIY
AYON
M. BILLINGSA
Ambarchik

VRANGELYA
(WRANGEL)

M. SHMIDTA

Arctic Circle

Nizhne-Kolymsk

Sredne-
Kolymsk

Zashiversk

Zyryanka

KHREBET
CHERSKOGO

Gora Chen
15194

Omyakon

Kolyma

Aldanskoye

Ust'-Maya

Amga

Aldan

Nel'kan

Ayan

Chumikan

STANOVOY KHREBET

Tyndinskiy

Skovorodino

Zeya

Svobodnyy

Belogorsk
Ust' Tyrma

Amur

Blagoveshchensk

Nenjiang

Goukou

Qiqihar

Suihua

Tao'an

Jarud Qi

Shuangliao

Fuyu

CHANGCHUN

MANCHURIA

Jilin

CHINA

Chifeng

Weichang

Chengde

SHENYANG FUSHUN

Lüshun Dalian

Bo
Hai

SHANDONG
BANDAO

YELLOW SEA

Korea Bay

CHUKOTSKOYE NAGORYE

M. SHELAGSKIY
CHUKOTSKIY
P-OV

Anadyr

M. OL'UTORSKIY

Penzhino

Grhiga

Tilichiki

KORYAKSKIY KHREBET

KARAGIN

M. OZERNOY

P-OV
KAMCHATKA

Klyuchevskaya
15584

Petropavlovsk-
Kamchatskiy

Ust'-Bol'sheretsk

SEA OF OKHOTSK

M. ALEVINA
Magadan
Okhotsk

DZHUGDZHUR KHREBET

M. TAYGONOS

Yamsk

M. YELIZAVETY

Okha

SAKHALIN

Aleksandrovsk

Poronaysk

Uglegorsk

M. TERPENIYA

Yuzhno-Sakhalinsk

Korsakov

Nikolayevsk-na-Amure

KHREBET BUREINSKIY

Komsomol'sk
na-Amure

Birobidzhan

Khabarovsk

Ussuriysk

Vladivostok

Nakhodka

Partizansk

Artëm

SIKHOTE ALIN'

Tatar Strait

Kholmsk

Wakkanai

HOKKAIDŌ

Otaru Sapporo

Asahi

J A P A N

HONSHŪ

Kanazawa

KYOTO
KŌBE
OSAKA

Kanazawa

Matsue Tottori

Hiroshima Okayama

Kōchi

PUSAN

NORTH
KOREA

Pyongyang

Kaesong

SEOUL SOUTH
KOREA

Taegu

SEA OF JAPAN

Najin

Chongjin

Hunchun

Dunhua

Mudanjiang

HARBIN

Longitude East of Greenwich

0   100 200 300 400 500 600 Miles
0 200 400 600 800 1000 Kilometers

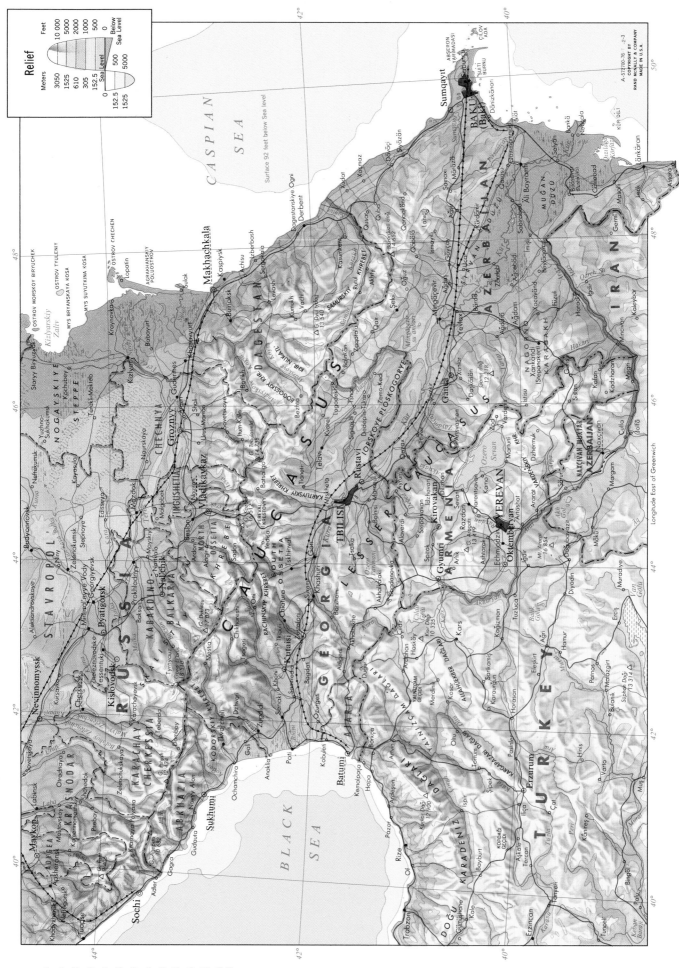

Relief

Meters	Feet
3050	10 000
1525	5000
610	2000
305	1000
152.5	500
0	Sea Level
	0 Below Sea Level

| 152.5 | 500 |
| 1525 | 5000 |

A-572700-76
COPYRIGHT BY
RAND McNALLY & COMPANY
MADE IN U.S.A.

CASPIAN SEA

Surface 92 feet below Sea level

BLACK SEA

Longitude East of Greenwich

Scale 1:4 000 000; one inch to 64 miles. Conic Projection
Elevations and depressions are given in feet

| 0 | 10 | 20 | 30 | 40 | 50 | 60 | 70 | 80 | 90 | 100 | 110 | 120 Miles |

| 0 | 20 | 40 | 60 | 80 | 100 | 120 | 140 | 160 | 180 | 200 Kilometers |

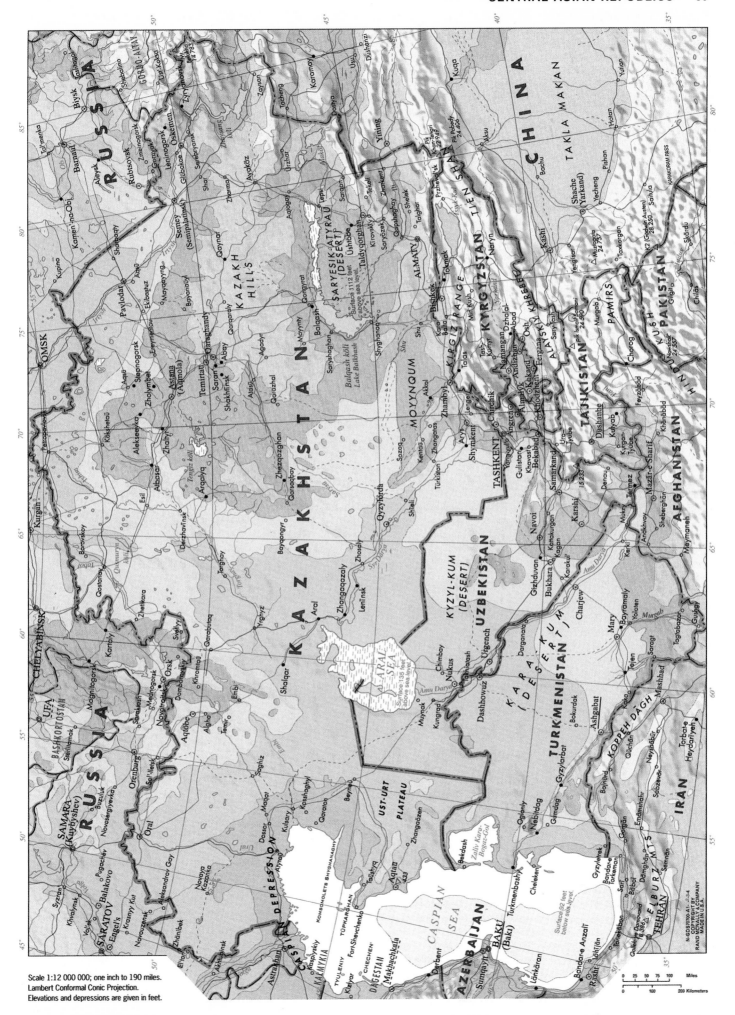

Scale 1:12 000 000; one inch to 190 miles.
Lambert Conformal Conic Projection.
Elevations and depressions are given in feet.

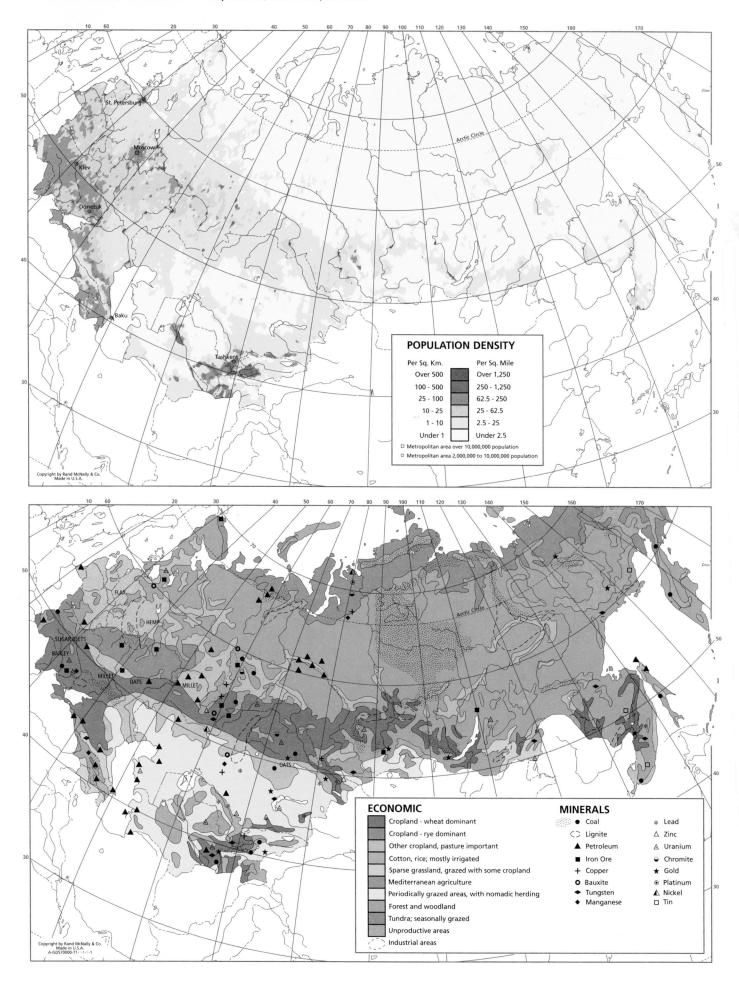

POPULATION DENSITY

Per Sq. Km.	Per Sq. Mile
Over 500	Over 1,250
100 - 500	250 - 1,250
25 - 100	62.5 - 250
10 - 25	25 - 62.5
1 - 10	2.5 - 25
Under 1	Under 2.5

□ Metropolitan area over 10,000,000 population
○ Metropolitan area 2,000,000 to 10,000,000 population

Copyright by Rand McNally & Co.
Made in U.S.A.

St. Petersburg
Moscow
Kiev
Donetsk
Baku
Tashkent

ECONOMIC

Cropland - wheat dominant
Cropland - rye dominant
Other cropland, pasture important
Cotton, rice; mostly irrigated
Sparse grassland, grazed with some cropland
Mediterranean agriculture
Periodically grazed areas, with nomadic herding
Forest and woodland
Tundra; seasonally grazed
Unproductive areas
⌐ Industrial areas

MINERALS

●	Coal	✳	Lead
◖	Lignite	△	Zinc
▲	Petroleum	△	Uranium
■	Iron Ore	◖	Chromite
+	Copper	★	Gold
○	Bauxite	◉	Platinum
◆	Tungsten	▲	Nickel
◆	Manganese	□	Tin

FLAX
HEMP
SUGAR BEETS
BARLEY
MILLET
OATS
MILLET
OATS

Copyright by Rand McNally & Co.
Made in U.S.A.
A-GDS70000-T1- -1-1-1

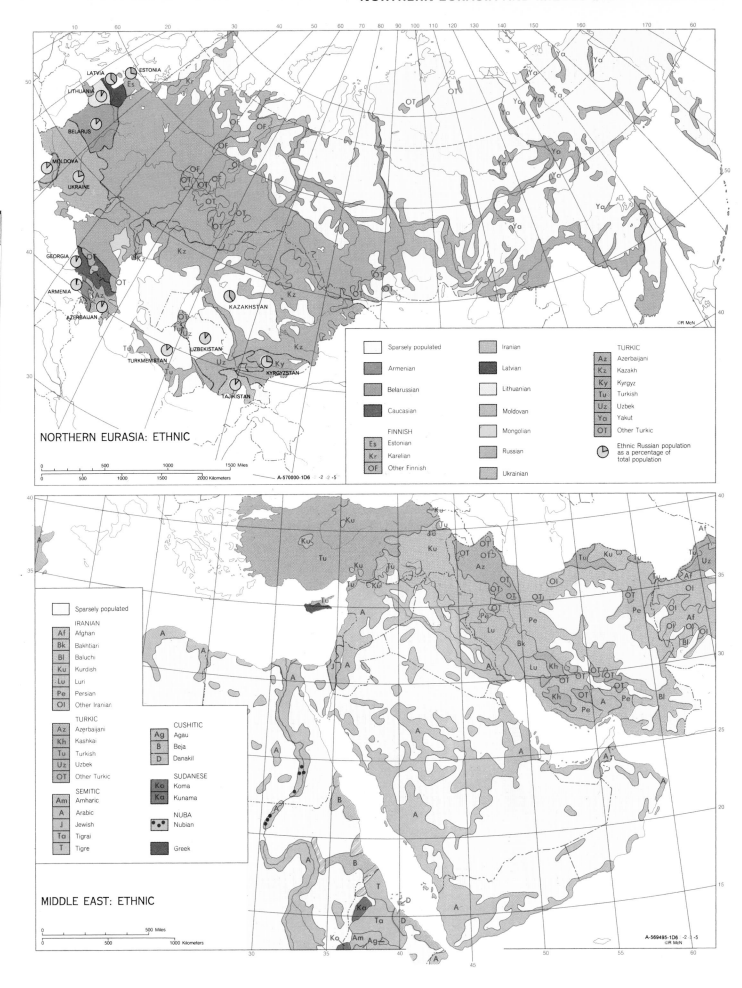

NORTHERN EURASIA: ETHNIC

MIDDLE EAST: ETHNIC

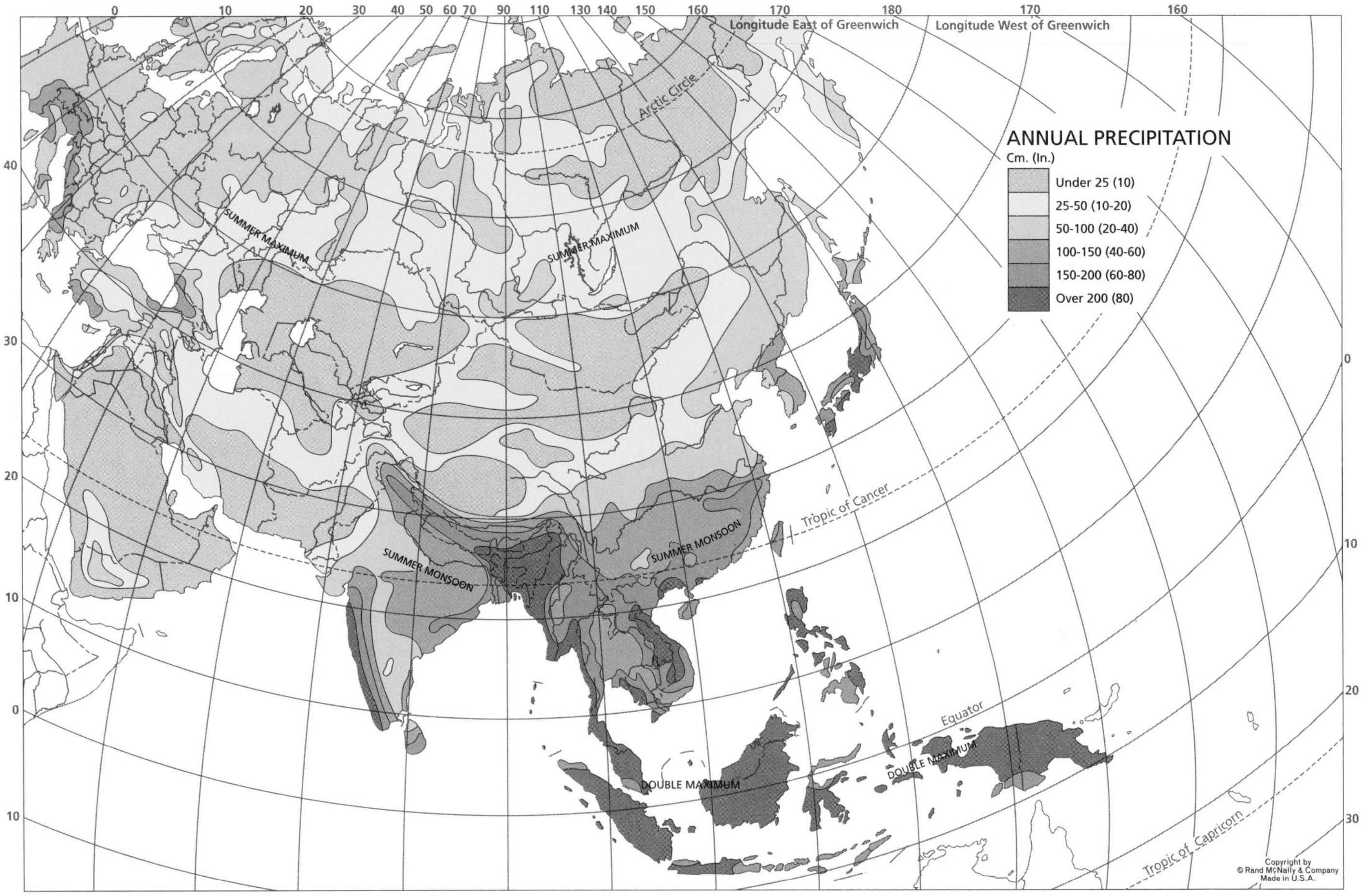

ANNUAL PRECIPITATION

Cm. (In.)

- Under 25 (10)
- 25-50 (10-20)
- 50-100 (20-40)
- 100-150 (40-60)
- 150-200 (60-80)
- Over 200 (80)

SUMMER MAXIMUM

SUMMER MAXIMUM

SUMMER MONSOON

SUMMER MONSOON

DOUBLE MAXIMUM

DOUBLE MAXIMUM

Longitude East of Greenwich   Longitude West of Greenwich

Arctic Circle

Tropic of Cancer

Equator

Tropic of Capricorn

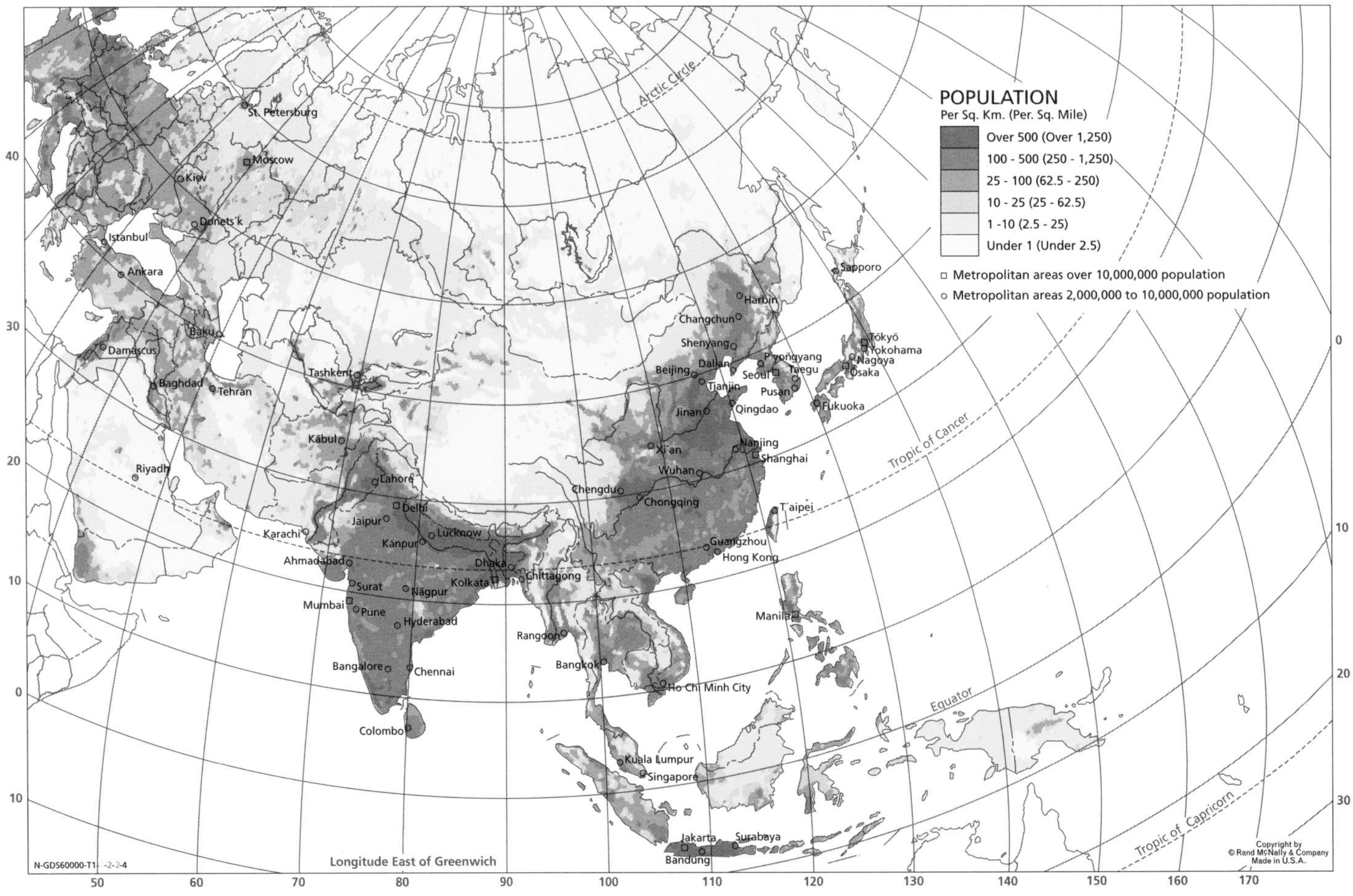

POPULATION

Per Sq. Km. (Per. Sq. Mile)

- Over 500 (Over 1,250)
- 100 - 500 (250 - 1,250)
- 25 - 100 (62.5 - 250)
- 10 - 25 (25 - 62.5)
- 1 -10 (2.5 - 25)
- Under 1 (Under 2.5)

□ Metropolitan areas over 10,000,000 population
○ Metropolitan areas 2,000,000 to 10,000,000 population

St. Petersburg
Kiev
Moscow
Donets'k
Istanbul
Ankara
Baku
Damascus
Baghdad
Tehran
Tashkent
Riyadh
Kābul
Karachi
Lahore
Jaipur
Delhi
Kanpur
Lucknow
Ahmadabad
Dhaka
Chittagong
Surat
Nagpur
Kolkata
Mumbai
Pune
Hyderabad
Rangoon
Bangalore
Chennai
Bangkok
Colombo
Ho Chi Minh City
Kuala Lumpur
Singapore
Jakarta
Surabaya
Bandung

Sapporo
Harbin
Changchun
Shenyang
Tōkyō
Yokohama
Nagoya
Beijing
Dalian
Pyongyang
Seoul
Taegu
Osaka
Tianjin
Pusan
Fukuoka
Jinan
Qingdao
Xi'an
Nanjing
Shanghai
Wuhan
Chengdu
Chongqing
T'aipei
Guangzhou
Hong Kong
Manila

Arctic Circle

Tropic of Cancer

Equator

Tropic of Capricorn

Longitude East of Greenwich

N-GDS60000-T1- -2-2-4

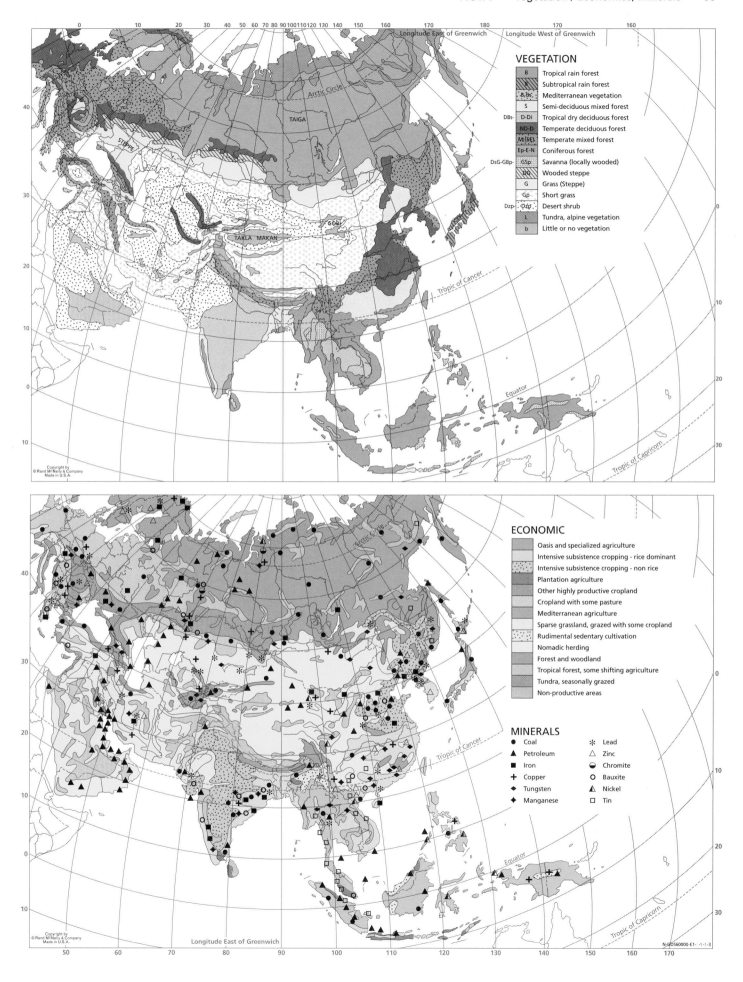

VEGETATION

B	Tropical rain forest
	Subtropical rain forest
B-Bs	Mediterranean vegetation
S	Semi-deciduous mixed forest
D-Di	Tropical dry deciduous forest
ND-D	Temperate deciduous forest
M-(SE)	Temperate mixed forest
Ep-E-N	Coniferous forest
GSp	Savanna (locally wooded)
DG	Wooded steppe
G	Grass (Steppe)
Gp	Short grass
Dzp	Desert shrub
L	Tundra, alpine vegetation
b	Little or no vegetation

DBs-
DsG-GBp-
Dzp-

ECONOMIC

	Oasis and specialized agriculture
	Intensive subsistence cropping - rice dominant
	Intensive subsistence cropping - non rice
	Plantation agriculture
	Other highly productive cropland
	Cropland with some pasture
	Mediterranean agriculture
	Sparse grassland, grazed with some cropland
	Rudimental sedentary cultivation
	Nomadic herding
	Forest and woodland
	Tropical forest, some shifting agriculture
	Tundra, seasonally grazed
	Non-productive areas

MINERALS

●	Coal	✳	Lead
▲	Petroleum	△	Zinc
■	Iron	◖	Chromite
+	Copper	○	Bauxite
◆	Tungsten	◭	Nickel
◆	Manganese	□	Tin

Copyright by
© Rand McNally & Company
Made in U.S.A.

N-GDS60000-E1- -1-1-3

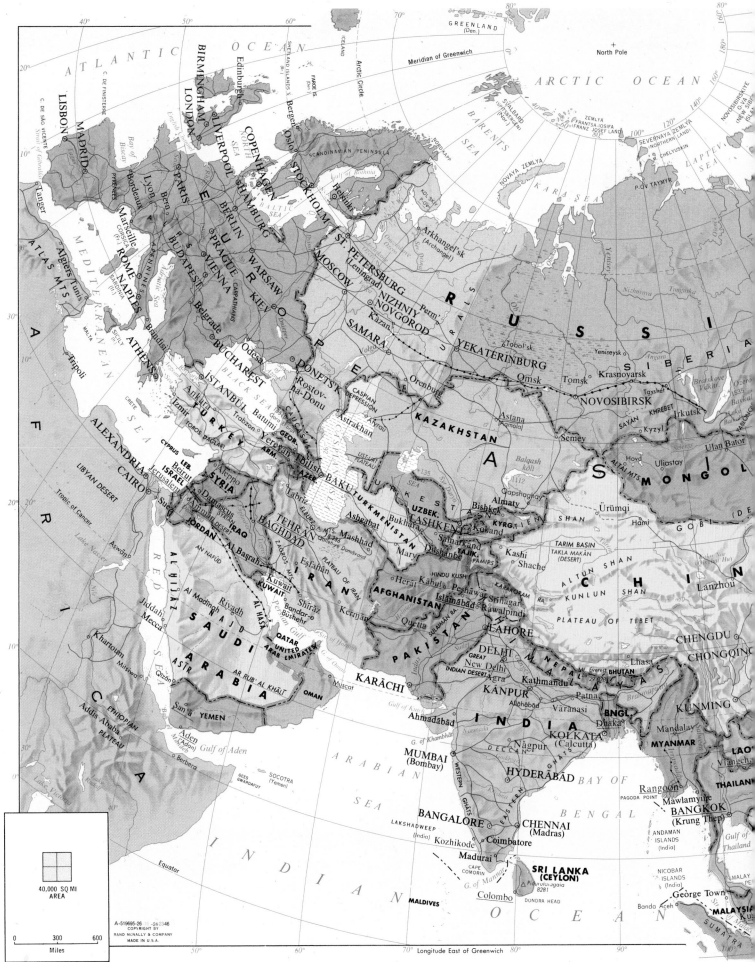

Scale 1:40 000 000; one inch to 630 miles. Lambert's Azimuthal, Equal Area Projection
Elevations and depressions are given in feet

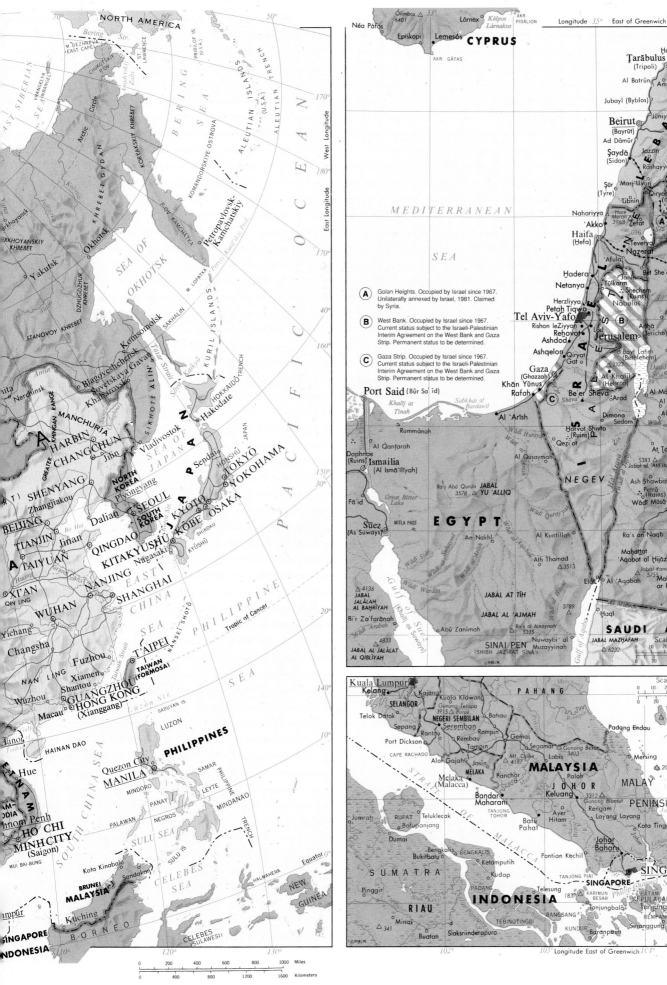
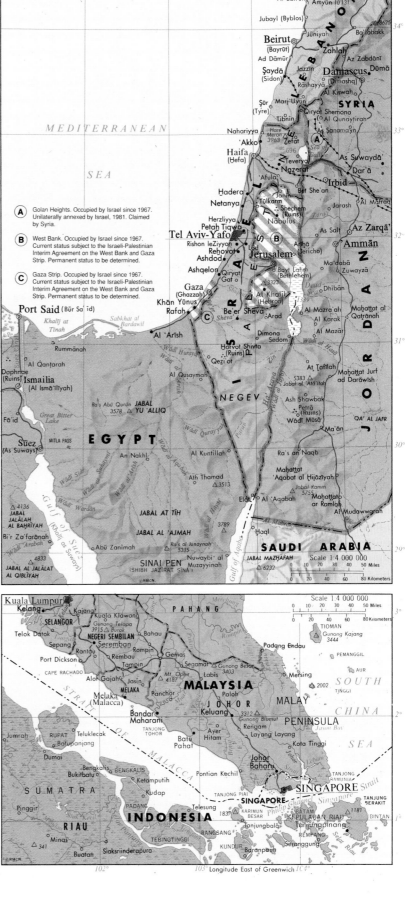

66

Scale 1:40 000 000; one inch to 630 miles. Lambert's Azimuthal, Equal Area Projection
Elevations and depressions are given in feet

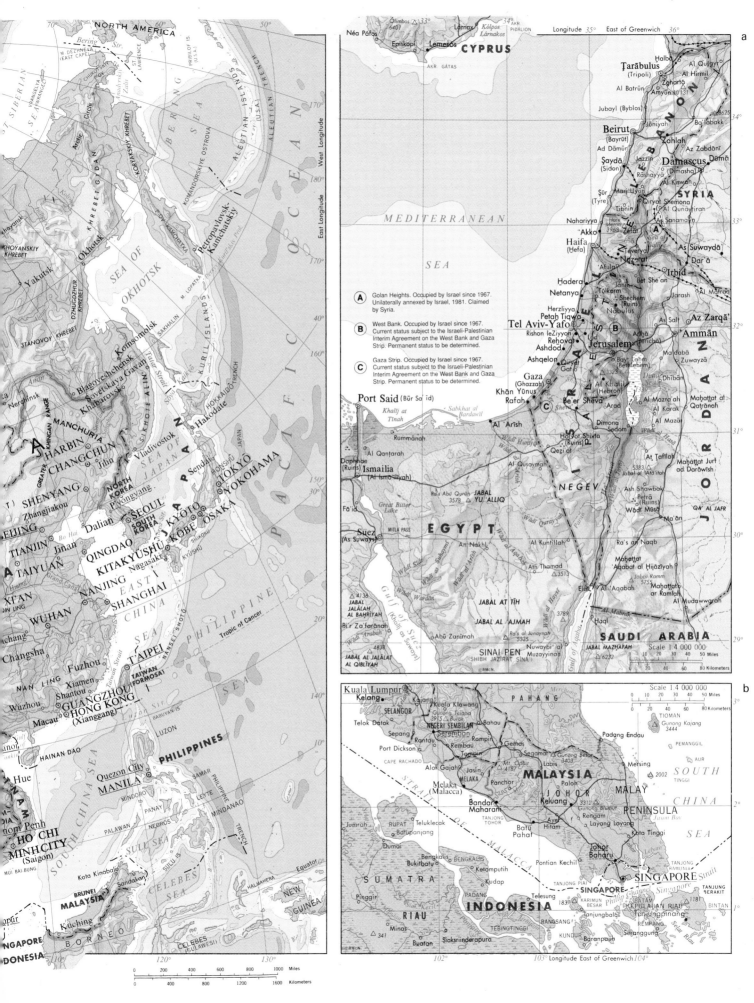

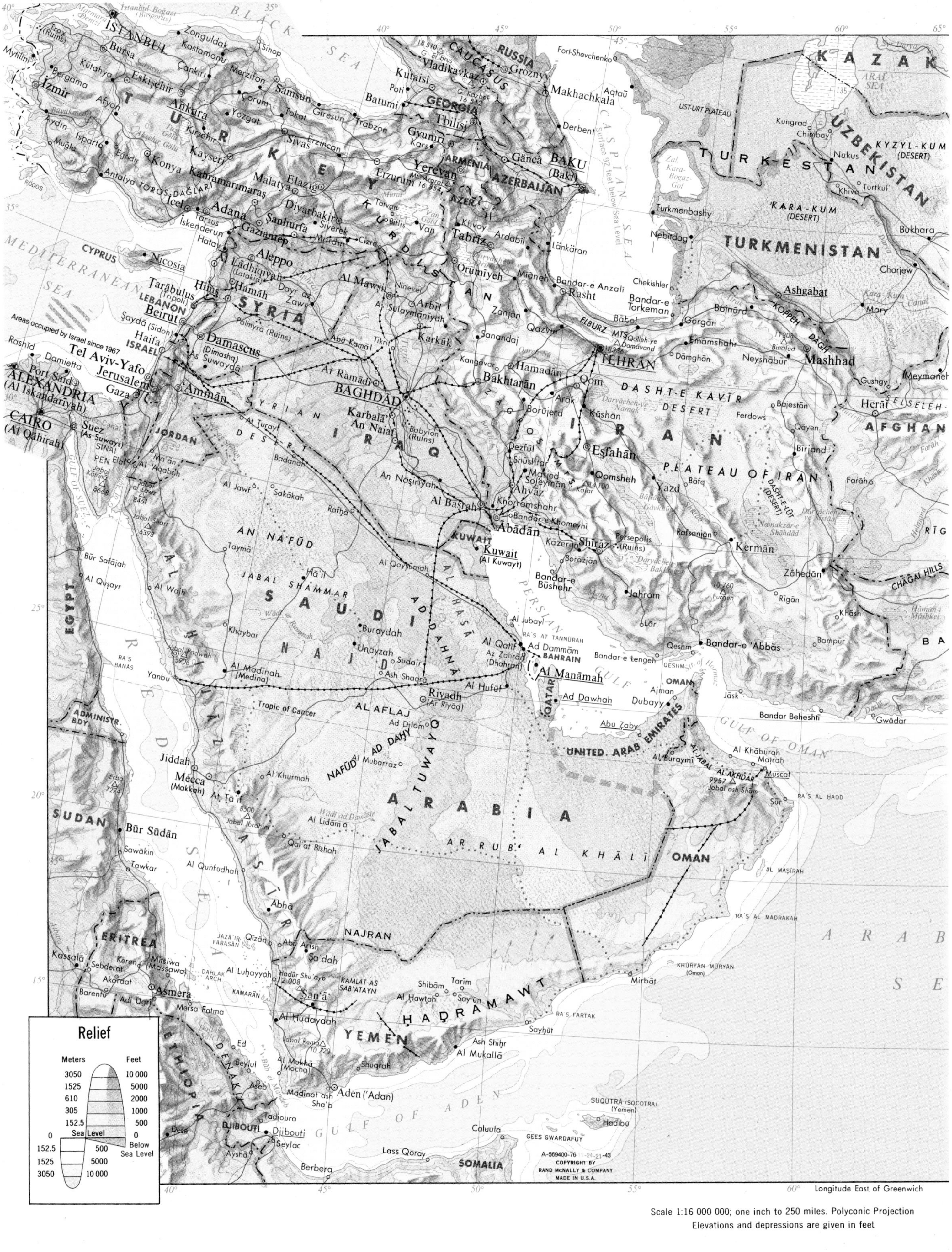

BLACK SEA

TURKEY

İstanbul Boğazı (Bosporus)
İstanbul
Marmara Denizi
Troy (Ruins)
Mytilini
İzmir
Bergama
Kütahya
Bursa
Aydın
Muğla
Afyon
Isparta
Eğridir Gölü
Akşehir Gölü
Konya
Antalya
TOROS DAĞLARI
Içel
Tarsus
İskenderun
Hatay
RODOS

Zonguldak
Kastamonu
Sinop
Çankırı
Merzifon
Samsun
Çorum
Yozgat
Sivas
Tokat
Giresun
Trabzon
Ankara
Kırşehir
Kayseri
Kahramanmaraş
Malatya
Elazığ
Diyarbakır
Siverek
Şanlıurfa
Gaziantep
Adana
Erzincan

RUSSIA
CAUCASUS
Grozny
Vladikavkaz
Kutaisi
Poti
Batumi
Elbrus 18 510
Kazbek
Makhachkala
Derbent

GEORGIA
Tbilisi
Gyumri
Kars
ARMENIA
Yerevan
Mt. Ararat 16 854
AZERBAIJAN
Gäncä (Bäk)
BAKU (Baki)
Länkäran

Erzurum 16 854
Murat
Tatvan
Van Gölü
Van
Bitlis
Cizre
KURDISTAN

Fort-Shevchenko
Aqtaū
CASPIAN SEA
Surface 92 feet below Sea Level
UST-URT PLATEAU

KAZAK
ARAL SEA
Kungrad
Chimbay
Nukus
Turtkul
Khiva

UZBEKISTAN
KYZYL-KUM (DESERT)
Bukhara

ZAL Kara-Bogaz-Gol
Turkmenbashy
Nebitdag
Chekishler
Bandar-e Torkeman
Bojnūrd
Gorgān
KARA-KUM (DESERT)
Kara-Kum Canal
Ashgabat
KÖPPEH DAĞ
Mary
Charjew

TURKMENISTAN

Nicosia
CYPRUS
MEDITERRANEAN SEA

Tarābulus (Tripoli)
Ladhiqiyah (Latakia)
Hims
Hamāh
LEBANON
Beirut
Şaydā (Sidon)
SYRIA
Damascus (Dimashq)
As Suwaydā'
Palmyra (Ruins)
Al Mawşil
Nineveh
Arbil
As Sulaymānīyah
Dayr az Zawr
Abū Kamāl
Tikrit
Karkūk
Kanaqin
Sanandaj
Hamadān

Tabriz
Orūmīyeh
Ardabil
Khvoy
Miāneh
Bandar-e Anzali
Rasht
Qazvin
Zanjān
TEHRĀN
Qolleh-ye Damāvand
Dāmghān
ELBURZ MTS
Bābol
Emāmshahr
Neyshābūr
Mashhad
Gushgy
Meymaneh
Herāt
EL SELEH
AFGHAN

Haifa
ISRAEL
Tel Aviv-Yafo
Jerusalem
Gaza
Rashīd
Damietta
Port Said
ALEXANDRIA (Al Iskandarīyah)
Areas occupied by Israel since 1967
CAIRO (Al Qāhirah)
Suez (As Suways)
SINAI
PEN Elat
Al 'Aqabah
Ma'ān
At Turayf
JORDAN
Amman
SYRIAN DESERT
Ar Ramādī
BAGHDAD
Karbalā'
An Najaf
Babylon (Ruins)
IRAQ
An Nāşirīyah
Al Basrah
Khorramshahr
Bandar-e Khomeyni
Abādān
KUWAIT (Al Kuwayt)

Bakhtarān
Qom
Arāk
Borūjerd
Kāshān
Dezfūl
Shūshtar
Masjed-e Soleymān
Ahvāz
Eşfahān
Qomsheh
Yazd
Bāfq
DASHT-E KAVIR DESERT
Darvācheh-ye Namak
Ferdows
Bājestān
Qāyen
Biriand
Farāh
IRAN
PLATEAU OF IRAN
DASHT-E LŪT (DESERT)
Namakzār-e Shāhdād

Shīrāz
Persepolis (Ruins)
Kāzerūn
Borāzjān
Jahrom
Lār
Kermān
Rafsanjān
Zāhedān
Khāsh
CHĀGAI HILLS
Būr Safājah
Al Quşayr
RA'S BANAS
Al Wajh
EGYPT

AN NAFŪD
Al Jawf
Sakākah
Rafhā'
Badanah
Taymā'
Ha'il
JABAL SHAMMAR
Khaybar
Buraydah
Wādī ar Rummah
'Unayzah
Sudayr
Ash Shaqrā'
SAUDI
NAJD
Al Qayşūmah
An Nafūd
AD DAHNĀ

Al Hasā
Al Qatif
Az Zahrān (Dhahran)
Al Hufūf
Ad Dammām
BAHRAIN
Al Manāmah
QATAR
Ad Dawhah
Al Jubayl
RA'S AT TANNŪRAH
Bandar-e Būshehr
PERSIAN GULF
Qeshm
Bandar-e Lengeh
Bandar-e 'Abbās
Bampūr
Rigān
10 760 Furang
OMAN
Ajman
Dubayy
Abū Žaby
UNITED ARAB EMIRATES
Jāsk
Bandar Beheshtī
GULF OF OMAN
Gwādar
JABAL AL AKHDAR
9957 Jabal ash Shām
Al Khābūrah
Matrah
Muscat
Al Buraymi
Sūr
RA'S AL HADD

RA'S BANAS
Yanbu'
Al Madinah (Medina)
Tropic of Cancer
Jiddah
Mecca (Makkah)
Al Tā'if
Jabal Ibrāhīm 8500
Al Khurmah
Mubarraz
Riyadh (Ar Riyād)
Ad Dilam
AL AFLAJ
AD DAHY
NAFŪD
Wādī ad Dawāsir
Al Lidām
Qal'at Bishah
JABAL AT TUWAYQ
ARABIA
AR RUB' AL KHĀLI
OMAN
RA'S AL MASIRAH
AL MAŞIRAH

Būr Sūdān
SUDAN
Sawākin
Tawkar
Qal'at Bishah
Al Qunfudhah
Abha
Najrān
Jāzā'IR FARASAN
Qizān
Abū 'Arīsh
Şa'dah
Hadūr Shu'ayb 3 008
Shibām
Tarīm
Say'ūn
Al Hawtah
HADRAMAWT
Mirbāt
RA'S FARTAK
Sayhūt
KHŪRYAN MŪRYĀN (Oman)
RA'S AL MADRAKAH

Kassalā
Sebderat
Keren
Mitsiwa (Massawa)
Akordat
ERITREA
Asmera
Adi Ugri
Barentu
Adi Ugri
Mersa Fatma
DAHLAK ARCH
KAMARAN
Al Luhayyah
San'ā'
Al Hudaydah
Jabal Remā 10 729
YEMEN
Al Mukhā (Mocha)
Shuqrah
Ash Shihr
Al Mukallā

ETHIOPIA
DANAKIL
Ed
Beylul
Aseb
Madinat ash Sha'b
Aden ('Adan)
GULF OF ADEN
SUQUTRA (SOCOTRA) (Yemen)
Hadībū

ADMINISTR. BDY
Tadjoura
DJIBOUTI
Djibouti
Aysha
Dese
Seylac
Berbera
Lass Qoray
GEES GWARDAFUY
Caluula
SOMALIA

A-569400-76 11-24-21-43
COPYRIGHT BY
RAND McNALLY & COMPANY
MADE IN U.S.A.

ARABIAN SEA

Longitude East of Greenwich

Relief

Meters	Feet
3050	10 000
1525	5000
610	2000
305	1000
152.5	500
0 Sea Level	0
152.5	500 Below Sea Level
1525	5000
3050	10 000

Scale 1:16 000 000; one inch to 250 miles. Polyconic Projection
Elevations and depressions are given in feet

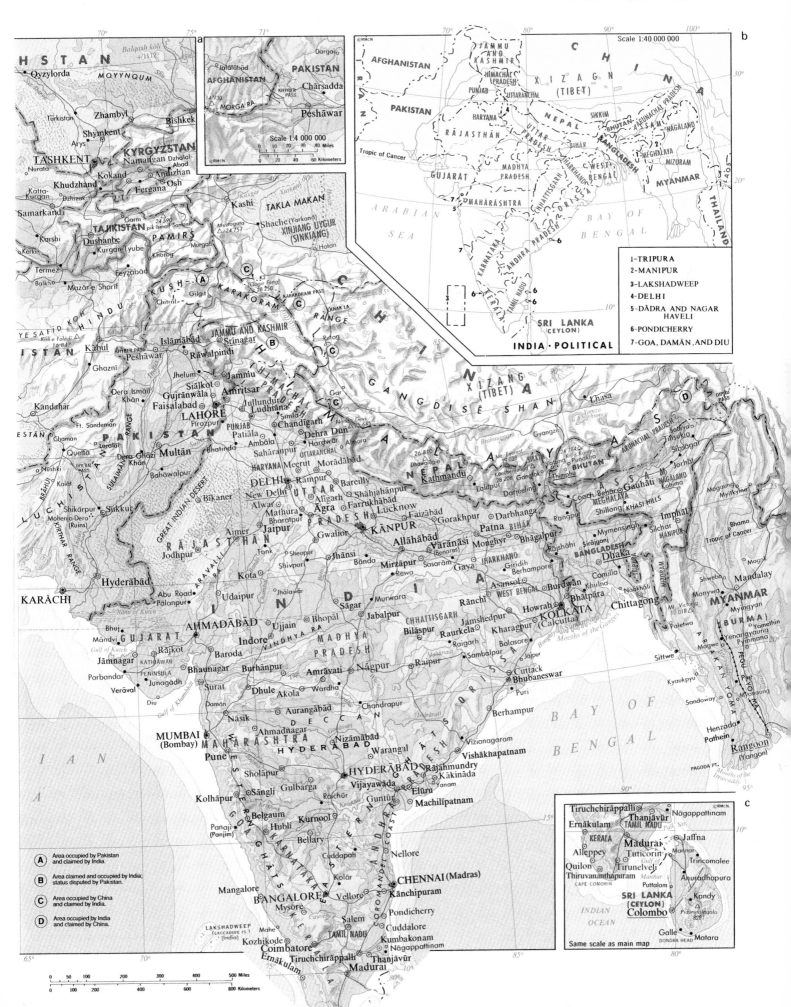

a

Dārgai
Jalālābād
AFGHANISTAN
PAKISTAN
KHYBER PASS
Chārsadda
MORGA RA.
Peshāwar

Scale 1:4 000 000
0   10   20   30   40 Miles
0   20   40   60 Kilometers

©RMcN.

b

Scale 1:40 000 000

©RMcN.

AFGHANISTAN
JAMMU AND KASHMIR
CHINA
IRAN
PAKISTAN
HIMACHAL PRADESH
XIZAGN (TIBET)
PUNJAB
UTTARANCHAL
HARYANA
NEPAL
SIKKIM
BHUTAN
ARUNACHAL PRADESH
ASSAM
NAGALAND
RAJASTHAN
UTTAR PRADESH
BIHAR
MEGHALAYA
MIZORAM
Tropic of Cancer
GUJARAT
MADHYA PRADESH
JHARKHAND
WEST BENGAL
BANGLADESH
MYANMAR
ARABIAN SEA
CHHATTISGARH
ORISSA
BAY OF BENGAL
MAHĀRĀSHTRA
KARNATAKA
ANDHRA PRADESH
KERALA
TAMIL NADU
SRI LANKA (CEYLON)

INDIA · POLITICAL

1-TRIPURA
2-MANIPUR
3-LAKSHADWEEP
4-DELHI
5-DĀDRA AND NAGAR HAVELI
6-PONDICHERRY
7-GOA, DAMĀN, AND DIU

c

©RMcN.

Tiruchchirāppalli
Nāgappattinam
Ernākulam
Thanjāvūr
TAMIL NADU
KERALA
Madurai
Jaffna
Alleppey
Tuticorin
Mannar
Quilon
Tirunelveli
Trincomalee
Thiruvananthapuram
Anuradhapura
CAPE COMORIN
Puttalam
SRI LANKA (CEYLON)
Kandy
INDIAN OCEAN
Colombo
Galle
DONDRA HEAD
Matara

Same scale as main map

Ⓐ Area occupied by Pakistan and claimed by India.
Ⓑ Area claimed and occupied by India; status disputed by Pakistan.
Ⓒ Area occupied by China and claimed by India.
Ⓓ Area claimed by India and occupied by China.

0   50   100   200   300   400   500 Miles
0   100   200   400   600   800 Kilometers

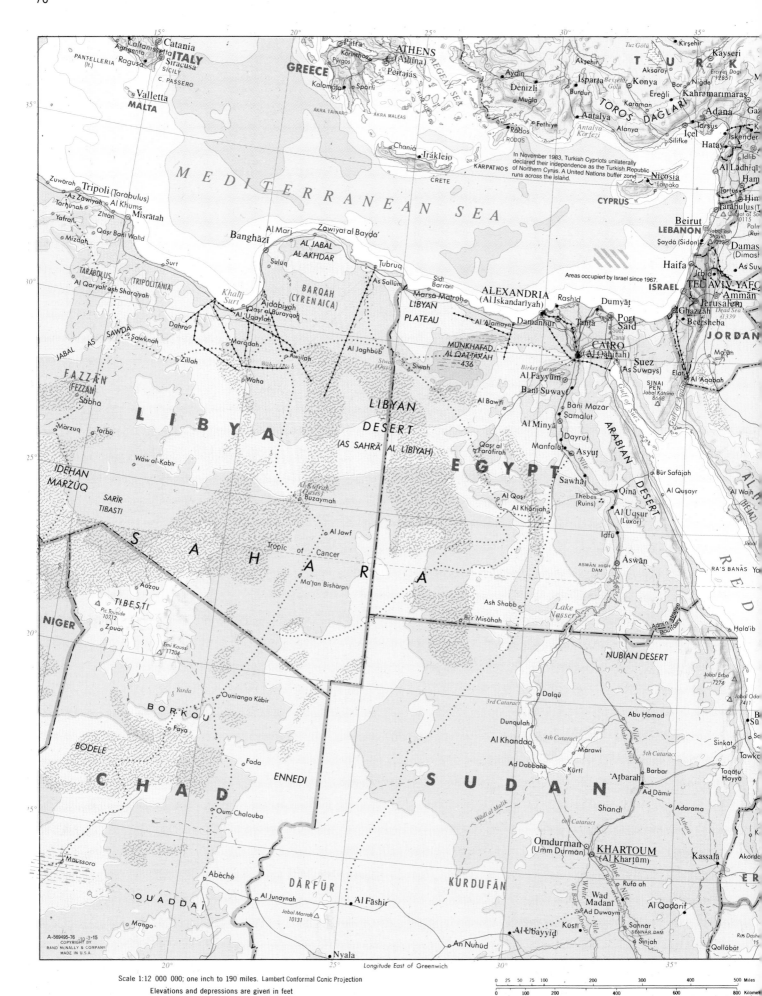

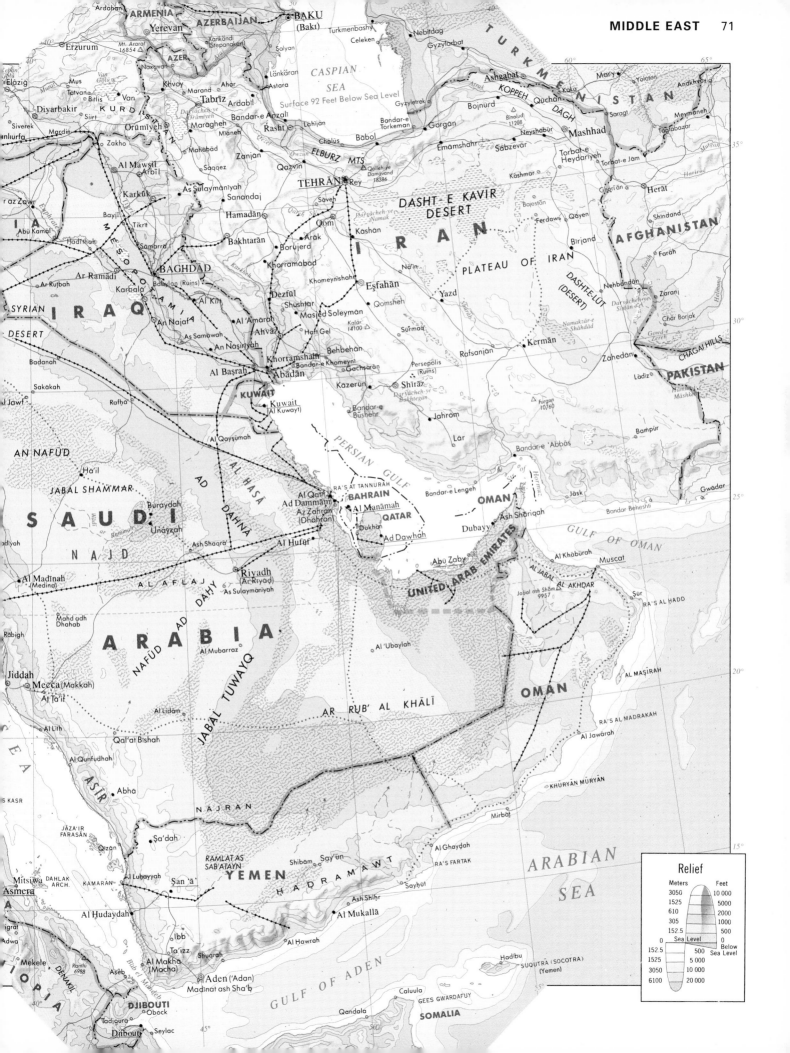

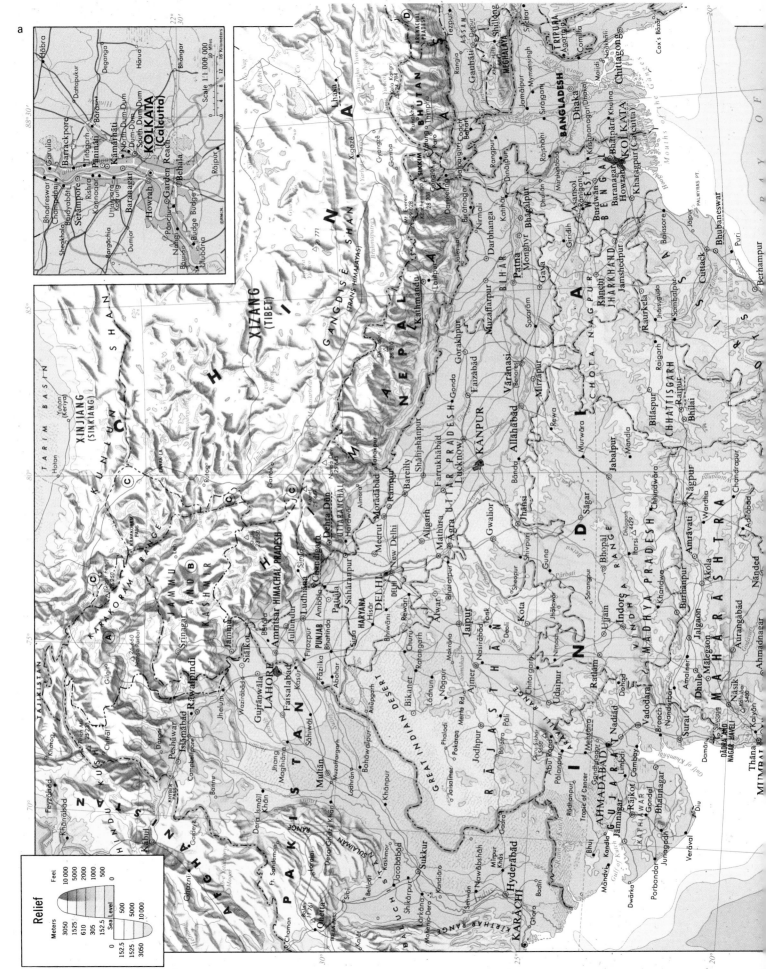

a

KOLKATA (Calcutta)

Scale 1:1 000 000

Relief

Meters	Feet
3050	10 000
1525	5000
610	2000
305	1000
152.5	500
0	Sea Level
152.5	500
1525	5000
3050	10 000

Scale 1:10 000 000; one inch to 160 miles. Lambert Conformal Conic Projection
Elevations and depressions are given in feet

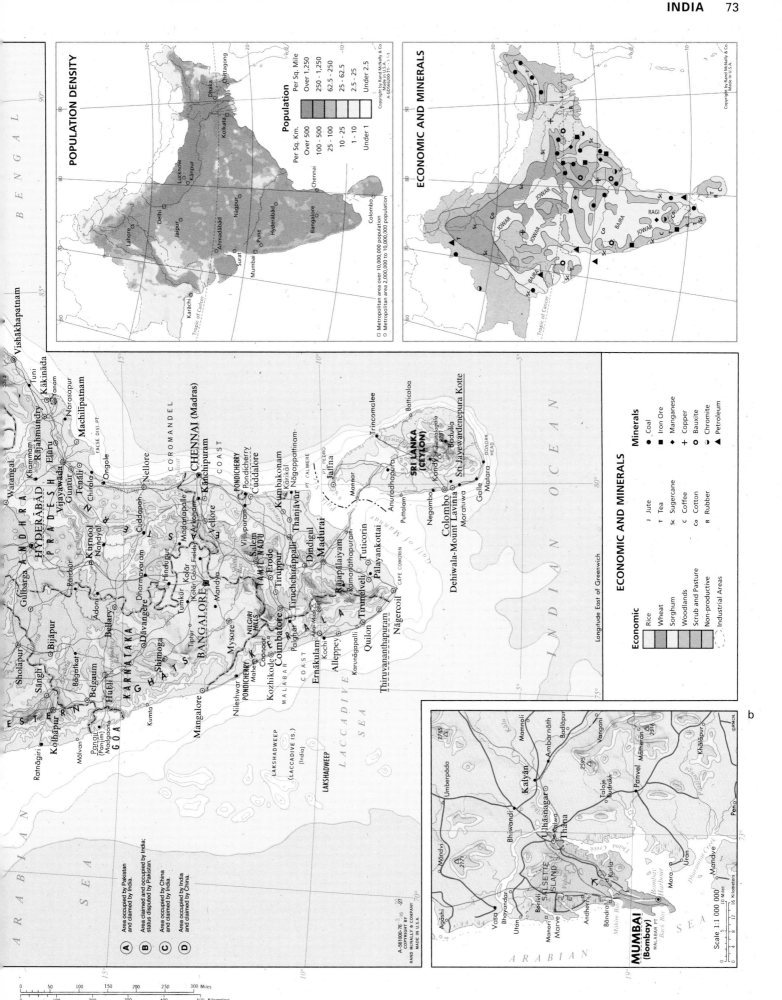

POPULATION DENSITY

Population

Per Sq. Km.	Per Sq. Mile
Over 500	Over 1,250
100 - 500	250 - 1,250
25 - 100	62.5 - 250
10 - 25	25 - 62.5
1 - 10	2.5 - 25
Under 1	Under 2.5

☐ Metropolitan area over 10,000,000 population
○ Metropolitan area 2,000,000 to 10,000,000 population

ECONOMIC AND MINERALS

Economic

Rice
Wheat
Sorghum
Woodlands
Scrub and Pasture
Non-productive
Industrial Areas

J	Jute	T	Tea
Sc	Sugarcane	C	Coffee
Co	Cotton	R	Rubber

Minerals

● Coal
● Iron Ore
◆ Manganese
+ Copper
○ Bauxite
◑ Chromite
◀ Petroleum

A Area occupied by Pakistan and claimed by India.
B Area claimed and occupied by India; status disputed by Pakistan.
C Area occupied by China and claimed by India.
D Area occupied by India and claimed by China.

A-561000-76
COPYRIGHT BY
RAND McNALLY & COMPANY
MADE IN U.S.A.

Copyright by Rand McNally & Co.
Made in U.S.A.
A-GD046200-T1-.-1.-1

MUMBAI
(Bombay)

Scale 1:1 000 000

b

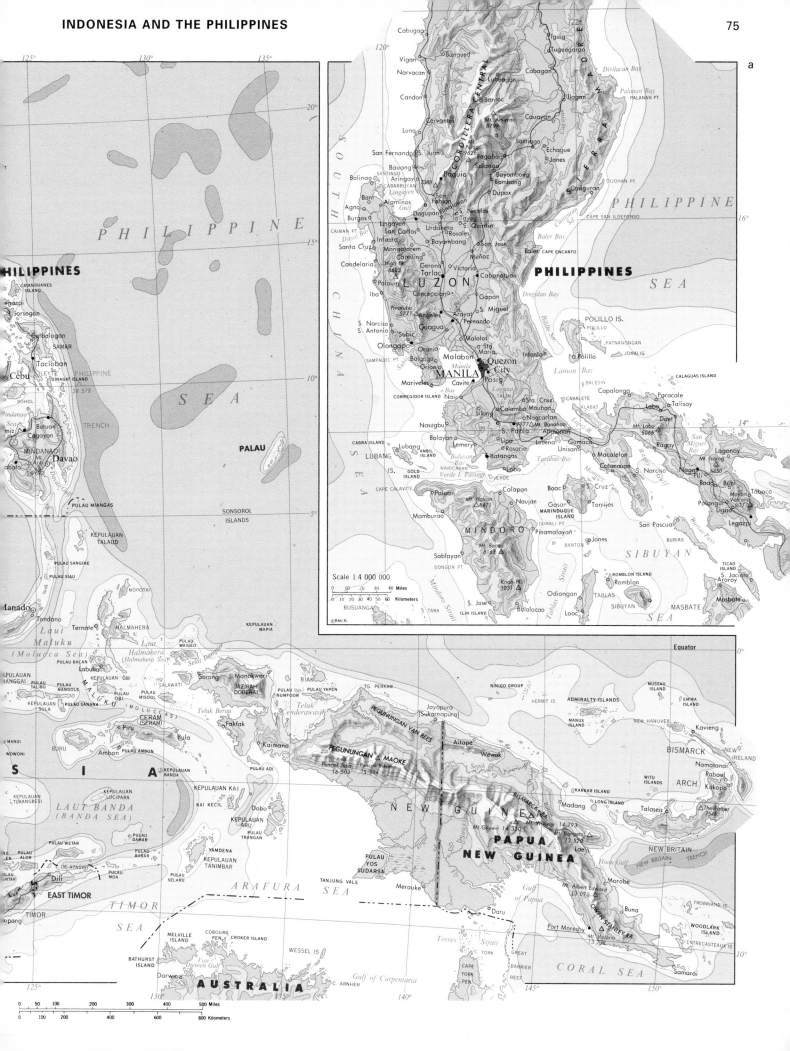

a

PHILIPPINES

PHILIPPINE

SEA

SOUTH CHINA SEA

PHILIPPINES

PHILIPPINE SEA

Cabuga
Iguig
Tuguegarao

Cabugao
Bangued
Vigan
Lubuagan
Cabagan

Narvacan
Bontoc
Cabagan
Divilacan Bay

Candon
Cervantes
Ilagan
Palaman Bay
PALANAN PT.

Luna
Mt. Amuyao
8799
Cauayan

San Fernando
S. Juan
Santiago
Echague

Bauang
Aringay
Baguio
Bayombong
Jones
Casiguran

Bolinao
CABARRUYAN
Lingayen
Bambang
Dupax
DIJOHAN PT.

Agno
Alaminos
T388
Dagupan
S. Nicolas
CAPE SAN ILDEFONSO
16°

Burgos
Lingayen
San Fabian
Bagabag

CAIMAN PT.
Dasol Bay
San Carlos
Urdaneta
S. Quintin
Baler Bay
Casiguran

Santa Cruz
Infanta
Rosales
Baler
CAPE ENCANTO

Candelaria
Mangatarem
Bayambang
San Jose
Dingalan Bay

High Pk.
6683
Camiling
Gerona
Muñoz

Palauig
Tarlac
Victoria
Cabanatuan

Iba
Concepcion
Gapan
POLILLO IS.

Pinatubo
5771
Angeles
S. Miguel
POLILLO

S. Narciso
Arayat
S. Fernando
Infanta
PATNANONGAN

S. Antonio
Guagua
Malolos
Polillo
JOMALIG

Olongapo
Orani
Sta. Maria
BALESIN
Capalonga
Paracale
Talisay

Bataan
Bolanao
Malabon
Quezon City
Labo
Daet

Orion
Pasig
CABALETE
ALABAT
Macalelon

CORREGIDOR ISLAND
Cavite
MANILA
Laguna de Bay
CALAGUAS ISLAND

Mariveles
Naic
Sta. Cruz
Mauban
S. Miguel Bay
Mt. Isarog
6450

Balayan
Silang
Calamba
Nagcarlan
Catanauan
Naga

LUBANG
Lemery
Lipa
S. Pablo
Gumaca
S. Narciso
Pili

AMBIL ISLAND
Rosario
Mt. Banahao
Atimonan
Iriga
Baao
Buhi

CABRA ISLAND
Balayan Bay
Batangas
Lucena
Unisan
Polangui
Mt. Mayon Volcano
8077

LUBANG IS.
GOLD ISLAND
Loba
Tarabas Bay
Ligao

MARICABAN
VERDE
Boac
S. Cruz
Legazpi

CAPE CALAVITE
Verde I. Passage
Calapan
Torrijos
BURIAS

Paluan
Naujan
MARINDUQUE ISLAND
SIBUYAN

Mamburao
Mt. Halcon
8471
Gasan
San Pascual
TICAO ISLAND
S. Jacinto
Aroroy

MINDORO
DUMALI PT.
Pinamalayan
BANTON
Jones
ROMBLON ISLAND
Masbate

Sablayan
Mt. Baco
8163
Romblon
TABLAS

DONGON PT.
Knob Pk.
3031
Odiongan
SIBUYAN
MASBATE SEA

BUSUANGA
S. Jose
Bulalacao
Looc
TARA
ILIN ISLAND
Mindoro Strait

Scale 1:4 000 000

0 10 20 30 40 Miles

0 10 20 30 40 50 60 Kilometers

©RMcN

Manado
Tondano
Ternate
HALMAHERA
MOROTAI
KEPULAUAN MAPIA

Laut Maluku
(Molucca Sea)
Laut Halmahera
(Halmahera Sea)
PULAU WAIGEO

Selat Dampier

PULAU BACAN
SALAWATI
Sorong
Manokwari
BIAK
PULAU YAPEN
TG. PERKAM
NINIGO GROUP
ADMIRALTY ISLANDS
MUSSAU ISLAND

PULAU OBI
JAZIRAH DOBERAI
PULAU NUMFOOR
HERMIT IS.
MANUS ISLAND
EMIRA ISLAND

KEPULAUAN OBI
PULAU MISOOL
Teluk Berau
Jayapura
(Sukarnapura)

A
KEPULAUAN SULA
PULAU SANANA
(MOLUCCAS) (SERAM)
Fakfak
Teluk Cenderawasih
Aitape
Wewak
NEW HANOVER

S
PULAU TALIBU
PULAU MANGOLE
CERAM (SERAM)
Piru
Bula
Kaimana
Sepik
Kavieng

I
PULAU BURU
Ambon
PULAU AMBON
PULAU ADI
PEGUNUNGAN VAN REES
BISMARCK

MANUI
WOWONI
BURU
PEGUNUNGAN MAOKE
Puncak Jaya
16 503
Puncak Trikora
15 584
Madang
LONG ISLAND
ARCH.

KEPULAUAN TUKANGBESI
KEPULAUAN BANDA
KEPULAUAN KAI
NEW GUINEA
Mt. Wilhelm 14 793
Mt. Bangeta
13 524
KARKAR ISLAND
WITU ISLANDS
Namatanai
Rabaul
Kokopo

LAUT BANDA
(BANDA SEA)
KEPULAUAN LUCIPARA
KAI KECIL
Dobo
KEPULAUAN ARU
Mt. Giluwe 14 330
PAPUA NEW GUINEA
Talasea
NEW IRELAND

PULAU DAMAR
PULAU TRANGAN
Lae
Huon Gulf
The Father
7546

PULAU WETAR
YAMDENA
KEPULAUAN TANIMBAR
Digul
Morobe
NEW BRITAIN
NEW BRITAIN TRENCH

PULAU ALOR
DE-ATAURO
PULAU BABAR
PULAU YOS SUDARSA
Merauke
Mt. Albert Edward
13 090
Gulf of Papua
TROBRIAND IS.

Dili
PULAU MOA
PULAU SELARU
TANJUNG VALS
Buna
WOODLARK ISLAND

EAST TIMOR
Kupang
ARAFURA SEA
Daru
Port Moresby
OWEN STANLEY RA.
Mt. Victoria
13 238
D'ENTRECASTEAUX IS.

TIMOR
TIMOR SEA
COBOURG PEN.
CROKER ISLAND
Torres Strait
C. YORK
GREAT BARRIER REEF
CORAL SEA
Samarai

MELVILLE ISLAND
WESSEL IS.
CAPE YORK PEN.

BATHURST ISLAND
Van Diemen Gulf
Darwin
Gulf of Carpentaria
C. ARNHEM

AUSTRALIA

0 50 100 200 300 400 500 Miles

0 100 200 400 600 800 Kilometers

Cabugao
Egazpi
Sorsogon
CATANDUANES ISLAND

Catbalogan
SAMAR
Tacloban

Cebu
LEYTE
DINAGAT ISLAND

BOHOL
PHILIPPINE TRENCH
34 578

Mindanao Sea
Butuan
Cagayan

MINDANAO
Mt. Apo
9692
Davao

Zamboanga

PULAU MIANGAS

KEPULAUAN TALAUD

PALAU

SONSOROL ISLANDS

PULAU SANGIHE

PULAU SIAU

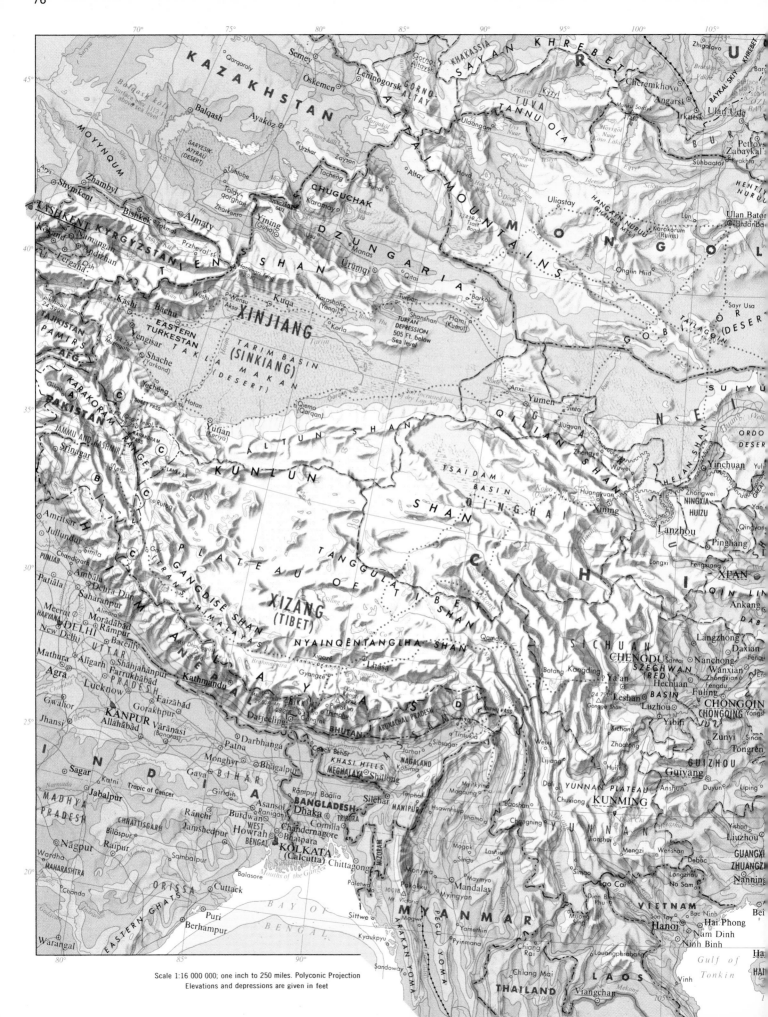

Scale 1:16 000 000; one inch to 250 miles. Polyconic Projection
Elevations and depressions are given in feet

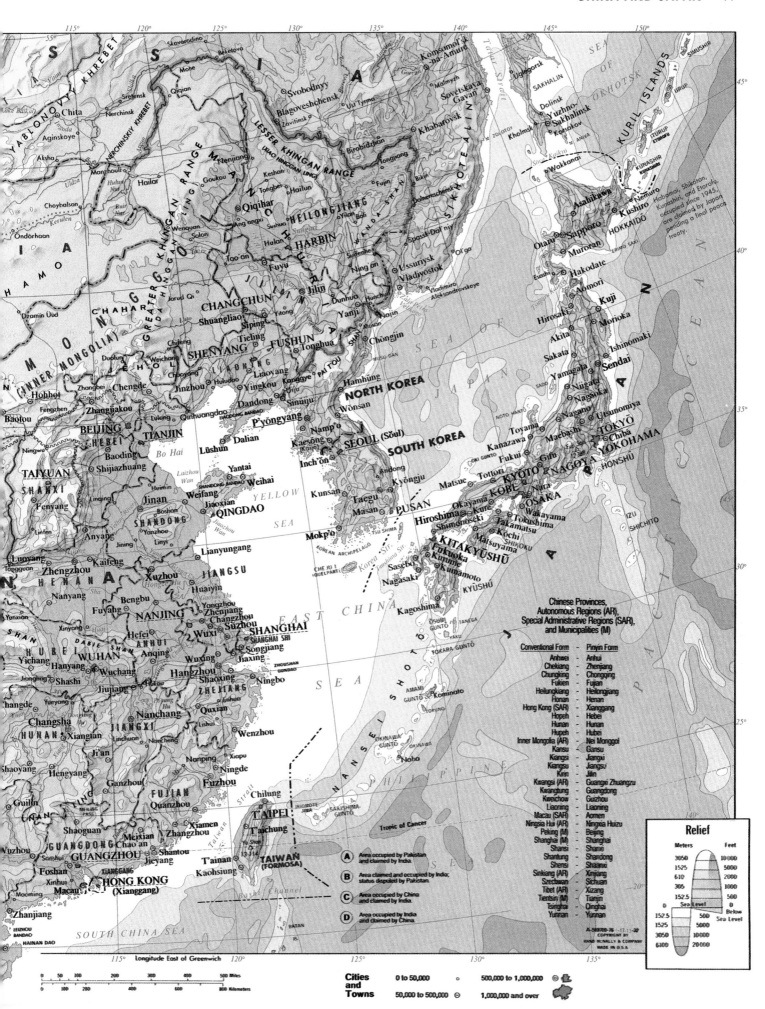

Chinese Provinces,
Autonomous Regions (AR),
Special Administrative Regions (SAR),
and Municipalities (M)

Conventional Form	Pinyin Form
Anhwei	Anhui
Chekiang	Zhejiang
Chungking	Chongqing
Fukien	Fujian
Heilungkiang	Heilongjiang
Honan	Henan
Hong Kong (SAR)	Xianggang
Hopeh	Hebei
Hunan	Hunan
Hupeh	Hubei
Inner Mongolia (AR)	Nei Monggol
Kansu	Gansu
Kiangsi	Jiangxi
Kiangsu	Jiangsu
Kirin	Jilin
Kwangsi (AR)	Guangxi Zhuangzu
Kwangtung	Guangdong
Kweichow	Guizhou
Liaoning	Liaoning
Macau (SAR)	Aomen
Ningsia Hui (AR)	Ningxia Huizu
Peking (M)	Beijing
Shanghai (M)	Shanghai
Shansi	Shanxi
Shantung	Shandong
Shensi	Shaanxi
Sinkiang (AR)	Xinjiang
Szechwan	Sichuan
Tibet (AR)	Xizang
Tientsin (M)	Tianjin
Tsinghai	Qinghai
Yunnan	Yunnan

A   Area occupied by Pakistan and claimed by India.
B   Area claimed and occupied by India; status disputed by Pakistan.
C   Area occupied by China and claimed by India.
D   Area occupied by India and claimed by China.

Relief

Meters	Feet
3050	10,000
1525	5000
610	2000
305	1000
152.5	500
0	Sea Level
152.5	500
1525	5000
3050	10,000
6100	20,000

115° Longitude East of Greenwich

0 50 100 200 300 400 500 Miles
0 100 200 400 600 800 Kilometers

Cities and Towns

0 to 50,000   500,000 to 1,000,000
50,000 to 500,000   1,000,000 and over

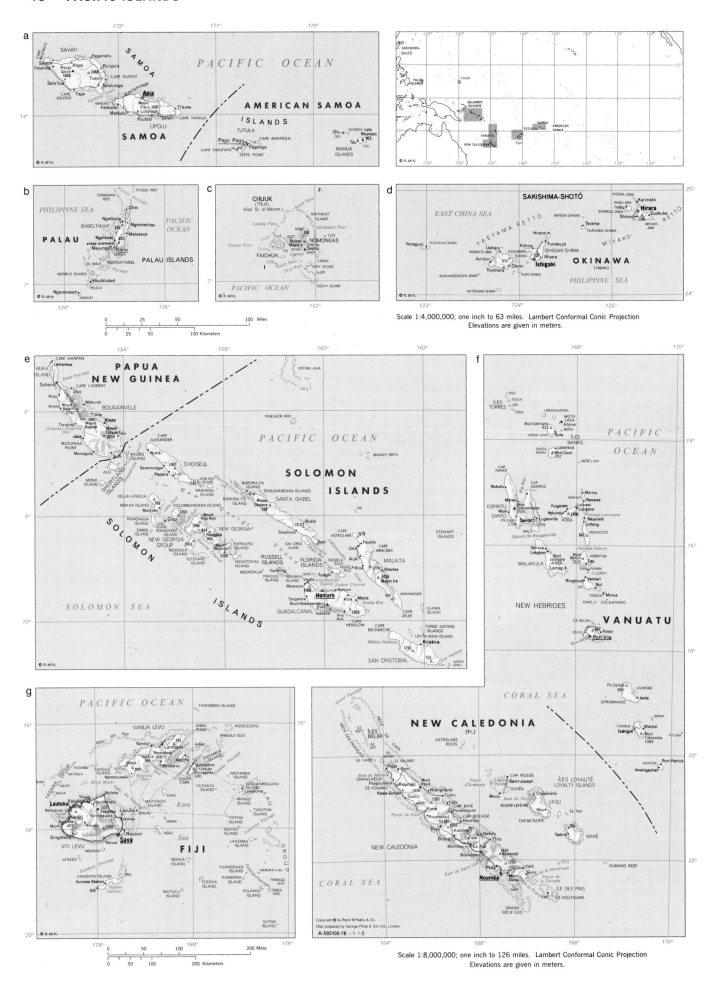

Scale 1:4,000,000; one inch to 63 miles. Lambert Conformal Conic Projection
Elevations are given in meters.

Scale 1:8,000,000; one inch to 126 miles. Lambert Conformal Conic Projection
Elevations are given in meters.

Copyright © by Rand M&Nally & Co.
Map prepared by George Philip & Son Ltd, London.
A-593100-76 :-1-1-5

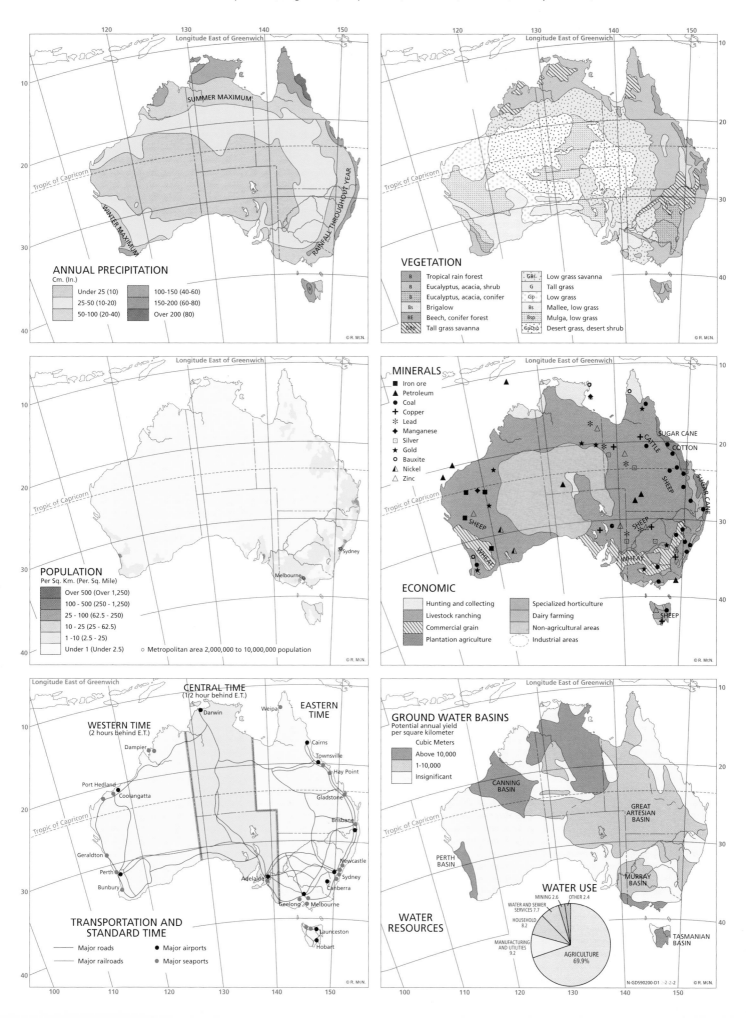

ANNUAL PRECIPITATION

Cm. (In.)

- Under 25 (10)
- 25-50 (10-20)
- 50-100 (20-40)
- 100-150 (40-60)
- 150-200 (60-80)
- Over 200 (80)

SUMMER MAXIMUM

WINTER MAXIMUM

RAINFALL THROUGHOUT YEAR

Tropic of Capricorn

Longitude East of Greenwich

© R. McN.

VEGETATION

- B — Tropical rain forest
- B — Eucalyptus, acacia, shrub
- Bc — Eucalyptus, acacia, conifer
- Bs — Brigalow
- BE — Beech, conifer forest
- Gbt — Tall grass savanna
- GBs — Low grass savanna
- G — Tall grass
- Gp — Low grass
- Bs — Mallee, low grass
- Bsp — Mulga, low grass
- GpDsb — Desert grass, desert shrub

© R. McN.

POPULATION

Per Sq. Km. (Per. Sq. Mile)

- Over 500 (Over 1,250)
- 100 - 500 (250 - 1,250)
- 25 - 100 (62.5 - 250)
- 10 - 25 (25 - 62.5)
- 1 - 10 (2.5 - 25)
- Under 1 (Under 2.5)

○ Metropolitan area 2,000,000 to 10,000,000 population

Sydney
Melbourne

© R. McN.

MINERALS

- ■ Iron ore
- ▲ Petroleum
- ● Coal
- ✛ Copper
- ✳ Lead
- ◆ Manganese
- ☐ Silver
- ★ Gold
- ○ Bauxite
- ◢ Nickel
- △ Zinc

SUGAR CANE
CATTLE
COTTON
SHEEP
SUGAR CANE
SHEEP
WHEAT
SHEEP
WHEAT
SHEEP

ECONOMIC

- Hunting and collecting
- Livestock ranching
- Commercial grain
- Plantation agriculture
- Specialized horticulture
- Dairy farming
- Non-agricultural areas
- Industrial areas

© R. McN.

TRANSPORTATION AND STANDARD TIME

WESTERN TIME
(2 hours behind E.T.)

CENTRAL TIME
(1/2 hour behind E.T.)

EASTERN TIME

Darwin
Weipa
Cairns
Townsville
Hay Point
Dampier
Port Hedland
Coolangatta
Gladstone
Brisbane
Geraldton
Newcastle
Perth
Sydney
Bunbury
Adelaide
Canberra
Geelong
Melbourne
Launceston
Hobart

- —— Major roads
- —— Major railroads
- ● Major airports
- ● Major seaports

© R. McN.

GROUND WATER BASINS

Potential annual yield per square kilometer
Cubic Meters

- Above 10,000
- 1-10,000
- Insignificant

CANNING BASIN
GREAT ARTESIAN BASIN
PERTH BASIN
MURRAY BASIN
TASMANIAN BASIN

WATER RESOURCES

WATER USE

- MINING 2.6
- OTHER 2.4
- WATER AND SEWER SERVICES 7.7
- HOUSEHOLD 8.2
- MANUFACTURING AND UTILITIES 9.2
- AGRICULTURE 69.9%

N-GD590200-D1 -2-2-2 © R. McN.

40,000 SQ MI
AREA

0   100   200

Miles

Longitude 115° East of Greenwich

Scale 1:16 000 000; one inch to 250 miles. Lambert's Azimuthal, Equal Area Projection
Elevations and depressions are given in feet

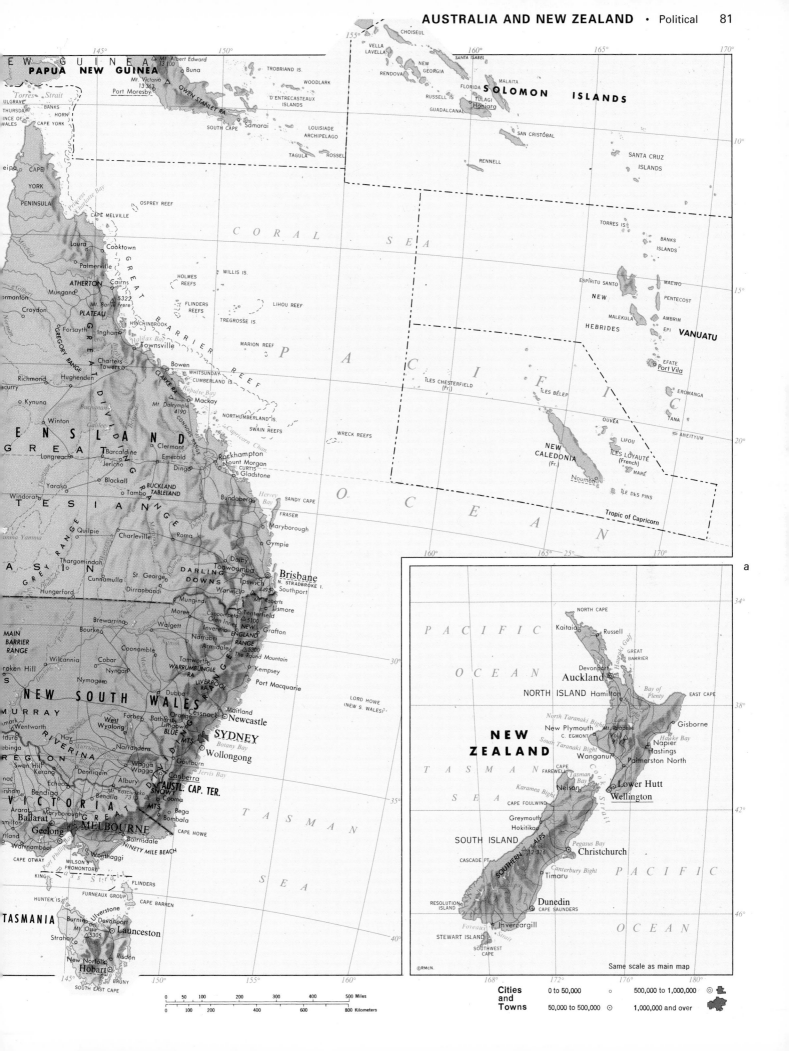

NEW GUINEA

PAPUA NEW GUINEA

Mt. Albert Edward
13,100

Port Moresby

Mt. Victoria
13,363

OWEN STANLEY RA.

Torres Strait

ULGRAVE

BANKS

THURSDAY

HORN

INCE OF

WALES

CAPE YORK

eipa

CAPE

YORK

PENINSULA

Buna

TROBRIAND IS.

WOODLARK

D'ENTRECASTEAUX
ISLANDS

SOUTH CAPE

Samarai

LOUISIADE
ARCHIPELAGO

TAGULA

ROSSEL

Princess
Charlotte Bay

CAPE MELVILLE

OSPREY REEF

CHOISEUL

VELLA
LAVELLA

NEW
GEORGIA

RENDOVA

RUSSELL IS.

GUADALCANAL

SANTA ISABEL

FLORIDA

TULAGI
Honiara

MALAITA

SOLOMON ISLANDS

SAN CRISTÓBAL

RENNELL

TORRES IS.

SANTA CRUZ
ISLANDS

BANKS
ISLANDS

CORAL SEA

ESPÍRITU SANTO

MAEWO

NEW

HEBRIDES

MALEKULA

AMBRIM

EPI

PENTECOST

AMBRIM

VANUATU

EFATE
Port Vila

EROMANGA

TANA

ANEITYUM

Laura

Cooktown

Palmerville

ATHERTON

Mungana

Cairns

5322

GREAT

Mt. Bartle Frere

PLATEAU

Croydon

Forsayth

Ingham

HINCHINBROOK I.

Townsville

ormanton

HOLMES
REEFS

WILLIS IS.

FLINDERS
REEFS

LIHOU REEF

TREGROSSE IS.

MARION REEF

PACIFIC

ÎLES CHESTERFIELD
(Fr.)

ÎLES BÉLEP

NEW
CALEDONIA
(Fr.)

Nouméa

ÎLE DES PINS

OUVÉA

LIFOU

ÎLES LOYAUTÉ
(French)

MARÉ

Richmond

Hughenden

Charters
Towers

Bowen

Mackay

NORTHUMBERLAND IS.

SWAIN REEFS

WRECK REEFS

Halifax Bay

Rapulse Bay

WHITSUNDAY
CUMBERLAND IS.

Mt. Dalrymple
4190

BARRIER

REEF

GREAT

Winton

QUEENSLAND

GREA

Longreach

Barcaldine

Jericho

Clermont

Emerald

Dingo

Mount Morgan
CURTIS

Rockhampton

Gladstone

CONNOR RANGE

Mt. Dairymple

Capricorn Chan.

PACIFIC

Yaraka

Blackall

Windorah

Tambo

BUCKLAND
TABLELAND

Bundaberg

Hervey Bay

SANDY CAPE

FRASER

OCEAN

Tropic of Capricorn

Thargomindah

Quilpie

Charleville

Roma

Maryborough

Gympie

Cunnamulla

Dalby

Toowoomba

DARLING
DOWNS

Ipswich

Brisbane

Southport

N. STRADBROKE I.

Warwick

4495

Mt. Roberts

Tenterfield

Lismore

Moree

Inverell

Glen Innes

5100

Grafton

NEW

ENGLAND

RANGE

Brewarrina

Walgett

Narrabri

Armidale

5300

Bourke

Coonamble

Tamworth

The Round Mountain

WARRUMBUNGLE
RA.

Kempsey

Wilcannia

Cobar

Nyngan

Port Macquarie

MAIN
BARRIER
RANGE

roken Hill

NEW SOUTH WALES

Nymagee

Dubbo

LIVERPOOL
RANGE

Maitland

Cessnock

Newcastle

MURRAY

Wentworth

Forbes

Bathurst

Orange

Lithgow

BLUE MTS.

SYDNEY

RIVERINA
REGION

Hay

West
Wyalong

Narrandera

Goulburn

Wollongong

Botany Bay

Jervis Bay

LORD HOWE
(NEW S. WALES)

Swan Hill

Wagga
Wagga

Albury

Kerang

Echuca

Benalla

Canberra

AUSTL. CAP. TER.

Mt. Kosciusko 7316

SNOWY
MTS.

Cooma

Bendigo

VICTORIA

GREAT

Bega

Bombala

Ararat

Maryborough

MELBOURNE

CAPE HOWE

Ballarat

Geelong

Bairnsdale

NINETY MILE BEACH

Warrnambool

Wonthaggi

CAPE OTWAY

WILSON'S
PROMONTORY

Port Phillip

KING

Bass Strait

FLINDERS

FURNEAUX GROUP

CAPE BARREN

HUNTER IS.

TASMAN

SEA

TASMANIA

Burnie

Ulverstone

Devonport

Mt. Ossa
5305

Launceston

Strahan

New Norfolk

Risdon

Hobart

BRUNY

SOUTH EAST CAPE

a

PACIFIC

OCEAN

NORTH CAPE

Kaitaia

Russell

Devonport

Auckland

NORTH ISLAND

Hamilton

GREAT
BARRIER

Bay of
Plenty

EAST CAPE

NEW
ZEALAND

New Plymouth

C. EGMONT

Wanganui

Gisborne

Hawke Bay

Napier

Hastings

Palmerston North

TASMAN

SEA

CAPE
FAREWELL

Nelson

CAPE FOULWIND

Greymouth

Hokitika

SOUTH ISLAND

SOUTHERN ALPS

CASCADE PT.

RESOLUTION
ISLAND

STEWART ISLAND

SOUTHWEST
CAPE

Lower Hutt

Wellington

Cook Strait

Pegasus Bay

Christchurch

Canterbury Bight

Timaru

Dunedin

CAPE SAUNDERS

Invercargill

Foveaux Strait

12,316

North Taranaki Bight

South Taranaki Bight

Karamea Bight

PACIFIC

OCEAN

Same scale as main map

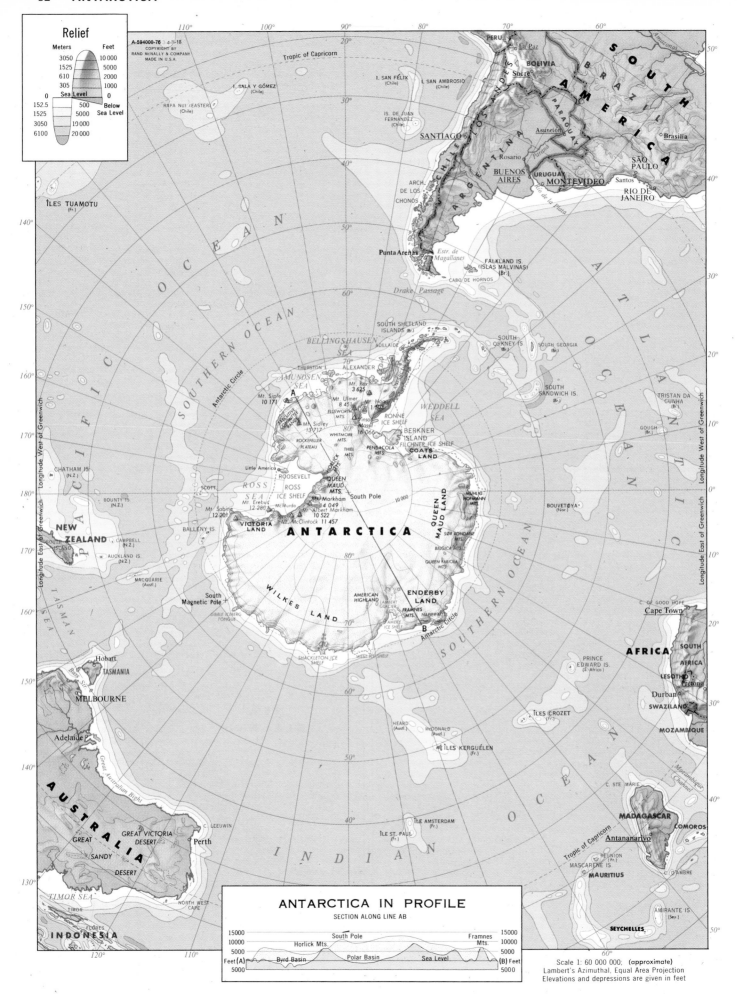

Relief

Meters		Feet
3050		10 000
1525		5000
610		2000
305		1000
	Sea Level	
0		0
	Below	
152.5	Sea Level	500
1525		5000
3050		10 000
6100		20 000

A-594000-76
COPYRIGHT BY
RAND McNALLY & COMPANY
MADE IN U.S.A.

ANTARCTICA IN PROFILE

SECTION ALONG LINE AB

Scale 1: 60 000 000; (approximate)
Lambert's Azimuthal, Equal Area Projection
Elevations and depressions are given in feet

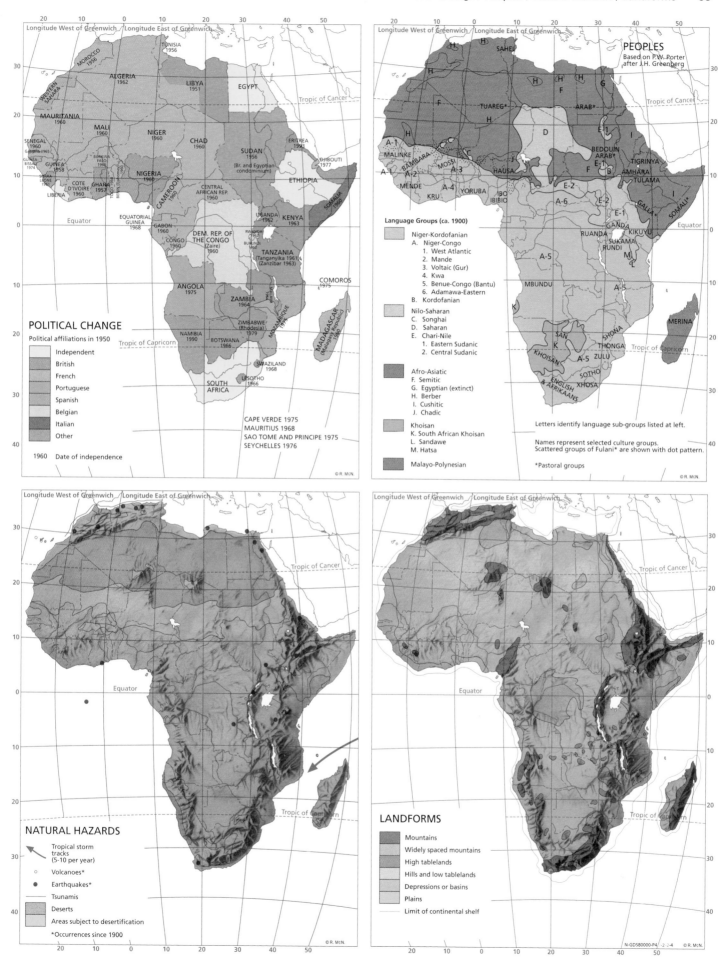

POLITICAL CHANGE

Political affiliations in 1950

- Independent
- British
- French
- Portuguese
- Spanish
- Belgian
- Italian
- Other

1960   Date of independence

CAPE VERDE 1975
MAURITIUS 1968
SAO TOME AND PRINCIPE 1975
SEYCHELLES 1976

© R. McN.

PEOPLES
Based on P.W. Porter
after J.H. Greenberg

Language Groups (ca. 1900)

- Niger-Kordofanian
  - A. Niger-Congo
    1. West Atlantic
    2. Mande
    3. Voltaic (Gur)
    4. Kwa
    5. Benue-Congo (Bantu)
    6. Adamawa-Eastern
  - B. Kordofanian

- Nilo-Saharan
  - C. Songhai
  - D. Saharan
  - E. Chari-Nile
    1. Eastern Sudanic
    2. Central Sudanic

- Afro-Asiatic
  - F. Semitic
  - G. Egyptian (extinct)
  - H. Berber
  - I. Cushitic
  - J. Chadic

- Khoisan
  - K. South African Khoisan
  - L. Sandawe
  - M. Hatsa

- Malayo-Polynesian

Letters identify language sub-groups listed at left.

Names represent selected culture groups.
Scattered groups of Fulani* are shown with dot pattern.

*Pastoral groups

© R. McN.

NATURAL HAZARDS

- ⟶ Tropical storm tracks (5–10 per year)
- ○ Volcanoes*
- ● Earthquakes*
- — Tsunamis
- Deserts
- Areas subject to desertification

*Occurrences since 1900

© R. McN.

LANDFORMS

- Mountains
- Widely spaced mountains
- High tablelands
- Hills and low tablelands
- Depressions or basins
- Plains

— Limit of continental shelf

N-GDS80000-P4/-2-2-4   © R. McN.

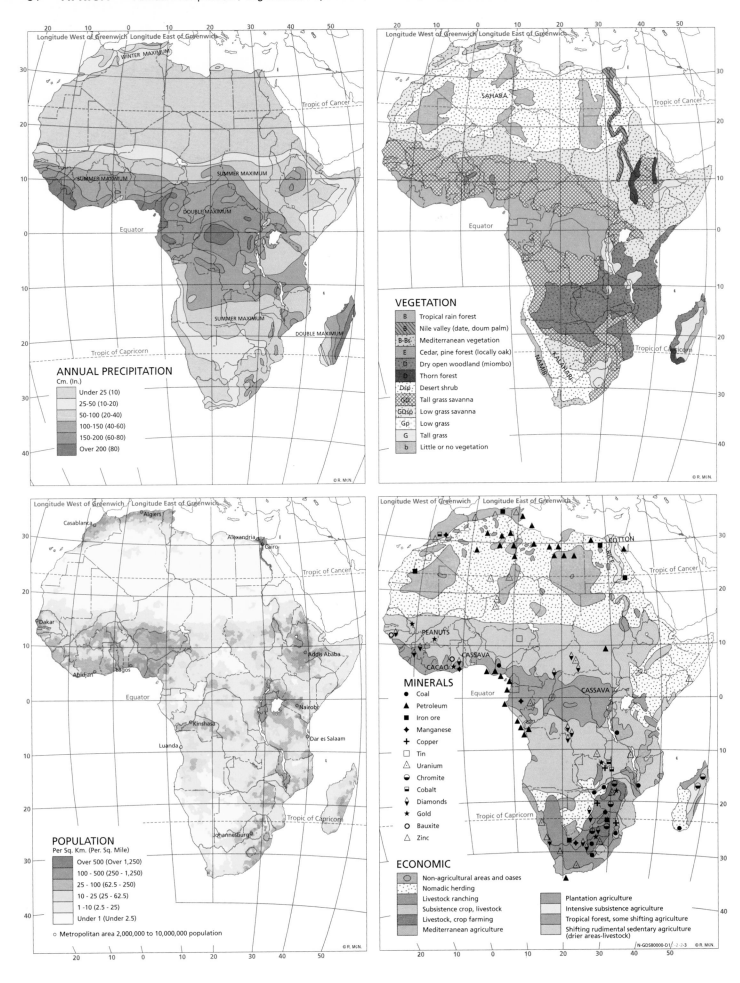

ANNUAL PRECIPITATION
Cm. (In.)
- Under 25 (10)
- 25-50 (10-20)
- 50-100 (20-40)
- 100-150 (40-60)
- 150-200 (60-80)
- Over 200 (80)

WINTER MAXIMUM

SUMMER MAXIMUM

SUMMER MAXIMUM

DOUBLE MAXIMUM

SUMMER MAXIMUM

DOUBLE MAXIMUM

Tropic of Cancer

Equator

Tropic of Capricorn

VEGETATION
B	Tropical rain forest
B	Nile valley (date, doum palm)
B-Bs	Mediterranean vegetation
E	Cedar, pine forest (locally oak)
D	Dry open woodland (miombo)
D	Thorn forest
Dsp	Desert shrub
GD	Tall grass savanna
GDsp	Low grass savanna
Gp	Low grass
G	Tall grass
b	Little or no vegetation

SAHARA

NAMIB

KALAHARI

POPULATION
Per Sq. Km. (Per. Sq. Mile)
- Over 500 (Over 1,250)
- 100 - 500 (250 - 1,250)
- 25 - 100 (62.5 - 250)
- 10 - 25 (25 - 62.5)
- 1 - 10 (2.5 - 25)
- Under 1 (Under 2.5)

○ Metropolitan area 2,000,000 to 10,000,000 population

Casablanca, Algiers, Alexandria, Cairo, Dakar, Addis Ababa, Abidjan, Lagos, Nairobi, Kinshasa, Dar es Salaam, Luanda, Johannesburg

MINERALS
- ● Coal
- ▲ Petroleum
- ■ Iron ore
- ◆ Manganese
- ✚ Copper
- □ Tin
- △ Uranium
- ◓ Chromite
- ▱ Cobalt
- ⬠ Diamonds
- ★ Gold
- ○ Bauxite
- △ Zinc

COTTON, RICE, PEANUTS, CACAO, CASSAVA, CASSAVA

ECONOMIC
- Non-agricultural areas and oases
- Nomadic herding
- Livestock ranching
- Subsistence crop, livestock
- Livestock, crop farming
- Mediterranean agriculture
- Plantation agriculture
- Intensive subsistence agriculture
- Tropical forest, some shifting agriculture
- Shifting rudimental sedentary agriculture (drier areas-livestock)

N-GDS80000-D1/ -2-2-3   © R. McN.

© R. McN.

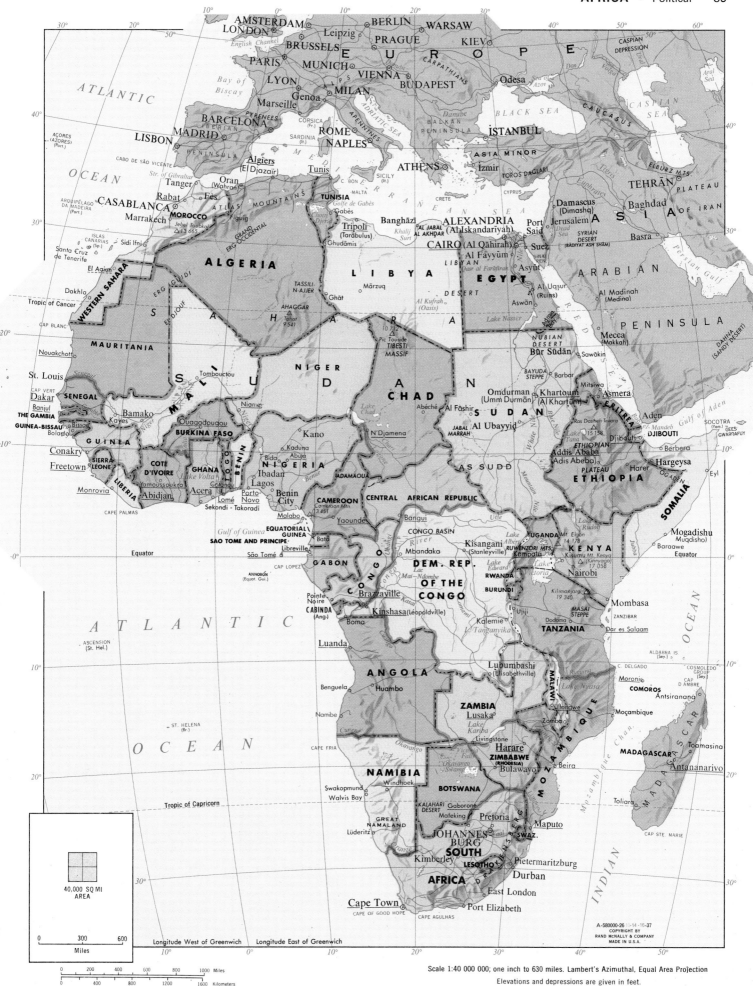

Scale 1:40 000 000; one inch to 630 miles. Lambert's Azimuthal, Equal Area Projection
Elevations and depressions are given in feet.

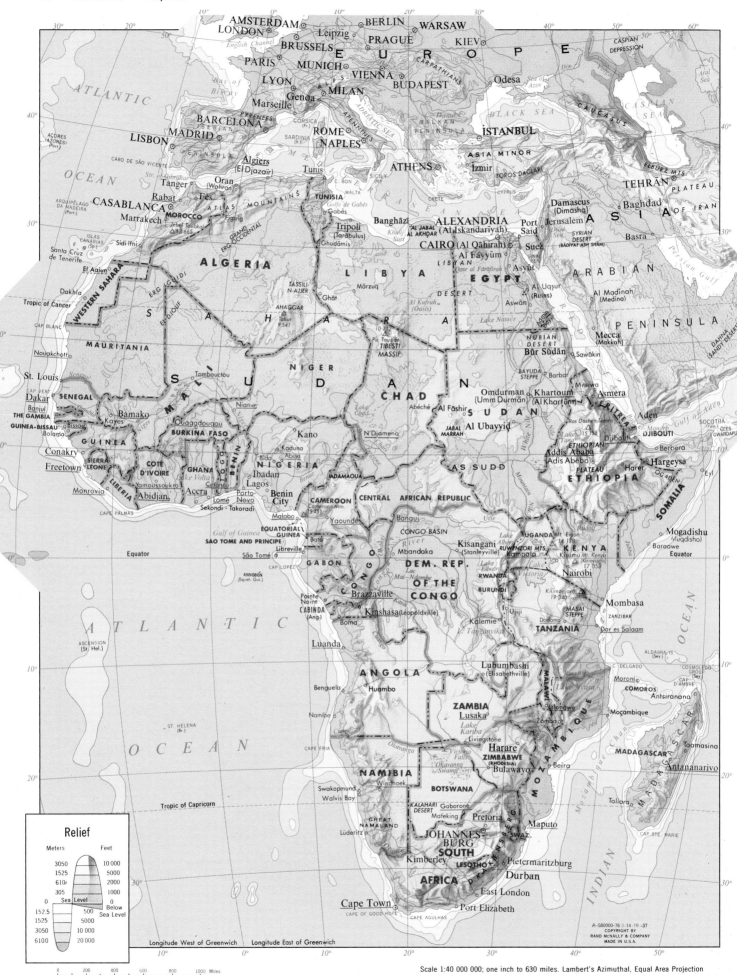

Relief

Meters	Feet
3050	10 000
1525	5000
610	2000
305	1000
0 Sea Level	Below Sea Level
152.5	500
1525	5000
3050	10 000
6100	20 000

Longitude West of Greenwich   Longitude East of Greenwich

0   200   400   600   800   1000 Miles
0   400   800   1200   1600 Kilometers

Scale 1:40 000 000; one inch to 630 miles. Lambert's Azimuthal, Equal Area Projection
Elevations and depressions are given in feet.

A-580000-76 8-14-16-37
COPYRIGHT BY
RAND McNALLY & COMPANY
MADE IN U.S.A.

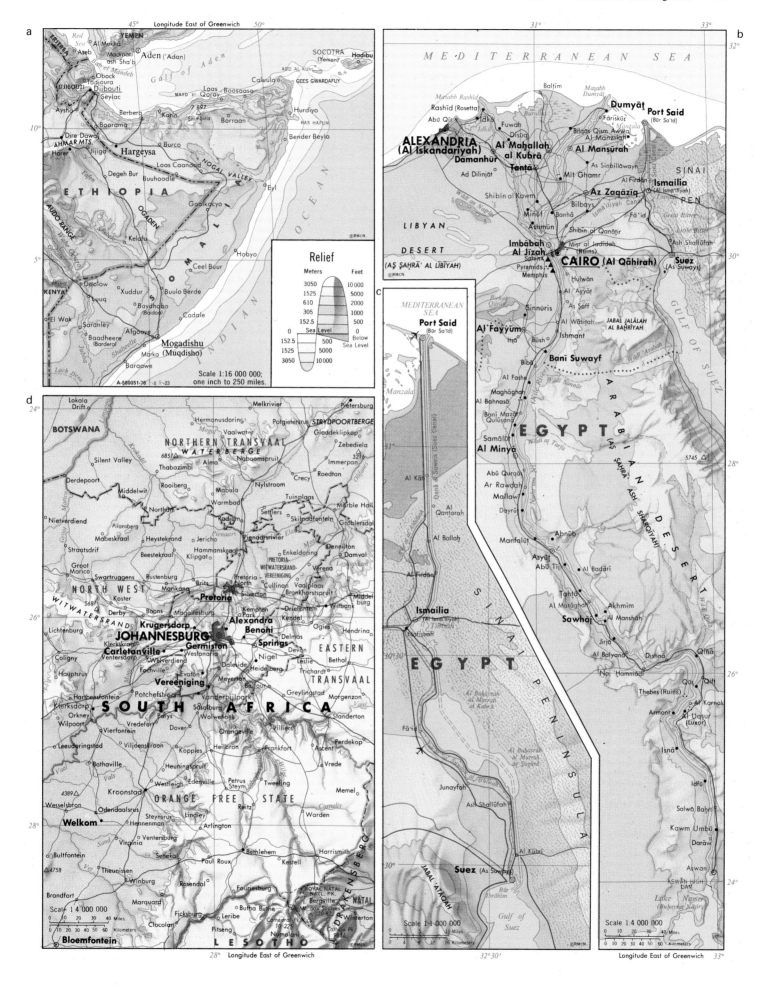

a

Red Sea
YEMEN
Al Mukha
Aden ('Adan)
Madinat ash Sha'b
ERITREA
Aseb
Bab el Mandeb
Gulf of Aden
SOCOTRA (Yemen)
Hadibu
ABD AL-KURI
GEES GWARDAFUY
Caluula
Obock
Tadjoura
DJIBOUTI
Djibouti
Seylac
Boosaaso
MAYD
Laas Qoray
Aysha
Berbera
Karin
Shimbiris 7 897
Borraan
Hurdiyo
RAS HAFUN
Dire Dawa
Jijiga
Hargeysa
Laas Caanood
Burco
Bender Beyla
AHMAR MTS
Harer
Degeh Bur
Buuhoodle
NOGAL VALLEY
Eyl
ETHIOPIA
OGADEN
AUDO RANGE
Gaalkacyo
Kelafa
Ceel Buur
Hobyo
KENYA
Doolow
Xuddur
Buulo Berde
Luuq
Baydhabo (Baidoa)
Cadale
El Wak
Saranley
Afgooye
Baadheere (Bardera)
SOMALIA
Mogadishu (Muqdisho)
Marka (Merca)
Baraawe
Lach Dera
INDIAN OCEAN

Relief

Meters		Feet
3050		10 000
1525		5000
610		2000
305		1000
152.5		500
0	Sea Level	0
152.5		500	Below Sea Level
1525		5000
3050		10 000

Scale 1:16 000 000;
one inch to 250 miles.
A-580051-76   -8 5 -23

b

MEDITERRANEAN SEA
Baltim
Masabb Dumyat
Masabb Rashid
Rashid (Rosetta)
Abu Qir
Idku
Fuwah
Burullus
Dumyat
Port Said (Bur Sa'id)
Fariskur
Bilqas Qism Awwal
Al Manzilah
Mangala
SINAI
ALEXANDRIA (Al Iskandariyah)
Damanhur
Al Mahallah al Kubra
Al Mansurah
As Sinbillawayn
Al Firdan
Ismailia (Al Isma'iliyah)
Tanta
Ad Dilinjat
Mit Ghamr
Az Zaqaziq
SINAI PEN.
Shibin al Kawm
Minuf
Bilbays
Isma'iliyah Can.
Fa'id
Great Bitter
LIBYAN
Banha
Ashmun
Shibin al Qanatir
Little Bitter
DESERT (AS SAHRA' AL LIBIYAH)
Imbabah
Al Jizah
Misr al Jadidah (Ruins)
Ash Shallufah
Pyramids
Sphinx
CAIRO (Al Qahirah)
Suez (As Suways)
Memphis
Hulwan
Birkat Qarun
Al 'Ayyat
Sinnuris
As Saff
Al Fayyum
Al Wasitah
JABAL JALALAH AL BAHRIYAH
GULF OF SUEZ
Ishmant
Itsa
Bush
Bani Suwayf
Biba
Wadi 'Arabah
Al Fashn
Maghaghah
Al Bahnasa
Bani Mazar
Quluṣana
EGYPT
Samalut
Al Minya
5745
Abu Qurqaş
Ar Rawdah
Mallawi
Dayrut
Wadi at Tarfa
Abnub
Manfalut
Asyut
Abu Tij
Al Badari
Tahta
Al Maraghah
Akhmim
Sawhaj
Al Manshah
Jirja
Al Balyana
Dishna
Naj Hammadi
Qina
Qus
Qift
Thebes (Ruins)
Al Karnak
Armant
Al Uqsur (Luxor)
Isna
Idfu
Salwa Bahri
Kawm Umbu
Daraw
Aswan
ASWAN HIGH DAM
Lake Nasser (Buhayrat Nasir)

ARABIAN DESERT (AS SAHRA' ASH SHARQIYAH)

c

MEDITERRANEAN SEA
Port Said (Bur Sa'id)
Manzala
Al Kab
Al Qantarah
Qana (Suwais (Suez Canal))
Al Ballah
Al Firdan
Ismailia (Al Isma'iliyah)
Timsah
Nafishah
Fa'id
Al Buhayrat al Murrah al Kubra
SINAI PENINSULA
EGYPT
Junayfah
Ash Shallufah
Al Buhayrat al Murrah aş Şughra
Al Kubri
Suez (As Suways)
JABAL 'ATAQAH
Bur Ibrahim
Gulf of Suez

Scale 1:1 000 000
0   10 Miles
0   4   8   12   16 Kilometers

Scale 1:4 000 000
0 10 20 30 40 Miles
0 10 20 30 40 50 60 Kilometers
Longitude East of Greenwich

d

Lokala Drift
Melkrivier
Pietersburg
BOTSWANA
Hermanusdoring
Potgietersrus
STRYDPOORTBERGE
Gladdeklipkop
NORTHERN TRANSVAAL
Vaalwater
Zebediela
6851 WATERBERGE
Alma
Naboomspruit
Silent Valley
Thabazimbi
Immerpan
3216
Derdepoort
Rooiberg
Crecy
Roedtan
Middelwit
Mabula
Nylstroom
Tuinplaas
Marble Hall
Northam
Warmbad
Settlers
Skilpadfontein
Groblersdal
Nietverdiend
Pilansberg
Radium
Jericho
Pienaarsrivier
Enkeldoring
Dennilton
Damval
Straatsdrif
Mabeskraal
Heystekrand
Klipgat
Hammanskraal
Verena
Groot Marico
Swartruggens
Rustenburg
Marikana
Brits
PRETORIA-WITWATERSRAND-VEREENIGING
Pretoria North
Cullinan
Vaalplaas
Bronkhorstspruit
NORTH WEST
568
Koster
Derby
Boons
Magaliesburg
Silverton
PRETORIA
Kempton Park
Driefontein
Witbank
Middelburg
Lichtenburg
Krugersdorp
Alexandra
Kendal
Hendrina
Coligny
Klerksdorp
JOHANNESBURG
Benoni
Ogies
EASTERN
Hauptsrus
Carletonville
Germiston
Springs
Delmas
Venterdorp
Fochville
Nigel
Devon
Leslie
Bethal
TRANSVAAL
Vereeniging
Evaton
Meyerton
Heidelberg
Trichardt
Hartbeesfontein
Potchefstroom
Vanderbijlpark
Balfour
Morgenzon
Standerton
Orkney
WITWATERSRAND
Wolwehoek
Greylingstad
Villiers
SOUTH AFRICA
Vredefort
Dover
Perdekop
Wilpoort
Vierfontein
Orangeville
Ascent
Klip
Leeudoringstad
Viljoenskroon
Koppies
Heilbron
Frankfort
Vrede
Bothaville
Westleigh
Edenville
Memel
4389
Kroonstad
Petrus Steyn
Tweeling
Vals
Reitz
ORANGE FREE STATE
Lindley
Warden
Cornelis
Wesselsbron
Steynsrus
Arlington
Welkom
Hennenman
Virginia
Ventersburg
Senekal
Bethlehem
Harrismith
Bultfontein
Winburg
Paul Roux
Kestell
4758
Theunissen
Rosendal
Brandfort
Marquard
Fouriesburg
DRAKENSBERG
NATAL
Ficksburg
Bergville
ROYAL NATAL NATL. PK.
aux Sources 10 822
Winterton
Clocolan
Leribe
Cathedral Pk
10 225
Bloemfontein
Butha Buthe
Pitseng
Cathkin Pk 9856
LESOTHO
Numolani
Clarens

Scale 1:4 000 000
0 10 20 30 40 Miles
0 10 20 30 40 Kilometers

28° Longitude East of Greenwich

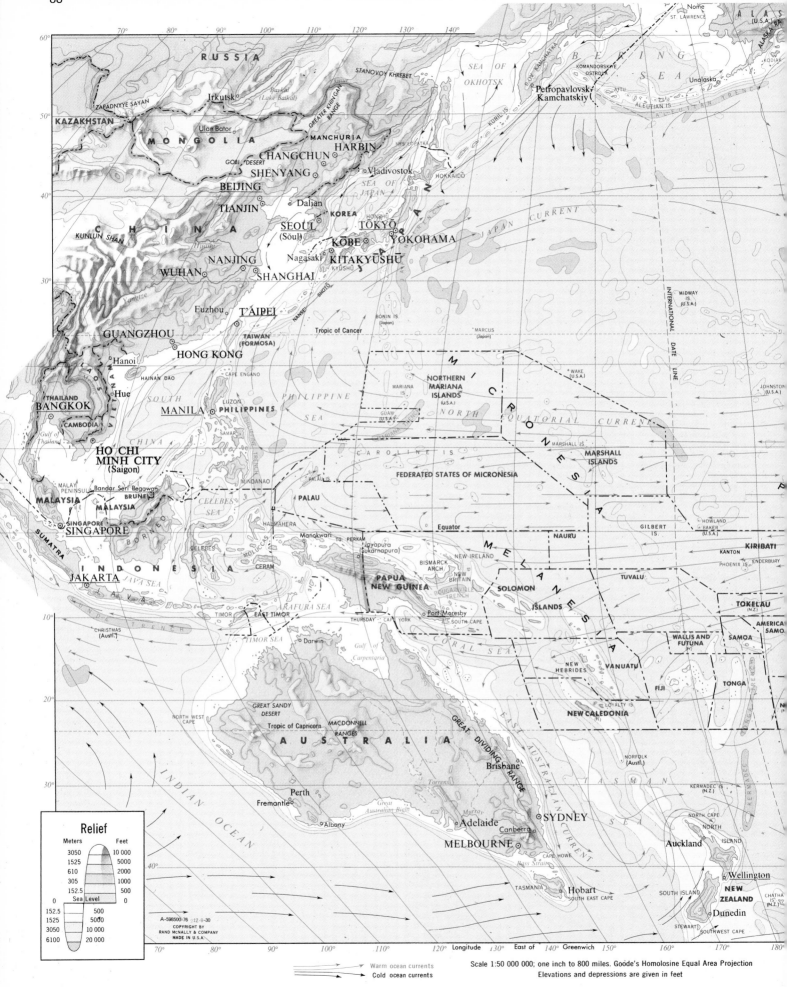

Relief

Meters	Feet
3050	10 000
1525	5000
610	2000
305	1000
152.5	500
0 Sea Level	0
152.5	500
1525	5000
3050	10 000
6100	20 000

A-598500-76 12-9-30
COPYRIGHT BY
RAND McNALLY & COMPANY
MADE IN U.S.A.

→ Warm ocean currents
→ Cold ocean currents

Scale 1:50 000 000; one inch to 800 miles. Goode's Homolosine Equal Area Projection
Elevations and depressions are given in feet

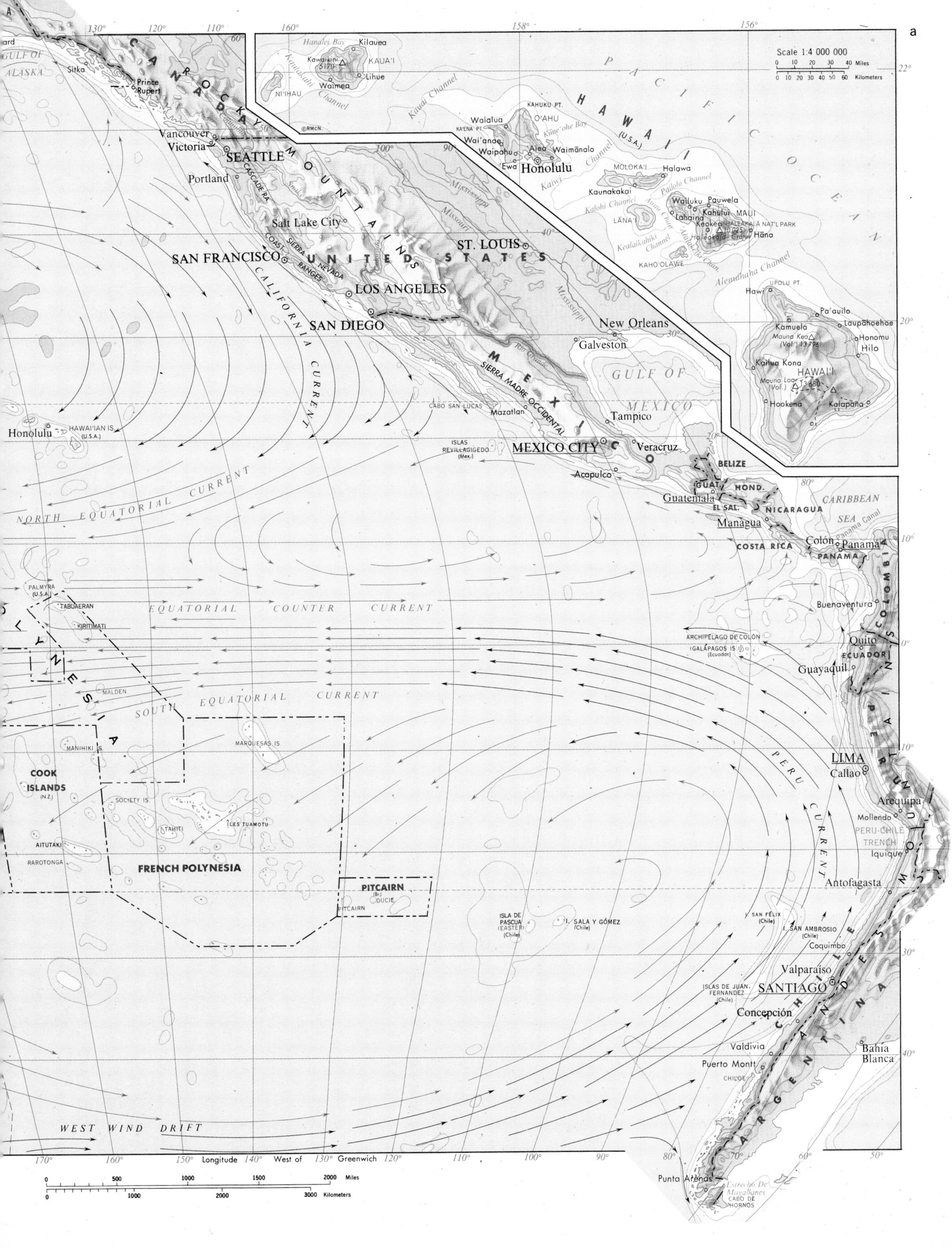

a

Scale 1:4 000 000
0 10 20 30 40 Miles
0 10 20 30 40 50 60 Kilometers

GULF OF ALASKA

Sitka
Prince Rupert
Vancouver
Victoria
SEATTLE
Portland
CANADA
ROCKY MOUNTAINS
CASCADE RA.
Salt Lake City
SAN FRANCISCO
COAST RANGES
SIERRA NEVADA
UNITED STATES
ST. LOUIS
LOS ANGELES
SAN DIEGO
CALIFORNIA CURRENT
MEXICO
SIERRA MADRE OCCIDENTAL
CABO SAN LUCAS
Mazatlan
Rio Grande
New Orleans
Galveston
GULF OF MEXICO
Tampico
ISLAS REVILLAGIGEDO (Mex.)
MEXICO CITY
Veracruz
Acapulco
BELIZE
GUAT. HOND.
Guatemala
EL SAL. NICARAGUA
Managua
COSTA RICA
Panama Canal
Colón Panamá
PANAMA
CARIBBEAN SEA

Honolulu
HAWAIIAN IS. (U.S.A.)
NORTH EQUATORIAL CURRENT
PALMYRA (U.S.A.)
TABUAERAN
KIRITIMATI
EQUATORIAL COUNTER CURRENT
POLYNESIA
MALDEN
SOUTH EQUATORIAL CURRENT
MANIHIKI IS.
MARQUESAS IS.
COOK ISLANDS (N.Z.)
SOCIETY IS.
ÎLES TUAMOTU
AITUTAKI
TAHITI
RAROTONGA
FRENCH POLYNESIA
PITCAIRN (Br.) DUCIE
PITCAIRN
ISLA DE PASCUA (EASTER) (Chile)
I. SALA Y GÓMEZ (Chile)

Buenaventura
ARCHIPIÉLAGO DE COLÓN (GALÁPAGOS IS.) (Ecuador)
Quito
ECUADOR
Guayaquil
COLOMBIA
PERU
LIMA
Callao
Arequipa
Mollendo
PERU-CHILE TRENCH
Iquique
PERU CURRENT
Antofagasta
SAN FÉLIX (Chile)
I. SAN AMBROSIO (Chile)
Coquimbo
Valparaíso
ISLAS DE JUAN FERNANDEZ (Chile)
SANTIAGO
Concepción
ANDES
CHILE
ARGENTINA
Valdivia
Bahía Blanca
Puerto Montt
CHILOÉ

WEST WIND DRIFT

Punta Arenas
Estrecho De Magallanes
CABO DE HORNOS

170° 160° 150° Longitude 140° West of 130° Greenwich 120° 110° 100° 90° 80° 70° 60° 50°

0 500 1000 1500 2000 Miles
0 1000 2000 3000 Kilometers

Handei Bay
Kilauea
Kawaikini 5170
KAUA'I
Lihue
NI'IHAU
Waimea
Kauai Channel
KAHUKU PT.
Waialua
O'AHU
KA'ENA PT.
Kāne'ohe Bay
Wai'anae
Waipahu
Aiea Waimānalo
Ewa
Honolulu
Kaiwi Channel
MOLOKA'I
Kaunakakai
Halawa
Kalohi Channel
LĀNA'I
Wailuku Pauwela
Kahului MAUI
Lahaina Keokea HALEAKALA NAT'L PARK
Haleakala Crater Hāna
KAHO'OLAWE
Kealaikahiki Channel
Alenuihaha Channel
HAWAII (U.S.A.)
UPOLU PT.
Hawi
Pa'auilo
Laupāhoehoe
Kamuela
Mauna Kea (Vol.) 13,796
Honomu
Hilo
Kailua Kona
HAWAI'I
Mauna Loa (Vol.)
Hookena
Kalapana
PACIFIC OCEAN

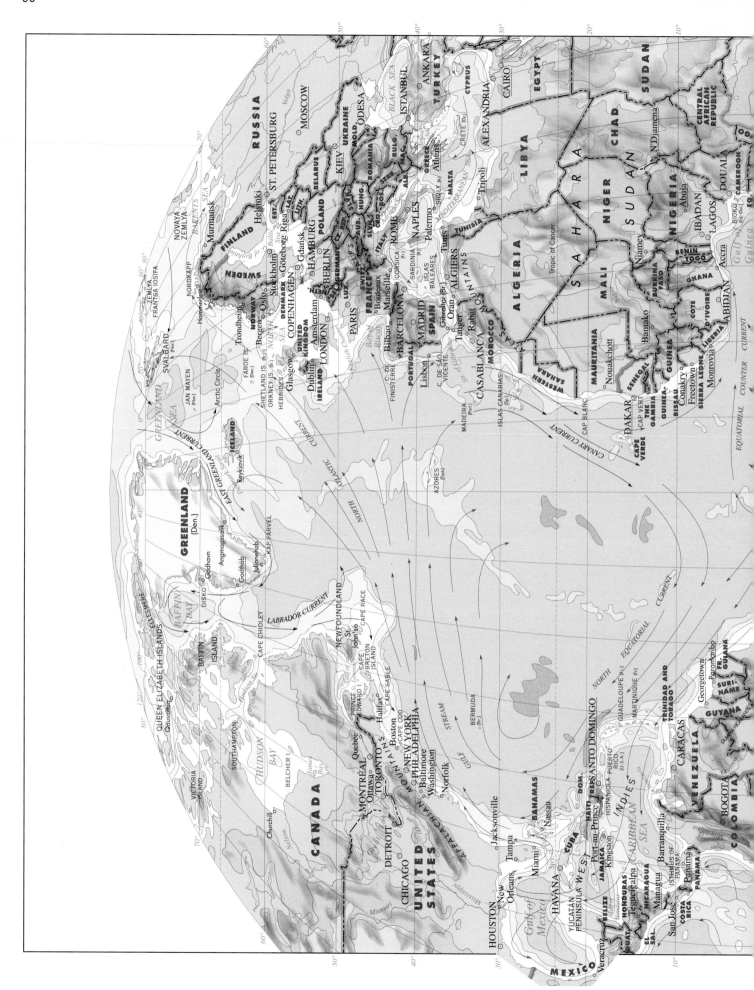

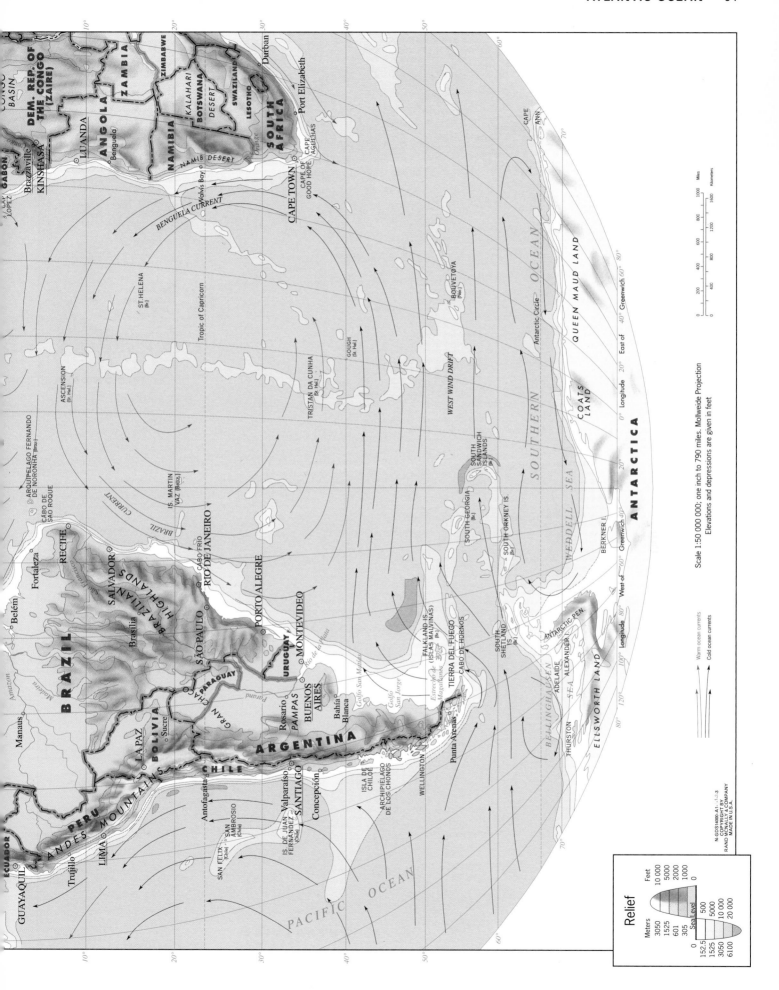

Scale 1:50 000 000; one inch to 790 miles. Mollweide Projection

Elevations and depressions are given in feet

→ Warm ocean currents

→ Cold ocean currents

N-GDS14000-A1--.1--3
COPYRIGHT BY
RAND McNALLY & COMPANY
MADE IN U.S.A.

Relief

Meters	Feet
3050	10 000
1525	5000
601	2000
305	1000
0	Sea Level
152.5	500
1525	5000
3050	10 000
6100	20 000

Miles
Kilometers
0 200 400 600 800 1000
0   400   800 1200 1600

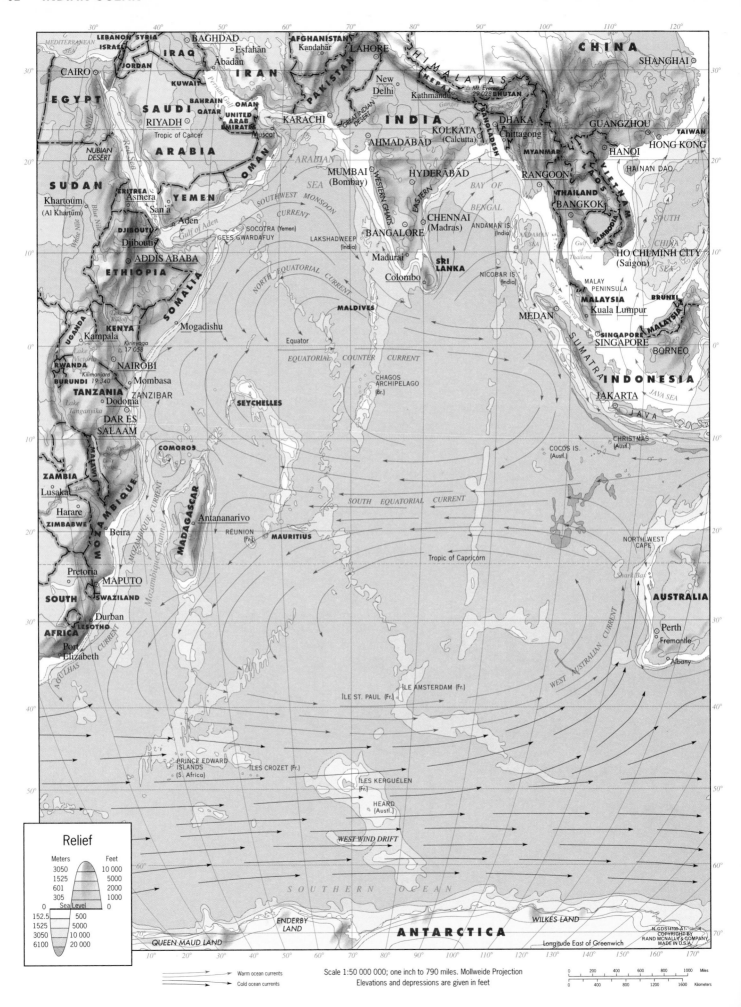

Relief

Meters		Feet
3050		10 000
1525		5000
601		2000
305		1000
0	Sea Level	0
152.5		500
1525		10 000
3050		10 000
6100		20 000

Warm ocean currents
Cold ocean currents

Scale 1:50 000 000; one inch to 790 miles. Mollweide Projection
Elevations and depressions are given in feet

| 0 | 200 | 400 | 600 | 800 | 1000 | Miles |
| 0 | 400 | 800 | 1200 | 1600 | | Kilometers |

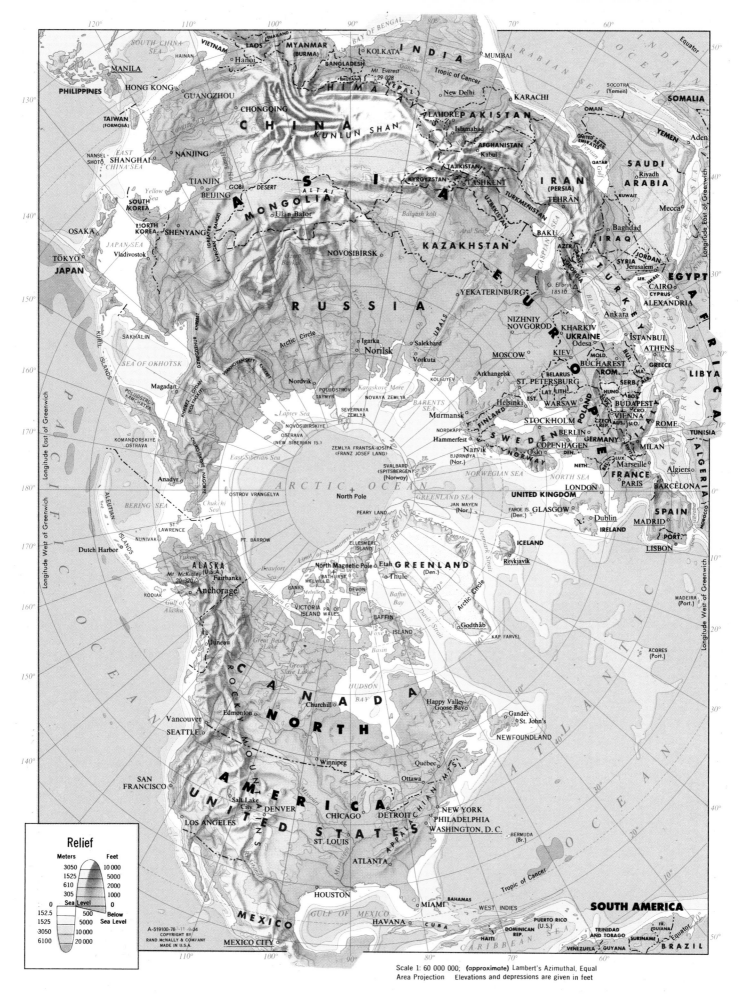

Relief

Meters	Feet
3050	10 000
1525	5000
610	2000
305	1000
0 Sea Level	0
152.5	500 Below Sea Level
1525	5000
3050	10 000
6100	20 000

A-519100-76 -11 -9-94
COPYRIGHT BY
RAND McNALLY & COMPANY
MADE IN U.S.A.

Scale 1: 60 000 000; (approximate) Lambert's Azimuthal, Equal
Area Projection   Elevations and depressions are given in feet

WORLD POLITICAL INFORMATION TABLE

This table gives the area, population, population density, political status, capital, and predominant languages for every country in the world. The political units listed are categorized by political status in the form of government column of the table, as follows: A—independent countries; B—internally independent political entities which are under the protection of another country in matters of defense and foreign affairs; C—colonies and other dependent political units; and D—the major administrative subdivisions of Australia, Canada, China, the United Kingdom, and the United States. For comparison, the table also includes the continents and the world. All footnotes appear at the end of the table.

The populations are estimates for January 1, 2004, made by Rand McNally on the basis of official data, United States Census Bureau estimates, and other available information. Area figures include inland water.

REGION OR POLITICAL DIVISION	Area Sq. Mi.	Est. Pop. 1/1/04	Pop. Per Sq. Mi.	Form of Government and Ruling Power	Capital	Predominant Languages	International Organizations
Afars and Issas see Djibouti							
Afghanistan	251,773	29,205,000	116	Transitional ... A	Kābul	Dari, Pashto, Uzbek, Turkmen	UN
Africa	11,700,000	866,305,000	74				
Alabama	52,419	4,515,000	86	State (U.S.) ... D	Montgomery	English	
Alaska	663,267	650,000	1.0	State (U.S.) ... D	Juneau	English, indigenous	
Albania	11,100	3,535,000	318	Republic ... A	Tiranë	Albanian, Greek	NATO/PP, UN
Alberta	255,541	3,215,000	13	Province (Canada) ... D	Edmonton	English	
Algeria	919,595	33,090,000	36	Republic ... A	Algiers (El Djazaïr)	Arabic, Berber dialects, French	AL, AU, OPEC, UN
American Samoa	77	58,000	753	Unincorporated territory (U.S.) ... C	Pago Pago	Samoan, English	
Andorra	181	70,000	387	Parliamentary co-principality (Spanish and French) ... B	Andorra	Catalan, Spanish (Castilian), French, Portuguese	UN
Angola	481,354	10,875,000	23	Republic ... A	Luanda	Portuguese, indigenous	AU, COMESA, UN
Anguilla	37	13,000	351	Overseas territory (U.K.) ... C	The Valley	English	
Anhui	53,668	61,215,000	1,141	Province (China) ... D	Hefei	Chinese (Mandarin)	
Antarctica	5,400,000	(¹)					
Antigua and Barbuda	171	68,000	398	Parliamentary state ... A	St. John's	English, local dialects	OAS, UN
Aomen (Macau)	6.9	445,000	64,493	Special administrative region (China) ... D	Macau (Aomen)	Chinese (Cantonese), Portuguese	
Argentina	1,073,519	38,945,000	36	Republic ... A	Buenos Aires	Spanish, English, Italian, German, French	MERCOSUR, OAS, UN
Arizona	113,998	5,600,000	49	State (U.S.) ... D	Phoenix	English	
Arkansas	53,179	2,735,000	51	State (U.S.) ... D	Little Rock	English	
Armenia	11,506	3,325,000	289	Republic ... A	Yerevan	Armenian, Russian	CIS, NATO/PP, UN
Aruba	75	71,000	947	Self-governing territory (Netherlands protection) ... B	Oranjestad	Dutch, Papiamento, English, Spanish	
Ascension	34	1,000	29	Dependency (St. Helena) ... C	Georgetown	English	
Asia	17,300,000	3,839,320,000	222				
Australia	2,969,910	19,825,000	6.7	Federal parliamentary state ... A	Canberra	English, indigenous	ANZUS, UN
Australian Capital Territory	911	325,000	357	Territory (Australia) ... D	Canberra	English	
Austria	32,378	8,170,000	252	Federal republic ... A	Vienna (Wien)	German	EU, NATO/PP, UN
Azerbaijan	33,437	7,850,000	235	Republic ... A	Baku (Bakı)	Azeri, Russian, Armenian	CIS, NATO/PP, UN
Bahamas	5,382	300,000	56	Parliamentary state ... A	Nassau	English, Creole	OAS, UN
Bahrain	267	675,000	2,528	Monarchy ... A	Al Manāmah	Arabic, English, Persian, Urdu	AL, UN
Bangladesh	55,598	139,875,000	2,516	Republic ... A	Dkaha (Dacca)	Bangla, English	UN
Barbados	166	280,000	1,687	Parliamentary state ... A	Bridgetown	English	OAS, UN
Beijing (Peking)	6,487	14,135,000	2,179	Autonomous city (China) ... D	Beijing (Peking)	Chinese (Mandarin)	
Belarus	80,155	10,315,000	129	Republic ... A	Minsk	Belarussian, Russian	CIS, NATO/PP, UN
Belau see Palau							
Belgium	11,787	10,340,000	877	Constitutional monarchy ... A	Brussels (Bruxelles)	Dutch (Flemish), French, German	EU, NATO, UN
Belize	8,867	270,000	30	Parliamentary state ... A	Belmopan	English, Spanish, Mayan, Garifuna, Creole	OAS, UN
Benin	43,484	7,145,000	164	Republic ... A	Porto-Novo and Cotonou	French, Fon, Yoruba, indigenous	AU, UN
Bermuda	21	65,000	3,095	Overseas territory (U.K. protection) ... B	Hamilton	English, Portuguese	
Bhutan	17,954	2,160,000	120	Monarchy (Indian protection) ... B	Thimphu	Dzongkha, Tibetan and Nepalese dialects	UN
Bolivia	424,165	8,655,000	20	Republic ... A	La Paz and Sucre	Aymara, Quechua, Spanish	OAS, UN
Bosnia and Herzegovina	19,767	4,000,000	202	Republic ... A	Sarajevo	Bosnian, Serbian, Croatian	UN
Botswana	224,607	1,570,000	7.0	Republic ... A	Gaborone	English, Tswana	AU, UN
Brazil	3,300,172	183,080,000	55	Federal republic ... A	Brasília	Portuguese, Spanish, English, French	MERCOSUR, OAS, UN
British Columbia	364,764	4,245,000	12	Province (Canada) ... D	Victoria	English	
British Indian Ocean Territory	23	(¹)		Overseas territory (U.K.) ... C		English	
British Virgin Islands	58	22,000	379	Overseas territory (U.K.) ... C	Road Town	English	
Brunei	2,226	360,000	162	Monarchy ... A	Bandar Seri Begawan	Malay, English, Chinese	ASEAN, UN
Bulgaria	42,855	7,550,000	176	Republic ... A	Sofia (Sofiya)	Bulgarian, Turkish	NATO, UN
Burkina Faso	105,869	13,400,000	127	Republic ... A	Ouagadougou	French, indigenous	AU, UN
Burma see Myanmar							
Burundi	10,745	6,165,000	574	Republic ... A	Bujumbura	French, Kirundi, Swahili	AU, COMESA, UN
California	163,696	35,590,000	217	State (U.S.) ... D	Sacramento	English	
Cambodia	69,898	13,245,000	189	Constitutional monarchy ... A	Phnom Penh (Phnum Pénh)	Khmer, French, English	ASEAN, UN
Cameroon	183,568	15,905,000	87	Republic ... A	Yaoundé	English, French, indigenous	AU, UN
Canada	3,855,103	32,360,000	8.4	Federal parliamentary state ... A	Ottawa	English, French, other	NAFTA, NATO, OAS, UN
Cape Verde	1,557	415,000	267	Republic ... A	Praia	Portuguese, Crioulo	AU, UN
Cayman Islands	102	43,000	422	Overseas territory (U.K.) ... C	George Town	English	
Central African Republic	240,536	3,715,000	15	Republic ... A	Bangui	French, Sango, indigenous	AU, UN
Ceylon see Sri Lanka							
Chad	495,755	9,395,000	19	Republic ... A	N'Djamena	Arabic, French, indigenous	AU, UN
Channel Islands	75	155,000	2,067	Two crown dependencies (U.K. protection)		English, French	
Chile	291,930	15,745,000	54	Republic ... A	Santiago	Spanish	OAS, UN
China (excl. Taiwan)	3,690,045	1,298,720,000	352	Socialist republic ... A	Beijing (Peking)	Chinese dialects	UN
Chongqing	31,815	31,600,000	993	Autonomous city (China) ... D	Chongqing (Chungking)	Chinese (Mandarin)	
Christmas Island	52	400	7.7	External territory (Australia) ... C	Settlement	English, Chinese, Malay	
Cocos (Keeling) Islands	5.4	600	111	External territory (Australia) ... C	West Island	English, Cocos-Malay	
Colombia	439,737	41,985,000	95	Republic ... A	Bogotá	Spanish	OAS, UN
Colorado	104,094	4,565,000	44	State (U.S.) ... D	Denver	English	
Comoros (excl. Mayotte)	863	640,000	742	Republic ... A	Moroni	Arabic, French, Shikomoro	AL, AU, COMESA, UN
Congo	132,047	2,975,000	23	Republic ... A	Brazzaville	French, Lingala, Monokutuba, indigenous	AU, UN
Congo, Democratic Republic of the (Zaire)	905,446	57,445,000	63	Republic ... A	Kinshasa	French, Lingala, indigenous	AU, COMESA, UN
Connecticut	5,543	3,495,000	631	State (U.S.) ... D	Hartford	English	

REGION OR POLITICAL DIVISION	Area Sq. Mi.	Est. Pop. 1/1/04	Pop. Per Sq. Mi.	Form of Government and Ruling Power	Capital	Predominant Languages	International Organizations
Cook Islands	91	21,000	231	Self-governing territory (New Zealand protection) ... B	Avarua	English, Maori	
Costa Rica	19,730	3,925,000	199	Republic ... A	San José	Spanish, English	OAS, UN
Cote d'Ivoire (Ivory Coast)	124,504	17,145,000	138	Republic ... A	Abidjan and Yamoussoukro	French, Dioula and other indigenous	AU, UN
Croatia	21,829	4,430,000	203	Republic ... A	Zagreb	Croatian	NATO/PP, UN
Cuba	42,804	11,290,000	264	Socialist republic ... A	Havana (La Habana)	Spanish	OAS, UN
Cyprus	3,572	775,000	217	Republic ... A	Nicosia	Greek, Turkish, English	EU, UN
Czech Republic	30,450	10,250,000	337	Republic ... A	Prague (Praha)	Czech	EU, NATO, UN
Delaware	2,489	820,000	329	State (U.S.) ... D	Dover	English	
Denmark	16,640	5,405,000	325	Constitutional monarchy ... A	Copenhagen (København)	Danish	EU, NATO, UN
District of Columbia	68	565,000	8,309	Federal district (U.S.) ... D	Washington	English	
Djibouti	8,958	460,000	51	Republic ... A	Djibouti	French, Arabic, Somali, Afar	AL, AU, COMESA, UN
Dominica	290	69,000	238	Republic ... A	Roseau	English, French	OAS, UN
Dominican Republic	18,730	8,775,000	468	Republic ... A	Santo Domingo	Spanish	OAS, UN
East Timor	5,743	1,010,000	176	Republic ... A	Dili	Portuguese, Tetum, Bahasa Indonesia (Malay), English	UN
Ecuador	109,484	13,840,000	126	Republic ... A	Quito	Spanish, Quechua, indigenous	OAS, UN
Egypt	386,662	75,420,000	195	Republic ... A	Cairo (Al Qāhirah)	Arabic	AL, AU, CAEU, COMESA, UN
Ellice Islands see Tuvalu							
El Salvador	8,124	6,530,000	804	Republic ... A	San Salvador	Spanish, Nahua	OAS, UN
England	50,356	50,360,000	1,000	Administrative division (U.K.) ... D	London	English	
Equatorial Guinea	10,831	515,000	48	Republic ... A	Malabo	French, Spanish, indigenous, English	AU, UN
Eritrea	45,406	4,390,000	97	Republic ... A	Asmera	Afar, Arabic, Tigre, Kunama, Tigrinya, other	AU, COMESA, UN
Estonia	17,462	1,405,000	80	Republic ... A	Tallinn	Estonian, Russian, Ukrainian, Finnish, other	EU, NATO, UN
Ethiopia	426,373	67,210,000	158	Federal republic ... A	Addis Ababa (Adis Abeba)	Amharic, Tigrinya, Orominga, Guaraginga, Somali, Arabic	AU, COMESA, UN
Europe	3,800,000	729,330,000	192				
Falkland Islands (²)	4,700	3,000	0.6	Overseas territory (U.K.) ... C	Stanley	English	
Faroe Islands	540	47,000	87	Self-governing territory (Danish protection) ... B	Tórshavn	Danish, Faroese	
Fiji	7,056	875,000	124	Republic ... A	Suva	English, Fijian, Hindustani	UN
Finland	130,559	5,210,000	40	Republic ... A	Helsinki (Helsingfors)	Finnish, Swedish, Sami, Russian	EU, NATO/PP, UN
Florida	65,755	17,070,000	260	State (U.S.) ... D	Tallahassee	English	
France (excl. Overseas Departments)	208,482	60,305,000	289	Republic ... A	Paris	French	EU, NATO, UN
French Guiana	32,253	190,000	5.9	Overseas department (France) ... C	Cayenne	French	
French Polynesia	1,544	265,000	172	Overseas territory (France) ... C	Papeete	French, Tahitian	
Fujian	46,332	35,495,000	766	Province (China) ... D	Fuzhou	Chinese dialects	
Gabon	103,347	1,340,000	13	Republic ... A	Libreville	French, Fang, indigenous	AU, UN
Gambia, The	4,127	1,525,000	370	Republic ... A	Banjul	English, Malinke, Wolof, Fula, indigenous	AU, UN
Gansu	173,746	26,200,000	151	Province (China) ... D	Lanzhou	Chinese (Mandarin), Mongolian, Tibetan dialects	
Gaza Strip	139	1,300,000	9,353	Israeli territory with limited self-government		Arabic, Hebrew	(⁴)
Georgia	59,425	8,710,000	147	State (U.S.) ... D	Atlanta	English	
Georgia	26,911	4,920,000	183	Republic ... A	Tbilisi	Georgian, Russian, Armenian, Azeri, other	NATO/PP, UN
Germany	137,847	82,415,000	598	Federal republic ... A	Berlin	German	EU, NATO, UN
Ghana	92,098	20,615,000	224	Republic ... A	Accra	English, Akan and other indigenous	AU, UN
Gibraltar (²)	2.3	28,000	12,174	Overseas territory (U.K.) ... C	Gibraltar	English, Spanish, Italian, Portuguese	
Gilbert Islands see Kiribati							
Golan Heights	454	37,000	81	Occupied by Israel		Arabic, Hebrew	
Great Britain see United Kingdom							
Greece	50,949	10,635,000	209	Republic ... A	Athens (Athina)	Greek, English, French	EU, NATO, UN
Greenland	836,331	56,000	0.07	Self-governing territory (Danish protection) ... B	Godthåb (Nuuk)	Danish, Greenlandic, English	
Grenada	133	89,000	669	Parliamentary state ... A	St. George's	English, French	OAS, UN
Guadeloupe (incl. Dependencies)	687	440,000	640	Overseas department (France) ... C	Basse-Terre	French, Creole	
Guam	212	165,000	778	Unincorporated territory (U.S.) ... C	Hagåtña (Agana)	English, Chamorro, Japanese	
Guangdong	68,649	88,375,000	1,287	Province (China) ... D	Guangzhou (Canton)	Chinese dialects, Miao-Yao	
Guangxi Zhuangzu	91,236	45,905,000	503	Autonomous region (China) ... D	Nanning	Chinese dialects, Thai, Miao-Yao	
Guatemala	42,042	14,095,000	335	Republic ... A	Guatemala	Spanish, indigenous	OAS, UN
Guernsey (incl. Dependencies)	30	65,000	2,167	Crown dependency (U.K. protection) ... B	St. Peter Port	English, French	
Guinea	94,926	9,135,000	96	Republic ... A	Conakry	French, indigenous	AU, UN
Guinea-Bissau	13,948	1,375,000	99	Republic ... A	Bissau	Portuguese, Crioulo, indigenous	AU, UN
Guizhou	65,637	36,045,000	549	Province (China) ... D	Guiyang	Chinese (Mandarin), Thai, Miao-Yao	
Guyana	83,000	705,000	8.5	Republic ... A	Georgetown	English, indigenous, Creole, Hindi, Urdu	OAS, UN
Hainan	13,205	8,050,000	610	Province (China) ... D	Haikou	Chinese, Min, Tai	
Haiti	10,714	7,590,000	708	Republic ... A	Port-au-Prince	Creole, French	OAS, UN
Hawaii	10,931	1,260,000	115	State (U.S.) ... D	Honolulu	English, Hawaiian, Japanese	
Hebei	73,359	68,965,000	940	Province (China) ... D	Shijiazhuang	Chinese (Mandarin)	
Heilongjiang	181,082	37,725,000	208	Province (China) ... D	Harbin	Chinese dialects, Mongolian, Tungus	
Henan	64,479	94,655,000	1,468	Province (China) ... D	Zhengzhou	Chinese (Mandarin)	
Holland see Netherlands							
Honduras	43,277	6,745,000	156	Republic ... A	Tegucigalpa	Spanish, indigenous	OAS, UN
Hubei	72,356	61,645,000	852	Province (China) ... D	Wuhan	Chinese dialects	
Hunan	81,082	65,855,000	812	Province (China) ... D	Changsha	Chinese dialects, Miao-Yao	
Hungary	35,919	10,045,000	280	Republic ... A	Budapest	Hungarian	EU, NATO, UN
Iceland	39,769	280,000	7.0	Republic ... A	Reykjavik	Icelandic, English, other	EFTA, NATO, UN
Idaho	83,570	1,370,000	16	State (U.S.) ... D	Boise	English	
Illinois	57,914	12,690,000	219	State (U.S.) ... D	Springfield	English	
India (incl. part of Jammu and Kashmir)	1,222,510	1,057,415,000	865	Federal republic ... A	New Delhi	English, Hindi, Telugu, Bengali, indigenous	UN
Indiana	36,418	6,215,000	171	State (U.S.) ... D	Indianapolis	English	
Indonesia	735,310	236,680,000	322	Republic ... A	Jakarta	Bahasa Indonesia (Malay), English, Dutch, indigenous	ASEAN, OPEC, UN
Iowa	56,272	2,955,000	53	State (U.S.) ... D	Des Moines	English	
Iran	636,372	68,650,000	108	Islamic republic ... A	Tehrān	Persian, Turkish dialects, Kurdish, other	OPEC, UN
Iraq	169,235	25,025,000	148	Republic ... A	Baghdād	Arabic, Kurdish, Assyrian, Armenian	AL, CAEU, OPEC, UN
Ireland	27,133	3,945,000	145	Republic ... A	Dublin (Baile Átha Cliath)	English, Irish Gaelic	EU, NATO/PP, UN
Isle of Man	221	74,000	335	Crown dependency (U.K. protection) ... B	Douglas	English, Manx Gaelic	

REGION OR POLITICAL DIVISION	Area Sq. Mi.	Est. Pop. 1/1/04	Pop. Per Sq. Mi.	Form of Government and Ruling Power	Capital	Predominant Languages	International Organizations
Israel (excl. Occupied Areas)	8,019	6,160,000	768	Republic ... A	Jerusalem (Yerushalayim)	Hebrew, Arabic	UN
Italy	116,342	58,030,000	499	Republic ... A	Rome (Roma)	Italian, German, French, Slovene	EU, NATO, UN
Ivory Coast see Cote d'Ivoire	...	...	...				
Jamaica	4,244	2,705,000	637	Parliamentary state ... A	Kingston	English, Creole	OAS, UN
Japan	145,850	127,285,000	873	Constitutional monarchy ... A	Tōkyō	Japanese	UN
Jersey	45	90,000	2,000	Crown dependency (U.K. protection) ... B	St. Helier	English, French	
Jiangsu	39,614	76,065,000	1,920	Province (China) ... D	Nanjing (Nanking)	Chinese dialects	
Jiangxi	64,325	42,335,000	658	Province (China) ... D	Nanchang	Chinese dialects	
Jilin	72,201	27,895,000	386	Province (China) ... D	Changchun	Chinese (Mandarin), Mongolian, Korean	
Jordan	34,495	5,535,000	160	Constitutional monarchy ... A	'Ammān	Arabic	AL, CAEU, UN
Kansas	82,277	2,730,000	33	State (U.S.) ... D	Topeka	English	
Kazakhstan	1,049,156	16,780,000	16	Republic ... A	Astana (Aqmola)	Kazakh, Russian	CIS, NATO/PP, UN
Kentucky	40,409	4,130,000	102	State (U.S.) ... D	Frankfort	English	
Kenya	224,961	31,840,000	142	Republic ... A	Nairobi	English, Swahili, indigenous	AU, COMESA, UN
Kiribati	313	100,000	319	Republic ... A	Bairiki	English, I-Kiribati	UN
Korea, North	46,540	22,585,000	485	Socialist republic ... A	P'yŏngyang	Korean	UN
Korea, South	38,328	48,450,000	1,264	Republic ... A	Seoul (Sŏul)	Korean	UN
Kuwait	6,880	2,220,000	323	Constitutional monarchy ... A	Kuwait (Al Kuwayt)	Arabic, English	AL, CAEU, OPEC, UN
Kyrgyzstan	77,182	4,930,000	64	Republic ... A	Bishkek	Kirghiz, Russian	CIS, NATO/PP, UN
Laos	91,429	5,995,000	66	Socialist republic ... A	Viangchan (Vientiane)	Lao, French, English	ASEAN, UN
Latvia	24,942	2,340,000	94	Republic ... A	Riga	Latvian, Lithuanian, Russian, other	EU, NATO, UN
Lebanon	4,016	3,755,000	935	Republic ... A	Beirut (Bayrūt)	Arabic, French, Armenian, English	AL, UN
Lesotho	11,720	1,865,000	159	Constitutional monarchy ... A	Maseru	English, Sesotho, Zulu, Xhosa	AU, UN
Liaoning	56,255	43,340,000	770	Province (China) ... D	Shenyang (Mukden)	Chinese (Mandarin), Mongolian	
Liberia	43,000	3,345,000	78	Republic ... A	Monrovia	English, indigenous	AU, UN
Libya	679,362	5,565,000	8.2	Socialist republic ... A	Tripoli (Tarābulus)	Arabic	AL, AU, CAEU, OPEC, UN
Liechtenstein	62	33,000	532	Constitutional monarchy ... A	Vaduz	German	EFTA, UN
Lithuania	25,213	3,590,000	142	Republic ... A	Vilnius	Lithuanian, Polish, Russian	EU, NATO, UN
Louisiana	51,840	4,510,000	87	State (U.S.) ... D	Baton Rouge	English	
Luxembourg	999	460,000	460	Constitutional monarchy ... A	Luxembourg	French, Luxembourgish, German	EU, NATO, UN
Macedonia	9,928	2,065,000	208	Republic ... A	Skopje	Macedonian, Albanian, other	NATO/PP, UN
Madagascar	226,658	17,235,000	76	Republic ... A	Antananarivo	French, Malagasy	AU, COMESA, UN
Maine	35,385	1,310,000	37	State (U.S.) ... D	Augusta	English	
Malawi	45,747	11,780,000	258	Republic ... A	Lilongwe	Chichewa, English, indigenous	AU, COMESA, UN
Malaysia	127,320	23,310,000	183	Federal constitutional monarchy ... A	Kuala Lumpur and Putrajaya (²)	Bahasa Melayu, Chinese dialects, English, other	ASEAN, UN
Maldives	115	335,000	2,913	Republic ... A	Male'	Dhivehi	UN
Mali	478,841	11,790,000	25	Republic ... A	Bamako	French, Bambara, indigenous	AU, UN
Malta	122	400,000	3,279	Republic ... A	Valletta	English, Maltese	EU, UN
Manitoba	250,116	1,190,000	4.8	Province (Canada) ... D	Winnipeg	English	
Marshall Islands	70	57,000	814	Republic (U.S. protection) ... B	Majuro (island)	English, indigenous, Japanese	UN
Martinique	425	430,000	1,012	Overseas department (France) ... C	Fort-de-France	French, Creole	
Maryland	12,407	5,525,000	445	State (U.S.) ... D	Annapolis	English	
Massachusetts	10,555	6,455,000	612	State (U.S.) ... D	Boston	English	
Mauritania	397,956	2,955,000	7.4	Republic ... A	Nouakchott	Arabic, Wolof, Pular, Soninke, French	AL, AU, CAEU, UN
Mauritius (incl. Dependencies)	788	1,215,000	1,542	Republic ... A	Port Louis	English, French, Creole, other	AU, COMESA, UN
Mayotte (²)	144	180,000	1,250	Departmental collectivity (France) ... C	Mamoutzou	French, Swahili (Mahorian)	
Mexico	758,452	104,340,000	138	Federal republic ... A	Mexico City (Ciudad de México)	Spanish, indigenous	NAFTA, OAS, UN
Michigan	96,716	10,110,000	105	State (U.S.) ... D	Lansing	English	
Micronesia, Federated States of	271	110,000	406	Republic (U.S. protection) ... B	Palikir	English, indigenous	UN
Midway Islands	2.0	(¹)	...	Unincorporated territory (U.S.) ... C		English	
Minnesota	86,939	5,075,000	58	State (U.S.) ... D	St. Paul	English	
Mississippi	48,430	2,890,000	60	State (U.S.) ... D	Jackson	English	
Missouri	69,704	5,720,000	82	State (U.S.) ... D	Jefferson City	English	
Moldova	13,070	4,440,000	340	Republic ... A	Chişinău (Kishinev)	Romanian (Moldovan), Russian, Gagauz	CIS, NATO/PP, UN
Monaco	0.8	32,000	40,000	Constitutional monarchy ... A	Monaco	French, English, Italian, Monegasque	UN
Mongolia	604,829	2,730,000	4.5	Republic ... A	Ulan Bator (Ulaanbaatar)	Khalkha Mongol, Turkish dialects, Russian	UN
Montana	4,095	920,000	225	State (U.S.) ... D	Helena	English	
Montserrat	39	9,000	231	Overseas territory (U.K.) ... C	Plymouth	English	
Morocco (excl. Western Sahara)	172,414	31,950,000	185	Constitutional monarchy ... A	Rabat	Arabic, Berber dialects, French	AL, UN
Mozambique	309,496	18,695,000	60	Republic ... A	Maputo	Portuguese, indigenous	AU, UN
Myanmar (Burma)	261,228	42,620,000	163	Provisional military government ... A	Rangoon (Yangon)	Burmese, indigenous	ASEAN, UN
Namibia	317,818	1,940,000	6.1	Republic ... A	Windhoek	English, Afrikaans, German, indigenous	AU, COMESA, UN
Nauru	8.1	13,000	1,605	Republic ... A	Yaren District	Nauruan, English	UN
Nebraska	77,354	1,745,000	23	State (U.S.) ... D	Lincoln	English	
Nei Mongol (Inner Mongolia)	456,759	24,295,000	53	Autonomous region (China) ... D	Hohhot	Mongolian	
Nepal	56,827	26,770,000	471	Constitutional monarchy ... A	Kathmandu	Nepali, indigenous	UN
Netherlands	16,164	16,270,000	1,007	Constitutional monarchy ... A	Amsterdam and The Hague ('s-Gravenhage)	Dutch, Frisian	EU, NATO, UN
Netherlands Antilles	309	215,000	696	Self-governing territory (Netherlands protection) ... B	Willemstad	Dutch, Papiamento, English, Spanish	
Nevada	110,561	2,250,000	20	State (U.S.) ... D	Carson City	English	
New Brunswick	28,150	770,000	27	Province (Canada) ... D	Fredericton	English, French	
New Caledonia	7,172	210,000	29	Territorial collectivity (France) ... C	Nouméa	French, indigenous	
Newfoundland and Labrador	156,453	535,000	3.4	Province (Canada) ... D	St. John's	English	
New Hampshire	9,350	1,290,000	138	State (U.S.) ... D	Concord	English	
New Hebrides see Vanuatu	...	...	...				
New Jersey	8,721	8,665,000	994	State (U.S.) ... D	Trenton	English	
New Mexico	121,590	1,880,000	15	State (U.S.) ... D	Santa Fe	English, Spanish	
New South Wales	309,130	6,665,000	22	State (Australia) ... D	Sydney	English	
New York	54,556	19,245,000	353	State (U.S.) ... D	Albany	English	
New Zealand	104,454	3,975,000	38	Parliamentary state ... A	Wellington	English, Maori	ANZUS, UN
Nicaragua	50,054	5,180,000	103	Republic ... A	Managua	Spanish, English, indigenous	OAS, UN
Niger	489,192	11,210,000	23	Republic ... A	Niamey	French, Hausa, Djerma, indigenous	AU, UN
Nigeria	356,669	135,570,000	380	Transitional military government ... A	Abuja	English, Hausa, Fulani, Yoruba, Ibo, indigenous	AU, OPEC, UN
Ningxia Huizu	25,637	5,745,000	224	Autonomous region (China) ... D	Yinchuan	Chinese (Mandarin)	
Niue	100	2,000	20	Self-governing territory (New Zealand protection) ... B	Alofi	Niuean, English	
Norfolk Island	14	2,000	143	External territory (Australia) ... C	Kingston	English, Norfolk	

REGION OR POLITICAL DIVISION	Area Sq. Mi.	Est. Pop. 1/1/04	Pop. Per Sq. Mi.	Form of Government and Ruling Power	Capital	Predominant Languages	International Organizations
North America	9,500,000	505,780,000	53				
North Carolina	53,819	8,430,000	157	State (U.S.) ... D	Raleigh	English	
North Dakota	70,700	635,000	9.0	State (U.S.) ... D	Bismarck	English	
Northern Ireland	5,242	1,725,000	329	Administrative division (U.K.) ... D	Belfast	English	
Northern Mariana Islands	179	77,000	430	Commonwealth (U.S. protection) ... B	Saipan (island)	English, Chamorro, Carolinian	
Northern Territory	520,902	200,000	0.4	Territory (Australia) ... D	Darwin	English, indigenous	
Northwest Territories	519,735	43,000	0.08	Territory (Canada) ... D	Yellowknife	English, indigenous	
Norway (incl. Svalbard and Jan Mayen)	125,050	4,565,000	37	Constitutional monarchy ... A	Oslo	Norwegian, Sami, Finnish	EFTA, NATO, UN
Nova Scotia	21,345	965,000	45	Province (Canada) ... D	Halifax	English	
Nunavut	808,185	30,000	0.04	Territory (Canada) ... D	Iqaluit	English, indigenous	
Oceania (incl. Australia)	3,300,000	32,170,000	9.7				
Ohio	44,825	11,470,000	256	State (U.S.) ... D	Columbus	English	
Oklahoma	69,898	3,520,000	50	State (U.S.) ... D	Oklahoma City	English	
Oman	119,499	2,855,000	24	Monarchy ... A	Muscat (Masqat)	Arabic, English, Baluchi, Urdu, Indian dialects	AL, UN
Ontario	415,599	12,495,000	30	Province (Canada) ... D	Toronto	English	
Oregon	98,381	3,570,000	36	State (U.S.) ... D	Salem	English	
Pakistan (incl. part of Jammu and Kashmir)	339,732	152,210,000	448	Federal Islamic republic ... A	Islāmābād	English, Urdu, Punjabi, Sindhi, Pashto, other	UN
Palau (Belau)	188	20,000	106	Republic (U.S. protection) ... B	Koror and Melekeok (¹)	Angaur, English, Japanese, Palauan, Sonsorolese, Tobi	UN
Panama	29,157	2,980,000	102	Republic ... A	Panamá	Spanish, English	OAS, UN
Papua New Guinea	178,704	5,360,000	30	Parliamentary state ... A	Port Moresby	English, Motu, Pidgin, indigenous	UN
Paraguay	157,048	6,115,000	39	Republic ... A	Asunción	Guarani, Spanish	MERCOSUR, OAS, UN
Pennsylvania	46,055	12,400,000	269	State (U.S.) ... D	Harrisburg	English	
Peru	496,225	28,640,000	58	Republic ... A	Lima	Quechua, Spanish, Aymara	OAS, UN
Philippines	115,831	85,430,000	738	Republic ... A	Manila	English, Filipino, indigenous	ASEAN, UN
Pitcairn Islands (incl. Dependencies)	19	100	5.3	Overseas territory (U.K.) ... C	Adamstown	English, Pitcairnese	
Poland	120,728	38,625,000	320	Republic ... A	Warsaw (Warszawa)	Polish	EU, NATO, UN
Portugal	35,516	10,110,000	285	Republic ... A	Lisbon (Lisboa)	Portuguese, Mirandese	EU, NATO, UN
Prince Edward Island	2,185	140,000	64	Province (Canada) ... D	Charlottetown	English	
Puerto Rico	3,515	3,890,000	1,107	Commonwealth (U.S. protection) ... B	San Juan	Spanish, English	
Qatar	4,412	830,000	188	Monarchy ... A	Ad Dawḥah (Doha)	Arabic	AL, OPEC, UN
Qinghai	277,994	5,295,000	19	Province (China) ... D	Xining	Tibetan dialects, Mongolian, Turkish dialects, Chinese (Mandarin)	
Quebec	595,391	7,675,000	13	Province (Canada) ... D	Québec	French, English	
Queensland	668,208	3,785,000	5.7	State (Australia) ... D	Brisbane	English	
Reunion	969	760,000	784	Overseas department (France) ... C	Saint-Denis	French, Creole	
Rhode Island	1,545	1,080,000	699	State (U.S.) ... D	Providence	English	
Rhodesia see Zimbabwe							
Romania	91,699	22,370,000	244	Republic ... A	Bucharest (Bucureşti)	Romanian, Hungarian, German	NATO, UN
Russia	6,592,849	144,310,000	22	Federal republic ... A	Moscow (Moskva)	Russian, other	CIS, NATO/PP, UN
Rwanda	10,169	7,880,000	775	Republic ... A	Kigali	English, French, Kinyarwanda, Kiswahili	AU, COMESA, UN
St. Helena (incl. Dependencies)	121	7,500	62	Overseas territory (U.K.) ... C	Jamestown	English	
St. Kitts and Nevis	101	39,000	386	Parliamentary state ... A	Basseterre	English	OAS, UN
St. Lucia	238	165,000	693	Parliamentary state ... A	Castries	English, French	OAS, UN
St. Pierre and Miquelon	93	7,000	75	Territorial collectivity (France) ... C	Saint-Pierre	French	
St. Vincent and the Grenadines	150	115,000	767	Parliamentary state ... A	Kingstown	English, French	OAS, UN
Samoa	1,093	180,000	165	Constitutional monarchy ... A	Apia	English, Samoan	UN
San Marino	24	28,000	1,167	Republic ... A	San Marino	Italian	UN
Sao Tome and Principe	372	180,000	484	Republic ... A	São Tomé	Portuguese	AU, UN
Saskatchewan	251,366	1,025,000	4.1	Province (Canada) ... D	Regina	English	
Saudi Arabia	830,000	24,690,000	30	Monarchy ... A	Riyadh (Ar Riyāḍ)	Arabic	AL, OPEC, UN
Scotland	30,167	5,135,000	170	Administrative division (U.K.) ... D	Edinburgh	English, Scots Gaelic	
Senegal	75,951	10,715,000	141	Republic ... A	Dakar	French, Wolof and other indigenous	AU, UN
Serbia and Montenegro (Yugoslavia)	39,449	10,660,000	270	Republic ... A	Belgrade (Beograd)	Serbian, Albanian	UN
Seychelles	176	81,000	460	Republic ... A	Victoria	English, French, Creole	AU, COMESA, UN
Shaanxi	79,151	36,865,000	466	Province (China) ... D	Xi'an (Sian)	Chinese (Mandarin)	
Shandong	59,074	92,845,000	1,572	Province (China) ... D	Jinan	Chinese (Mandarin)	
Shanghai	2,394	17,120,000	7,151	Autonomous city (China) ... D	Shanghai	Chinese (Wu)	
Shanxi	60,232	33,715,000	560	Province (China) ... D	Taiyuan	Chinese (Mandarin)	
Sichuan	188,263	85,175,000	452	Province (China) ... D	Chengdu	Chinese (Mandarin), Tibetan dialects, Miao-Yao	
Sierra Leone	27,699	5,815,000	210	Republic ... A	Freetown	English, Krio, Mende, Temne, indigenous	AU, UN
Singapore	264	4,685,000	17,746	Republic ... A	Singapore	Chinese (Mandarin), English, Malay, Tamil	ASEAN, UN
Slovakia	18,924	5,420,000	286	Republic ... A	Bratislava	Slovak, Hungarian	EU, NATO, UN
Slovenia	7,821	1,935,000	247	Republic ... A	Ljubljana	Slovenian, Croatian, Serbian	EU, NATO, UN
Solomon Islands	10,954	515,000	47	Parliamentary state ... A	Honiara	English, indigenous	UN
Somalia	246,201	8,165,000	33	Transitional ... A	Mogadishu (Muqdisho)	Arabic, Somali, English, Italian	AL, AU, CAEU, UN
South Africa	470,693	42,770,000	91	Republic ... A	Pretoria, Cape Town, and Bloemfontein	Afrikaans, English, Xhosa, Zulu, other indigenous	AU, UN
South America	6,900,000	366,600,000	53				
South Australia	379,724	1,525,000	4.0	State (Australia) ... D	Adelaide	English	
South Carolina	32,020	4,160,000	130	State (U.S.) ... D	Columbia	English	
South Dakota	77,117	765,000	9.9	State (U.S.) ... D	Pierre	English	
South Georgia and the South Sandwich Islands (²)	1,450	(¹)	.....	Overseas territory (U.K.) ... C		English	
South West Africa see Namibia							
Spain	194,885	40,250,000	207	Constitutional monarchy ... A	Madrid	Spanish (Castilian), Catalan, Galician, Basque	EU, NATO, UN
Spanish North Africa (³)	12	140,000	11,667	Five possessions (Spain) ... C		Spanish, Arabic, Berber dialects	
Spanish Sahara see Western Sahara							
Sri Lanka	25,332	19,825,000	783	Socialist republic ... A	Colombo and Sri Jayewardenepura Kotte	English, Sinhala, Tamil	UN
Sudan	967,500	38,630,000	40	Provisional military government ... A	Khartoum (Al Kharṭūm)	Arabic, Nubian, and other indigenous, English	AL, AU, CAEU, COMESA, UN
Suriname	63,037	435,000	6.9	Republic ... A	Paramaribo	Dutch, Sranan Tongo, English, Hindustani, Javanese	OAS, UN

REGION OR POLITICAL DIVISION	Area Sq. Mi.	Est. Pop. 1/1/04	Pop. Per Sq. Mi.	Form of Government and Ruling Power	Capital	Predominant Languages	International Organizations
Swaziland	6,704	1,165,000	174	Monarchy ... A	Mbabane and Lobamba	English, siSwati	AU, COMESA, UN
Sweden	173,732	8,980,000	52	Constitutional monarchy ... A	Stockholm	Swedish, Sami, Finnish	EU, NATO/PP, UN
Switzerland	15,943	7,430,000	466	Federal republic ... A	Bern (Berne)	German, French, Italian, Romansch	EFTA, NATO/PP, UN
Syria	71,498	17,800,000	249	Republic ... A	Damascus (Dimashq)	Arabic, Kurdish, Armenian, Aramaic, Circassian	AL, CAEU, UN
Taiwan	13,901	22,675,000	1,631	Republic ... A	T'aipei	Chinese (Mandarin), Taiwanese (Min), Hakka	
Tajikistan	55,251	6,935,000	126	Republic ... A	Dushanbe	Tajik, Russian	CIS, NATO/PP, UN
Tanzania	364,900	36,230,000	99	Republic ... A	Dar es Salaam and Dodoma	English, Swahili, indigenous	AU, UN
Tasmania	26,409	475,000	18	State (Australia) ... D	Hobart	English	
Tennessee	42,143	5,860,000	139	State (U.S.) ... D	Nashville	English	
Texas	268,581	22,185,000	83	State (U.S.) ... D	Austin	English, Spanish	
Thailand	198,115	64,570,000	326	Constitutional monarchy ... A	Bangkok (Krung Thep)	Thai, indigenous	ASEAN, UN
Tianjin (Tientsin)	4,363	10,235,000	2,346	Autonomous city (China) ... D	Tianjin (Tientsin)	Chinese (Mandarin)	
Togo	21,925	5,495,000	251	Republic ... A	Lomé	French, Ewe, Mina, Kabye, Dagomba	AU, UN
Tokelau	4.6	1,500	326	Island territory (New Zealand) ... C		English, Tokelauan	
Tonga	251	110,000	438	Constitutional monarchy ... A	Nuku'alofa	Tongan, English	UN
Trinidad and Tobago	1,980	1,100,000	556	Republic ... A	Port of Spain	English, Hindi, French, Spanish, Chinese	OAS, UN
Tristan da Cunha	40	300	7.5	Dependency (St. Helena) ... C	Edinburgh	English	
Tunisia	63,170	9,980,000	158	Republic ... A	Tunis	Arabic, French	AL, AU, UN
Turkey	302,541	68,505,000	226	Republic ... A	Ankara	Turkish, Kurdish, Arabic, Armenian, Greek	NATO, UN
Turkmenistan	188,457	4,820,000	26	Republic ... A	Ashgabat (Ashkhabad)	Turkmen, Russian, Uzbek	CIS, NATO/PP, UN
Turks and Caicos Islands	166	20,000	120	Overseas territory (U.K.) ... C	Grand Turk	English	
Tuvalu	10	11,000	1,100	Parliamentary state ... A	Funafuti	Tuvaluan, English, Samoan, I-Kiribati	UN
Uganda	93,065	26,010,000	279	Republic ... A	Kampala	English, Luganda, Swahili, indigenous, Arabic	AU, COMESA, UN
Ukraine	233,090	47,890,000	205	Republic ... A	Kiev (Kyïv)	Ukrainian, Russian, Romanian, Polish, Hungarian	CIS, NATO/PP, UN
United Arab Emirates	32,278	2,505,000	78	Federation of monarchs ... A	Abū Ẓaby (Abu Dhabi)	Arabic, Persian, English, Hindi, Urdu	AL, CAEU, OPEC, UN
United Kingdom	93,788	60,185,000	642	Constitutional monarchy ... A	London	English, Welsh, Scots Gaelic	EU, NATO, UN
United States	3,794,083	291,680,000	77	Federal republic ... A	Washington	English, Spanish	ANZUS, NAFTA, NATO, OAS, UN
Upper Volta see Burkina Faso							
Uruguay	67,574	3,425,000	51	Republic ... A	Montevideo	Spanish	MERCOSUR, OAS, UN
Utah	84,899	2,360,000	28	State (U.S.) ... D	Salt Lake City	English	
Uzbekistan	172,742	26,195,000	152	Republic ... A	Tashkent (Toshkent)	Uzbek, Russian, Tajik	CIS, NATO/PP, UN
Vanuatu	4,707	200,000	42	Republic ... A	Port Vila	Bislama, English, French	UN
Vatican City	0.2	900	4,500	Ecclesiastical state ... A	Vatican City	Italian, Latin, French, other	
Venezuela	352,145	24,835,000	71	Federal republic ... A	Caracas	Spanish, indigenous	OAS, OPEC, UN
Vermont	9,614	620,000	64	State (U.S.) ... D	Montpelier	English	
Victoria	87,807	4,905,000	56	State (Australia) ... D	Melbourne	English	
Vietnam	128,066	82,150,000	641	Socialist republic ... A	Hanoi	Vietnamese, English, French, Chinese, Khmer, indigenous	ASEAN, UN
Virginia	42,774	7,410,000	173	State (U.S.) ... D	Richmond	English	
Virgin Islands (U.S.)	134	110,000	821	Unincorporated territory (U.S.) ... C	Charlotte Amalie	English, Spanish, Creole	
Wake Island	3.0	(¹)		Unincorporated territory (U.S.) ... C		English	
Wales	8,023	2,965,000	370	Administrative division (U.K.) ... D	Cardiff	English, Welsh Gaelic	
Wallis and Futuna	99	16,000	162	Overseas territory (France) ... C	Mata-Utu	French, Wallisian	
Washington	71,300	6,150,000	86	State (U.S.) ... D	Olympia	English	
West Bank (incl. Jericho and East Jerusalem)	2,263	2,275,000	1,005	Israeli territory with limited self-government		Arabic, Hebrew	(⁴)
Western Australia	976,792	1,945,000	2.0	State (Australia) ... D	Perth	English	
Western Sahara	102,703	265,000	2.6	Occupied by Morocco ... C		Arabic	
West Virginia	24,230	1,815,000	75	State (U.S.) ... D	Charleston	English	
Wisconsin	65,498	5,490,000	84	State (U.S.) ... D	Madison	English	
Wyoming	97,814	505,000	5.2	State (U.S.) ... D	Cheyenne	English	
Xianggang (Hong Kong)	425	7,440,000	17,506	Special administrative region (China) ... D	Hong Kong (Xianggang)	Chinese (Cantonese), English	
Xinjiang Uygur (Sinkiang)	617,764	19,685,000	32	Autonomous region (China) ... D	Ürümqi	Turkish dialects, Mongolian, Tungus, English	
Xizang (Tibet)	471,045	2,680,000	5.7	Autonomous region (China) ... D	Lhasa	Tibetan dialects	
Yemen	203,850	19,680,000	97	Republic ... A	San'a' (Sanaa)	Arabic	AL, CAEU, UN
Yugoslavia see Serbia and Montenegro							
Yukon Territory	186,272	32,000	0.2	Territory (Canada) ... D	Whitehorse	English, Inuktitut, indigenous	
Yunnan	152,124	43,850,000	288	Province (China) ... D	Kunming	Chinese (Mandarin), Tibetan dialects, Khmer, Miao-Yao	
Zaire see Congo, Democratic Republic of the							
Zambia	290,586	10,385,000	36	Republic ... A	Lusaka	English, indigenous	AU, COMESA, UN
Zhejiang	39,305	47,830,000	1,217	Province (China) ... D	Hangzhou	Chinese dialects	
Zimbabwe	150,873	12,630,000	84	Republic ... A	Harare (Salisbury)	English, indigenous	AU, COMESA, UN
WORLD	57,900,000	6,339,505,000	109				

... None, or not applicable
(1) No permanent population
(2) Claimed by Argentina
(3) Claimed by Spain
(4) The Palestinian Liberation Organization (PLO) is a member of AL and CAEU
(5) Future capital
(6) Claimed by Comoros
(7) Comprises Ceuta, Melilla, and several small islands

AL	Arab League (League of Arab States)
ANZUS	Australia-New Zealand-U.S. Security Treaty
ASEAN	Association of Southeast Asian Nations
AU	African Union
CAEU	Council of Arab Unity
CIS	Commonwealth of Independent States
COMESA	Common Market for Eastern and Southern Africa
EFTA	European Free Trade Association
EU	European Union
MERCOSUR	Southern Common Market
NAFTA	North American Free Trade Agreement
NATO	North Atlantic Treaty Organization
NATO/PP	NATO-Partnership for Peace Program
OAS	Organization of American States
OPEC	Organization of Petroleum Exporting Countries

WORLD DEMOGRAPHIC TABLE

CONTINENT/Country	Population Estimate 2004	Pop. Per Sq. Mile 2004	Percent Urban[1] 2001	Crude Birth Rate per 1,000[2] 2003	Crude Death Rate per 1,000[2] 2003	Natural Increase Percent[2] 2003	Fertility Rate (Children born/Woman)[3] 2003	Infant Mortality Rate per 1,000[3] 2003	Median Age[2] 2002	Life Expectancy Male[2] 2003	Life Expectancy Female[6] 2003
NORTH AMERICA											
Bahamas	300,000	56	64.7	19	9	1.0%	2	26	27	62	69
Belize	270,000	30	48.1	30	6	2.4%	4	27	19	65	70
Canada	32,360,000	8	78.9	11	8	0.3%	2	5	38	76	83
Costa Rica	3,925,000	199	59.5	19	4	1.5%	2	11	25	74	79
Cuba	11,290,000	264	75.5	12	7	0.5%	2	7	35	75	79
Dominica	69,000	238	71.4	17	7	1.0%	2	15	28	71	77
Dominican Republic	8,775,000	468	66.0	24	7	1.7%	3	34	24	66	70
El Salvador	6,530,000	804	61.5	28	6	2.2%	3	27	21	67	74
Guatemala	14,095,000	335	39.9	35	7	2.8%	5	38	18	64	66
Haiti	7,590,000	708	36.3	34	13	2.1%	5	76	18	50	53
Honduras	6,745,000	156	53.7	32	6	2.5%	4	30	19	65	68
Jamaica	2,705,000	637	56.6	17	5	1.2%	2	13	27	74	78
Mexico	104,340,000	138	74.6	22	5	1.7%	3	22	24	72	78
Nicaragua	5,180,000	103	56.5	26	5	2.2%	3	31	20	68	72
Panama	2,980,000	102	56.5	21	6	1.5%	3	21	26	70	75
St. Lucia	165,000	693	38.0	21	5	1.6%	2	14	24	70	77
Trinidad and Tobago	1,100,000	556	74.5	13	9	0.4%	2	25	30	67	72
United States	291,680,000	77	77.4	14	8	0.6%	2	7	36	74	80
SOUTH AMERICA											
Argentina	38,945,000	36	88.3	17	8	1.0%	2	16	29	72	79
Bolivia	8,655,000	20	62.9	26	8	1.8%	3	56	21	62	67
Brazil	183,080,000	55	81.7	18	6	1.2%	2	32	27	67	75
Chile	15,745,000	54	86.1	16	6	1.0%	2	9	30	73	80
Colombia	41,985,000	95	75.5	22	6	1.6%	3	22	26	67	75
Ecuador	13,840,000	126	63.4	25	5	2.0%	3	32	23	69	75
Guyana	705,000	9	36.7	18	9	0.9%	2	38	26	61	66
Paraguay	6,115,000	39	56.7	30	5	2.6%	4	28	21	72	77
Peru	28,640,000	58	73.1	23	6	1.7%	3	37	24	68	73
Suriname	435,000	7	74.8	19	7	1.3%	2	25	26	67	72
Uruguay	3,425,000	51	92.1	17	9	0.8%	2	14	32	73	79
Venezuela	24,835,000	71	87.2	20	5	1.5%	2	24	25	71	77
EUROPE											
Albania	3,535,000	318	42.9	15	5	1.0%	2	23	27	74	80
Austria	8,170,000	252	67.4	9	9	0%	1	5	39	76	82
Belarus	10,315,000	129	69.6	10	14	-0.4%	1	14	37	63	75
Belgium	10,340,000	877	97.4	11	10	0.1%	2	5	40	75	82
Bosnia and Herzegovina	4,000,000	202	43.4	13	8	0.4%	2	23	36	70	75
Bulgaria	7,550,000	176	67.4	10	14	-0.5%	1	22	41	68	75
Croatia	4,430,000	203	58.1	13	11	0.2%	2	7	39	71	78
Czech Republic	10,250,000	337	74.5	9	11	-0.1%	1	4	38	72	79
Denmark	5,405,000	325	85.1	12	11	0.1%	2	5	39	75	80
Estonia	1,405,000	80	69.4	9	13	-0.4%	1	12	38	64	77
Finland	5,210,000	40	58.5	11	10	0.1%	2	4	40	75	82
France	60,305,000	289	75.5	13	9	0.3%	2	4	38	76	83
Germany	82,415,000	598	87.7	9	10	-0.2%	1	4	41	75	82
Greece	10,635,000	209	60.3	10	10	0%	1	6	40	76	81
Hungary	10,045,000	280	64.8	10	13	-0.3%	1	9	38	68	77
Iceland	280,000	7	92.7	14	7	0.7%	2	4	34	78	82
Ireland	3,945,000	145	59.3	14	8	0.6%	2	6	33	75	80
Italy	58,030,000	499	67.1	9	10	-0.1%	1	6	41	76	83
Latvia	2,340,000	94	59.8	9	15	-0.6%	1	15	39	63	75
Lithuania	3,590,000	142	68.6	10	13	-0.2%	1	14	37	64	76
Luxembourg	460,000	460	91.9	12	8	0.4%	2	5	38	75	82
Macedonia	2,065,000	208	59.4	13	8	0.5%	2	12	33	72	77
Moldova	4,440,000	340	41.4	14	13	0.2%	2	42	32	61	69
Netherlands	16,270,000	1,007	89.6	12	9	0.3%	2	5	39	76	81
Norway	4,565,000	37	75.0	12	10	0.3%	2	4	38	77	82
Poland	38,625,000	320	62.5	10	10	0.1%	1	9	36	70	78
Portugal	10,110,000	285	65.8	11	10	0.1%	1	6	38	73	80
Romania	22,370,000	244	55.2	11	12	-0.1%	1	28	35	67	75
Serbia and Montenegro	10,660,000	270	51.7	13	11	0.2%	2	17	36	71	77
Slovakia	5,420,000	286	57.6	10	10	0.1%	1	8	35	70	78
Slovenia	1,935,000	247	49.1	9	10	-0.1%	1	4	39	72	80
Spain	40,250,000	207	77.8	10	9	0.1%	1	5	39	76	83
Sweden	8,980,000	52	83.3	11	10	0%	2	3	40	78	83
Switzerland	7,430,000	466	67.3	10	8	0.1%	1	4	40	77	83
Ukraine	47,890,000	205	68.0	10	16	-0.7%	1	21	38	61	72
United Kingdom	60,185,000	642	89.5	11	10	0.1%	2	5	38	76	81
Russia	144,310,000	22	72.9	10	14	-0.4%	1	20	38	62	73
ASIA											
Afghanistan	29,205,000	116	22.3	41	17	2.3%	6	142	19	48	46
Armenia	3,325,000	289	67.2	13	10	0.2%	2	41	32	62	71
Azerbaijan	7,850,000	235	51.8	19	10	1.0%	2	82	27	59	68
Bahrain	675,000	2,528	92.5	19	4	1.5%	3	19	29	71	76
Bangladesh	139,875,000	2,516	25.6	30	9	2.1%	3	66	21	61	61
Brunei	360,000	162	72.8	20	3	1.6%	2	14	26	72	77
Cambodia	13,245,000	189	17.5	27	9	1.8%	4	76	19	55	60
China	1,298,720,000	352	37.1	13	7	0.6%	2	25	32	70	74
Cyprus	775,000	217	70.2	13	8	0.5%	2	8	34	75	80
East Timor	1,010,000	176	7.5	28	6	2.1%	4	50	20	63	68
Georgia	4,920,000	183	56.5	12	15	-0.3%	2	51	35	61	68
India	1,057,415,000	865	27.9	23	8	1.5%	3	60	24	63	64
Indonesia	236,680,000	322	42.1	21	6	1.5%	3	38	26	67	71
Iran	68,650,000	108	64.7	17	6	1.2%	2	44	23	68	71
Iraq	25,025,000	148	67.4	34	6	2.8%	5	55	19	67	69
Israel	6,160,000	768	91.8	19	6	1.2%	3	7	29	77	81
Japan	127,285,000	873	78.9	10	9	0.1%	1	3	42	78	84
Jordan	5,535,000	160	78.7	24	3	2.1%	3	19	22	75	81
Kazakhstan	16,780,000	16	55.8	18	11	0.8%	2	59	28	58	69
Korea, North	22,585,000	485	60.5	18	7	1.1%	2	26	31	68	74
Korea, South	48,450,000	1,264	82.5	13	6	0.7%	2	7	33	72	79
Kuwait	2,220,000	323	96.1	22	2	1.9%	3	11	26	76	78

CONTINENT/Country	Population Estimate 2004	Pop. Per Sq. Mile 2004	Percent Urban[1] 2001	Crude Birth Rate per 1,000[2] 2003	Crude Death Rate per 1,000[2] 2003	Natural Increase Percent[2] 2003	Fertility Rate (Children born/Woman)[3] 2003	Infant Mortality Rate per 1,000[3] 2003	Median Age[2] 2002	Life Expectancy Male[2] 2003	Life Expectancy Female[2] 2003
Kyrgyzstan	4,930,000	64	34.3	26	9	1.7%	3	75	23	59	68
Laos	5,995,000	66	19.7	37	12	2.5%	5	89	19	52	56
Lebanon	3,755,000	935	90.1	20	6	1.3%	2	26	26	70	75
Malaysia	23,310,000	183	58.1	24	5	1.9%	3	19	24	69	75
Mongolia	2,730,000	5	56.6	21	7	1.4%	2	57	24	62	66
Myanmar	42,620,000	163	28.1	19	12	0.7%	2	70	25	54	58
Nepal	26,770,000	471	12.2	32	10	2.3%	4	71	20	59	59
Oman	2,855,000	24	76.5	37	4	3.4%	6	21	19	70	75
Pakistan	152,210,000	448	33.4	30	9	2.1%	4	77	20	61	63
Philippines	85,430,000	738	59.4	26	6	2.1%	3	25	22	66	72
Qatar	830,000	188	92.9	16	4	1.1%	3	20	31	71	76
Saudi Arabia	24,690,000	30	86.7	37	6	3.1%	6	48	19	67	71
Singapore	4,685,000	17,746	100.0	13	4	0.8%	1	4	35	77	84
Sri Lanka	19,825,000	783	23.1	16	6	1.0%	2	15	29	70	75
Syria	17,800,000	249	51.8	30	5	2.5%	4	32	20	68	71
Taiwan	22,675,000	1,631	[5]	13	6	0.7%	2	7	33	74	80
Tajikistan	6,935,000	126	27.7	33	8	2.4%	4	113	19	61	68
Thailand	64,570,000	326	20.0	16	7	1.0%	2	22	30	69	74
Turkey	68,505,000	226	66.2	18	6	1.2%	2	44	27	69	74
Turkmenistan	4,820,000	26	44.9	28	9	1.9%	4	73	21	58	65
United Arab Emirates	2,505,000	78	87.2	18	4	1.4%	3	16	28	72	77
Uzbekistan	26,195,000	152	36.6	26	8	1.8%	3	72	22	61	68
Vietnam	82,150,000	641	24.5	20	6	1.3%	2	31	25	68	73
Yemen	19,680,000	97	25.0	43	9	3.4%	7	65	16	59	63
AFRICA											
Algeria	33,090,000	36	57.7	22	5	1.7%	3	38	23	69	72
Angola	10,875,000	23	34.9	46	26	2.0%	6	194	18	36	38
Benin	7,145,000	164	43.0	43	14	3.0%	6	87	16	50	52
Botswana	1,570,000	7	49.4	26	31	-0.6%	3	67	19	32	32
Burkina Faso	13,400,000	127	16.9	45	19	2.6%	6	100	17	43	46
Burundi	6,165,000	574	9.3	40	18	2.2%	6	72	16	43	44
Cameroon	15,905,000	87	49.7	35	15	2.0%	5	70	18	47	49
Cape Verde	415,000	267	63.5	27	7	2.0%	4	51	19	67	73
Central African Republic	3,715,000	15	41.7	36	20	1.6%	5	93	18	40	43
Chad	9,395,000	19	24.1	47	16	3.1%	6	96	16	47	50
Comoros	640,000	742	33.8	39	9	3.0%	5	80	19	59	64
Congo	2,975,000	23	66.1	29	14	1.5%	4	95	20	49	51
Congo, Democratic Republic of the	57,445,000	63	30.7	45	15	3.0%	7	97	16	47	51
Cote d'Ivoire	17,145,000	138	44.0	40	18	2.2%	6	98	17	40	45
Djibouti	460,000	51	84.2	41	19	2.1%	6	107	18	42	44
Egypt	75,420,000	195	42.7	24	5	1.9%	3	35	23	68	73
Equatorial Guinea	515,000	48	49.3	37	13	2.4%	5	89	19	53	57
Eritrea	4,390,000	97	19.1	39	13	2.6%	6	76	18	51	55
Ethiopia	67,210,000	158	15.9	40	20	2.0%	6	103	17	40	42
Gabon	1,340,000	13	82.3	37	11	2.5%	5	55	19	55	59
Gambia, The	1,525,000	370	31.3	41	12	2.8%	6	75	17	52	56
Ghana	20,615,000	224	36.4	26	11	1.5%	3	53	20	56	57
Guinea	9,135,000	96	27.9	43	16	2.7%	6	93	18	48	51
Guinea-Bissau	1,375,000	99	32.3	38	17	2.2%	5	110	19	45	49
Kenya	31,840,000	142	34.4	29	16	1.3%	3	63	18	45	45
Lesotho	1,865,000	159	28.8	27	25	0.3%	4	86	20	37	37
Liberia	3,345,000	78	45.5	45	18	2.7%	6	132	18	47	49
Libya	5,565,000	8	88.0	27	3	2.4%	3	27	22	74	78
Madagascar	17,235,000	76	30.1	42	12	3.0%	6	80	17	54	59
Malawi	11,780,000	258	15.1	45	23	2.2%	6	105	16	38	38
Mali	11,790,000	25	30.9	48	19	2.9%	7	119	16	45	46
Mauritania	2,955,000	7	59.1	42	13	2.9%	6	74	17	50	54
Mauritius	1,215,000	1,542	41.6	16	7	0.9%	2	16	30	68	76
Morocco	31,950,000	185	56.1	23	6	1.7%	3	45	23	68	72
Mozambique	18,695,000	60	33.3	37	23	1.4%	5	138	19	39	37
Namibia	1,940,000	6	31.4	34	19	1.5%	5	68	18	44	41
Niger	11,210,000	23	21.1	50	22	2.8%	7	124	16	42	42
Nigeria	135,570,000	380	44.9	39	14	2.5%	5	71	18	51	51
Rwanda	7,880,000	775	6.3	40	22	1.8%	6	103	18	39	40
Sao Tome and Principe	180,000	484	47.7	42	7	3.5%	6	46	16	65	68
Senegal	10,715,000	141	48.2	36	11	2.5%	5	58	18	55	58
Sierra Leone	5,815,000	210	37.3	44	21	2.3%	6	147	18	40	45
Somalia	8,165,000	33	27.9	46	18	2.9%	7	120	18	46	49
South Africa	42,770,000	91	57.7	19	18	0%	2	61	25	47	47
Sudan	38,630,000	40	37.1	36	10	2.7%	5	66	18	57	59
Swaziland	1,165,000	174	26.7	29	21	0.8%	4	67	19	41	38
Tanzania	36,230,000	99	33.3	40	17	2.2%	5	104	18	43	46
Togo	5,495,000	251	33.9	35	12	2.4%	5	69	17	51	55
Tunisia	9,980,000	158	66.2	17	5	1.2%	2	27	26	73	76
Uganda	26,010,000	279	14.5	47	17	3.0%	7	88	15	43	46
Zambia	10,385,000	36	39.8	40	24	1.5%	5	99	17	35	35
Zimbabwe	12,630,000	84	36.0	30	22	0.8%	4	66	19	40	38
OCEANIA											
Australia	19,825,000	7	91.2	13	7	0.5%	2	5	36	77	83
Fiji	875,000	124	50.2	23	6	1.7%	3	13	24	66	71
Kiribati	100,000	319	38.6	31	9	2.3%	4	51	20	58	64
Micronesia, Federated States of	110,000	406	28.6	26	5	2.1%	4	32	19[4]	67	71
New Zealand	3,975,000	38	85.9	14	8	0.7%	2	6	33	75	81
Papua New Guinea	5,360,000	30	17.6	31	8	2.3%	4	55	21	62	66
Samoa	180,000	165	22.3	15	6	0.9%	3	30	24	67	73
Solomon Islands	515,000	47	20.2	32	4	2.8%	4	23	18	70	75
Tonga	110,000	438	33.0	25	6	1.9%	3	13	20	66	71
Vanuatu	200,000	42	22.1	24	8	1.6%	3	58	22	60	63

This table presents data for most independent nations having an area greater than 200 square miles
(1) Source: United Nations World Urbanization Prospects
(2) Source: United States Census Bureau International Database
(3) Source: United States Central Intelligence Agency World Factbook
(4) 2000 Census preliminary count from www.fsmgov.org/info/people.html
(5) Data for Taiwan is included with China

WORLD AGRICULTURE TABLE

CONTINENT/Country	Total Area Sq. Miles	Cropland Area[1] Sq. Miles	Cropland Area[1] %	Pasture Area[1] Sq. Miles	Pasture Area[1] %	Wheat[1] 1,000 metric tons	Rice[1] 1,000 metric tons	Corn[1] 1,000 metric tons	Cattle[1] 1,000	Pigs[1] 1,000	Sheep[1] 1,000
NORTH AMERICA											
Bahamas	5,382	46	0.9%	8	0.1%	-	-	-	1	5	6
Belize	8,867	402	4.5%	193	2.2%	-	12	36	52	25	4
Canada	3,855,103	177,144	4.6%	111,970	2.9%	24,676	-	8,168	13,340	12,970	819
Costa Rica	19,730	2,027	10.3%	9,035	45.8%	-	267	20	1,358	438	3
Cuba	42,804	17,239	40.3%	8,494	19.8%	-	342	207	4,305	2,600	310
Dominica	290	77	26.6%	8	2.7%	-	-	-	13	5	8
Dominican Republic	18,730	6,162	32.9%	8,108	43.3%	-	615	30	2,026	548	106
El Salvador	8,124	3,514	43.2%	3,066	37.7%	-	47	605	1,190	195	5
Guatemala	42,042	7,355	17.5%	10,046	23.9%	9	46	1,057	2,500	1,417	270
Haiti	10,714	4,247	39.6%	1,892	17.7%	-	111	211	1,390	934	147
Honduras	43,277	5,514	12.7%	5,822	13.5%	1	9	509	1,737	474	14
Jamaica	4,244	1,097	25.8%	884	20.8%	-	-	2	400	180	1
Mexico	758,452	105,406	13.9%	308,882	40.7%	3,263	324	18,466	30,428	16,112	6,048
Nicaragua	50,054	8,382	16.7%	18,591	37.1%	-	234	374	2,008	402	4
Panama	29,157	2,683	9.2%	5,927	20.3%	-	237	71	1,348	279	-
St. Lucia	238	69	29.2%	8	3.2%	-	-	-	12	15	13
Trinidad and Tobago	1,980	471	23.8%	42	2.1%	-	13	5	36	41	12
United States	3,794,083	684,401	18.0%	903,479	23.8%	58,862	9,222	244,296	98,197	60,229	7,071
SOUTH AMERICA											
Argentina	1,073,519	135,136	12.6%	548,265	51.1%	15,642	1,140	15,217	49,299	4,200	13,588
Bolivia	424,165	11,973	2.8%	130,618	30.8%	121	281	607	6,715	2,786	8,743
Brazil	3,300,172	256,623	7.8%	760,621	23.0%	2,461	10,998	35,119	170,295	30,608	14,728
Chile	291,930	8,880	3.0%	49,942	17.1%	1,490	113	685	4,117	2,395	4,153
Colombia	439,737	16,405	3.7%	161,391	36.7%	37	2,262	1,128	25,274	2,726	2,247
Ecuador	109,484	11,525	10.5%	19,653	18.0%	19	1,340	483	5,261	2,654	2,214
Guyana	83,000	1,969	2.4%	4,749	5.7%	-	560	3	220	20	130
Paraguay	157,048	12,008	7.6%	83,784	53.3%	256	112	804	9,758	2,633	402
Peru	496,225	16,255	3.3%	104,634	21.1%	180	1,963	1,205	4,936	2,795	14,414
Suriname	63,037	259	0.4%	81	0.1%	-	178	-	128	22	8
Uruguay	67,574	5,174	7.7%	52,290	77.4%	284	1,189	190	10,446	375	13,257
Venezuela	352,145	13,158	3.7%	70,425	20.0%	1	696	1,547	14,620	5,555	780
EUROPE											
Albania	11,100	2,699	24.3%	1,699	15.3%	298	-	203	719	96	1,929
Austria	32,378	5,676	17.5%	7,413	22.9%	1,412	-	1,774	2,166	3,556	357
Belarus	80,155	24,151	30.1%	11,564	14.4%	903	-	13	4,411	3,565	96
Belgium	11,787	3,344[2]	26.2%[2]	2,618[2]	20.5%[2]	1,535	-	420	3,165	7,462	150
Bosnia and Herzegovina	19,767	3,243	16.4%	4,633	23.4%	289	-	656	448	345	645
Bulgaria	42,855	17,900	41.8%	6,236	14.6%	3,071	8	1,137	664	1,459	2,536
Croatia	21,829	6,124	28.1%	6,035	27.6%	852	-	1,958	435	1,276	519
Czech Republic	30,450	12,788	42.0%	3,730	12.2%	4,196	-	324	1,604	3,761	87
Denmark	16,640	8,880	53.4%	1,452	8.7%	4,683	-	-	1,887	12,052	147
Estonia	17,462	2,691	15.4%	745	4.3%	123	-	-	276	304	29
Finland	130,559	8,490	6.5%	77	0.1%	427	-	-	1,060	1,303	101
France	208,482	75,618	36.3%	38,788	18.6%	35,327	110	15,928	20,377	14,693	9,754
Germany	137,847	46,409	33.7%	19,355	14.0%	21,358	-	3,362	14,723	26,021	2,746
Greece	50,949	14,873	29.2%	17,954	35.2%	2,111	153	2,007	584	925	8,977
Hungary	35,919	18,548	51.6%	4,097	11.4%	3,843	9	6,664	845	5,216	991
Iceland	39,769	27	0.1%	8,780	22.1%	-	-	-	72	44	477
Ireland	27,133	4,050	14.9%	12,934	47.7%	688	-	-	6,613	1,765	5,311
Italy	116,342	42,379	36.4%	16,907	14.5%	7,239	1,310	10,222	7,167	8,356	11,000
Latvia	24,942	7,220	28.9%	2,355	9.4%	410	-	-	393	407	28
Lithuania	25,213	11,541	45.8%	1,923	7.6%	1,062	-	-	856	984	14
Luxembourg	999	[3]	[3]	[3]	[3]	-	-	2	134	-	-
Macedonia	9,928	2,363	23.8%	2,432	24.5%	308	20	135	267	209	1,285
Moldova	13,070	8,398	64.3%	1,483	11.3%	902	-	1,096	423	646	929
Netherlands	16,164	3,622	22.4%	3,834	23.7%	995	-	148	4,108	13,253	1,335
Norway	125,050	3,398	2.7%	625	0.5%	265	-	-	1,017	414	2,342
Poland	120,728	55,267	45.8%	15,745	13.0%	8,946	-	962	6,124	17,588	366
Portugal	35,516	10,444	29.4%	5,548	15.6%	295	146	907	1,415	2,346	4,337
Romania	91,699	38,305	41.8%	19,039	20.8%	5,610	3	8,317	3,021	5,946	8,062
Serbia and Montenegro	39,449	14,394	36.5%	7,197	18.2%	2,207	-	5,013	1,550	4,012	1,853
Slovakia	18,924	6,085	32.2%	3,375	17.8%	1,445	-	612	671	1,548	344
Slovenia	7,821	784	10.0%	1,185	15.2%	153	-	283	473	585	80
Spain	194,885	69,298	35.6%	44,209	22.7%	5,785	844	4,208	6,140	22,079	24,185
Sweden	173,732	10,413	6.0%	1,726	1.0%	2,135	-	-	1,683	1,975	440
Switzerland	15,943	1,683	10.6%	4,417	27.7%	535	-	214	1,603	1,499	421
Ukraine	233,090	129,321	55.5%	30,541	13.1%	15,043	74	3,075	10,591	9,270	1,074
United Kingdom	93,788	22,019	23.5%	43,440	46.3%	14,380	-	-	11,052	6,537	41,205
Russia	6,592,849	485,400	7.4%	351,905	5.3%	37,455	509	1,133	27,936	17,076	12,954
ASIA											
Afghanistan	251,773	31,097	12.4%	115,831	46.0%	1,821	205	172	2,600	-	12,762
Armenia	11,506	2,162	18.8%	3,089	26.8%	211	-	9	478	75	515
Azerbaijan	33,437	7,471	22.3%	10,039	30.0%	1,172	19	107	1,965	21	5,321
Bahrain	267	23	8.7%	15	5.8%	-	-	-	12	-	17
Bangladesh	55,598	32,761	58.9%	2,317	4.2%	1,807	36,909	8	23,817	-	1,128
Brunei	2,226	27	1.2%	23	1.0%	-	4,035	146	2	6	2
Cambodia	69,898	14,699	21.0%	5,792	8.3%	-	4,035	146	2,896	2,079	-
China	3,690,045	599,520[4]	16.2%[4]	1,544,412[4]	41.9%[4]	102,463[4]	189,840[4]	116,240[4]	104,179[4]	440,384[4]	130,536[4]
Cyprus	3,572	436	12.2%	15	0.4%	12	-	-	55	419	240
East Timor	5,743	309	5.4%	579	10.1%	-	33	93	173	300	36
Georgia	26,911	4,104	15.3%	7,490	27.8%	207	-	358	1,117	433	541
India	1,222,510	655,987	53.7%	42,124	3.4%	72,140	132,818	12,285	217,773	17,000	57,900
Indonesia	735,310	129,730	17.6%	43,155	5.9%	-	50,953	9,409	11,370	6,098	7,316
Iran	636,372	63,892	10.0%	169,885	26.7%	8,740	2,103	1,113	8,273	-	53,900
Iraq	169,235	23,514	13.9%	15,444	9.1%	667	110	73	1,342	-	6,770
Israel	8,019	1,637	20.4%	548	6.8%	94	-	73	393	138	373
Japan	145,850	18,510	12.7%	1,564	1.1%	657	11,551	-	4,592	9,823	11
Jordan	34,495	1,544	4.5%	2,865	8.3%	18	-	13	66	-	1,900
Kazakhstan	1,049,156	83,672	8.0%	714,667	68.1%	10,938	225	256	4,021	984	8,785
Korea, North	46,540	10,811	23.2%	193	0.4%	88	2,031	1,253	575	3,076	186
Korea, South	38,328	7,293	19.0%	208	0.5%	4	7,204	67	2,191	8,266	1
Kuwait	6,880	58	0.8%	525	7.6%	-	-	-	19	-	543

CONTINENT/Country	Agricultural Area 2001 Total Area Sq. Miles	Cropland Area[1] Sq. Miles	Cropland Area[1] %	Pasture Area[1] Sq. Miles	Pasture Area[1] %	Average Production 1999-2001 Wheat[1] 1,000 metric tons	Rice[1] 1,000 metric tons	Corn[1] 1,000 metric tons	Average 1999-2001 Cattle[1] 1,000	Pigs[1] 1,000	Sheep[1] 1,000
Kyrgyzstan	77,182	5,664	7.3%	35,873	46.5%	1,113	17	363	942	98	3,101
Laos	91,429	3,699	4.0%	3,390	3.7%	-	2,213	108	1,106	1,390	-
Lebanon	4,016	1,208	30.1%	62	1.5%	60	-	4	76	63	354
Malaysia	127,320	29,286	23.0%	1,100	0.9%	-	2,170	63	744	1,943	167
Mongolia	604,829	4,633	0.8%	499,230	82.5%	148	-	-	2,997	17	14,587
Myanmar	261,228	41,023	15.7%	1,212	0.5%	105	20,683	413	10,974	3,923	390
Nepal	56,827	12,324	21.7%	6,784	11.9%	1,143	4,137	1,528	7,012	872	852
Oman	119,499	313	0.3%	3,861	3.2%	1	-	-	299	-	342
Pakistan	339,732	85,560	25.2%	19,305	5.7%	19,319	6,920	1,653	22,007	-	24,067
Philippines	115,831	41,120	35.5%	4,942	4.3%	-	12,377	4,540	2,467	10,724	30
Qatar	4,412	81	1.8%	193	4.4%	-	-	1	15	-	214
Saudi Arabia	830,000	14,649	1.8%	656,373	79.1%	1,871	-	5	304	-	7,848
Singapore	264	4	1.5%	-	0.0%	-	-	-	-	190	-
Sri Lanka	25,332	7,378	29.1%	1,699	6.7%	-	2,804	30	1,580	71	12
Syria	71,498	21,043	29.4%	31,942	44.7%	3,514	-	196	933	-	13,288
Taiwan	13,901	(5)	(5)	(5)	(5)	(5)	(5)	(5)	(5)	(5)	(5)
Tajikistan	55,251	4,093	7.4%	13,514	24.5%	375	67	38	1,045	1	1,481
Thailand	198,115	70,657	35.7%	3,089	1.6%	1	25,578	4,405	4,973	6,539	40
Turkey	302,541	101,757	33.6%	47,792	15.8%	19,341	350	2,266	10,949	4	29,394
Turkmenistan	188,457	7,008	3.7%	118,533	62.9%	1,472	33	9	863	46	5,750
United Arab Emirates	32,278	919	2.8%	1,178	3.6%	-	-	-	94	-	504
Uzbekistan	172,742	18,649	10.8%	88,031	51.0%	3,637	219	133	5,279	83	7,980
Vietnam	128,066	32,579	25.4%	2,479	1.9%	-	31,964	1,961	4,029	20,273	-
Yemen	203,850	6,158	3.0%	62,027	30.4%	145	-	48	1,320	-	4,758
AFRICA											
Algeria	919,595	31,861	3.5%	122,780	13.4%	1,414	-	1	1,667	6	19,000
Angola	481,354	12,741	2.6%	208,495	43.3%	4	16	417	3,995	800	345
Benin	43,484	8,745	20.1%	2,124	4.9%	-	46	740	1,486	463	650
Botswana	224,607	1,440	0.6%	98,842	44.0%	1	-	8	2,035	6	347
Burkina Faso	105,869	15,444	14.6%	23,166	21.9%	-	102	500	4,767	621	6,722
Burundi	10,745	4,865	45.3%	3,610	33.6%	7	57	124	321	67	215
Cameroon	183,568	27,645	15.1%	7,722	4.2%	-	69	759	5,761	1,232	3,734
Cape Verde	1,557	158	10.2%	97	6.2%	-	-	27	22	195	9
Central African Republic	240,536	7,799	3.2%	12,066	5.0%	-	23	101	3,096	669	218
Chad	495,755	14,016	2.8%	173,746	35.0%	3	114	88	5,852	22	2,374
Comoros	863	510	59.1%	58	6.7%	-	17	4	51	-	21
Congo	132,047	849	0.6%	38,610	29.2%	-	1	6	87	46	102
Congo, Democratic Republic of the	905,446	30,425	3.4%	57,915	6.4%	9	338	1,184	823	1,050	925
Cote d'Ivoire	124,504	28,958	23.3%	50,193	40.3%	-	1,217	693	1,398	333	1,439
Djibouti	8,958	4	0.0%	5,019	56.0%	-	-	-	269	-	465
Egypt	386,662	12,888	3.3%	-	0.0%	6,388	5,681	6,487	3,583	29	4,510
Equatorial Guinea	10,831	888	8.2%	402	3.7%	-	-	-	5	6	37
Eritrea	45,406	1,942	4.3%	26,900	59.2%	32	-	13	2,150	-	1,570
Ethiopia	426,373	44,255	10.4%	77,220	18.1%	1,340	-	2,938	35,025	25	22,333
Gabon	103,347	1,911	1.8%	18,012	17.4%	-	1	26	36	213	197
Gambia, The	4,127	985	23.9%	1,772	42.9%	-	28	24	350	12	115
Ghana	92,098	22,780	24.7%	32,240	35.0%	-	244	988	1,297	327	2,715
Guinea	94,926	5,888	6.2%	41,313	43.5%	-	830	96	2,576	93	824
Guinea-Bissau	13,948	2,116	15.2%	4,170	29.9%	-	95	26	509	347	283
Kenya	224,961	19,923	8.9%	82,240	36.6%	184	58	2,419	13,229	311	7,000
Lesotho	11,720	1,290	11.0%	7,722	65.9%	39	-	128	547	63	839
Liberia	43,000	2,317	5.4%	7,722	18.0%	-	188	-	36	127	210
Libya	679,362	8,301	1.2%	51,352	7.6%	128	-	-	207	-	5,100
Madagascar	226,658	13,707	6.0%	92,664	40.9%	9	2,412	175	10,339	1,267	793
Malawi	45,747	9,035	19.7%	7,143	15.6%	2	86	2,190	741	450	110
Mali	478,841	18,147	3.8%	115,831	24.2%	8	801	378	6,594	72	6,282
Mauritania	397,956	1,931	0.5%	151,545	38.1%	-	65	7	1,470	-	7,437
Mauritius	788	409	51.9%	27	3.4%	-	-	-	27	12	10
Morocco	172,414	37,529	21.8%	81,081	47.0%	2,284	33	95	2,629	8	17,059
Mozambique	309,496	16,351	5.3%	169,885	54.9%	1	168	1,136	1,317	179	125
Namibia	317,818	3,166	1.0%	146,719	46.2%	4	-	26	2,436	21	2,330
Niger	489,192	17,375	3.6%	46,332	9.5%	10	66	5	2,217	39	4,386
Nigeria	356,669	120,464	33.8%	151,352	42.4%	75	3,109	4,734	19,677	5,000	20,833
Rwanda	10,169	5,019	49.4%	2,124	20.9%	6	13	66	766	172	264
Sao Tome and Principe	372	205	55.0%	4	1.0%	-	-	2	4	2	3
Senegal	75,951	9,653	12.7%	21,815	28.7%	-	229	84	3,076	263	4,619
Sierra Leone	27,699	2,178	7.9%	8,494	30.7%	-	215	9	413	52	365
Somalia	246,201	4,135	1.7%	166,024	67.4%	1	2	188	5,133	4	13,100
South Africa	470,693	60,664	12.9%	324,048	68.8%	2,200	3	9,147	13,594	1,542	28,677
Sudan	967,500	64,298	6.6%	452,434	46.8%	230	8	48	37,081	-	45,980
Swaziland	6,704	734	10.9%	4,633	69.1%	-	-	94	613	32	27
Tanzania	364,900	19,112	5.2%	135,136	37.0%	87	509	2,567	17,350	449	3,513
Togo	21,925	10,154	46.3%	3,861	17.6%	-	69	480	277	287	1,528
Tunisia	63,170	18,954	30.0%	15,792	25.0%	1,111	-	-	760	6	6,862
Uganda	93,065	27,799	29.9%	19,738	21.2%	12	106	1,108	5,977	1,540	1,065
Zambia	290,586	20,386	7.0%	115,831	39.9%	80	11	768	2,709	324	137
Zimbabwe	150,873	12,934	8.6%	66,410	44.0%	282	-	1,698	5,840	494	602
OCEANIA											
Australia	2,969,910	195,368	6.6%	1,563,327	52.6%	23,654	1,417	363	27,645	2,607	116,736
Fiji	7,056	1,100	15.6%	676	9.6%	-	16	1	335	139	7
Kiribati	313	151	48.1%	-	0.0%	-	-	-	-	10	-
Micronesia, Federated States of	271	139	51.3%	42	15.7%	-	-	-	14	32	-
New Zealand	104,454	13,019	12.5%	53,525	51.2%	337	-	185	9,025	364	45,114
Papua New Guinea	178,704	3,320	1.9%	676	0.4%	-	1	7	87	1,583	6
Samoa	1,093	498	45.6%	8	0.7%	-	-	-	28	179	-
Solomon Islands	10,954	286	2.6%	154	1.4%	-	5	-	11	63	-
Tonga	251	185	73.8%	15	6.2%	-	-	-	11	81	-
Vanuatu	4,707	463	9.8%	162	3.4%	-	-	1	151	62	-

This table presents data for most independent nations having an area greater than 200 square miles
- Zero, insignificant, or not available
(1) Source: United Nations Food and Agriculture Organization
(2) Includes data for Luxembourg
(3) Data for Luxembourg is included with Belgium
(4) Includes data for Taiwan
(5) Data for Taiwan is included with China

WORLD ECONOMIC TABLE

CONTINENT/Country	GDP 2002 Total GDP[1]	GDP 2002 GDP Per Capita[1]	Trade Value of Exports[1]	Trade Value of Imports[1]	Commercial Energy Production Avg. 2000[2] Total (1,000 Metric Tons of Coal Equiv.)	Solid %	Liquid %	Gas %	Hydro & Nuclear %	Average Production 1999-2001 in Metric Tons Coal[3]	Petroleum[3]	Iron Ore[4]	Bauxite[4]
NORTH AMERICA													
Bahamas	$4,590,000,000	$17,000	$560,700,000	$1,860,000,000	-	-	-	-	-	-	-	-	-
Belize	$1,280,000,000	$4,900	$290,000,000	$430,000,000	12	-	-	-	100%	-	-	-	-
Canada	$934,100,000,000	$29,400	$260,500,000,000	$229,000,000,000	507,218	10%	33%	43%	14%	70,711,084	97,834,913	20,527,000	-
Costa Rica	$32,000,000,000	$8,500	$5,100,000,000	$6,400,000,000	1,937	-	-	-	100%	-	-	-	-
Cuba	$30,690,000,000	$2,300	$1,800,000,000	$4,800,000,000	4,626	-	83%	17%	-	-	2,134,520	-	-
Dominica	$380,000,000	$5,400	$50,000,000	$135,000,000	4	-	-	-	100%	-	-	-	-
Dominican Republic	$53,780,000,000	$6,100	$5,300,000,000	$8,700,000,000	115	-	-	-	100%	-	-	-	-
El Salvador	$29,410,000,000	$4,700	$3,000,000,000	$4,900,000,000	1,110	-	-	-	100%	-	-	-	-
Guatemala	$53,200,000,000	$3,700	$2,700,000,000	$5,600,000,000	1,822	-	81%	1%	18%	-	1,076,526	9,000	-
Haiti	$10,600,000,000	$1,700	$298,000,000	$1,140,000,000	33	-	-	-	100%	-	-	-	-
Honduras	$16,290,000,000	$2,600	$1,300,000,000	$2,700,000,000	347	-	-	-	100%	-	-	-	-
Jamaica	$10,080,000,000	$3,900	$1,400,000,000	$3,100,000,000	18	-	-	-	100%	-	-	-	11,728,000
Mexico	$924,400,000,000	$9,000	$158,400,000,000	$168,400,000,000	340,594	1%	79%	16%	4%	11,097,943	150,165,451	6,860,000	-
Nicaragua	$11,160,000,000	$2,500	$637,000,000	$1,700,000,000	706	-	-	-	100%	-	-	-	-
Panama	$18,060,000,000	$6,000	$5,800,000,000	$6,700,000,000	418	-	-	-	100%	-	-	-	-
St. Lucia	$866,000,000	$5,400	$68,300,000	$319,400,000	-	-	-	-	-	-	-	-	-
Trinidad and Tobago	$11,070,000,000	$9,500	$4,200,000,000	$3,800,000,000	22,768	-	39%	61%	-	-	5,964,991	-	-
United States	$10,450,000,000,000	$37,600	$733,900,000,000	$1,194,100,000,000	2,342,228	33%	22%	30%	14%	996,498,186	289,640,487	35,178,000	-
SOUTH AMERICA													
Argentina	$403,800,000,000	$10,200	$25,300,000,000	$9,000,000,000	118,739	-	50%	45%	5%	260,299	38,783,798	-	-
Bolivia	$21,150,000,000	$2,500	$1,300,000,000	$1,600,000,000	7,732	-	33%	64%	3%	-	1,599,401	-	-
Brazil	$1,376,000,000,000	$7,600	$59,400,000,000	$46,200,000,000	143,640	3%	63%	6%	28%	4,446,477	61,155,586	124,667,000	13,654,000
Chile	$156,100,000,000	$10,000	$17,800,000,000	$15,600,000,000	6,180	6%	11%	45%	38%	475,484	349,201	5,523,000	-
Colombia	$251,600,000,000	$6,500	$12,900,000,000	$12,500,000,000	99,513	36%	52%	9%	4%	38,112,136	34,896,672	348,000	-
Ecuador	$42,650,000,000	$3,100	$4,900,000,000	$6,000,000,000	32,171	-	94%	3%	3%	-	19,520,185	-	-
Guyana	$2,628,000,000	$4,000	$500,000,000	$575,000,000	1	-	-	-	100%	-	-	-	2,272,000
Paraguay	$25,190,000,000	$4,200	$2,000,000,000	$2,400,000,000	6,577	-	-	-	100%	-	-	-	-
Peru	$138,800,000,000	$4,800	$7,600,000,000	$7,300,000,000	10,933	-	73%	9%	18%	52,297	4,932,561	2,701,000	-
Suriname	$1,469,000,000	$3,500	$445,000,000	$300,000,000	1,022	-	84%	-	16%	-	496,400	-	3,946,000
Uruguay	$26,820,000,000	$7,800	$2,100,000,000	$1,870,000,000	867	-	-	-	100%	-	-	-	-
Venezuela	$131,700,000,000	$5,500	$28,600,000,000	$18,800,000,000	311,899	3%	81%	14%	2%	7,482,998	146,621,238	10,497,000	4,309,000
EUROPE													
Albania	$15,690,000,000	$4,500	$340,000,000	$1,500,000,000	1,089	1%	42%	2%	55%	32,666	284,321	-	-
Austria	$227,700,000,000	$27,700	$70,000,000,000	$74,000,000,000	9,611	5%	15%	24%	56%	1,197,660	921,120	525,000	-
Belarus	$90,190,000,000	$8,200	$7,700,000,000	$8,800,000,000	3,644	18%	73%	9%	-	-	1,830,872	-	-
Belgium	$299,700,000,000	$29,000	$162,000,000,000	$152,000,000,000	18,451	2%	-	-	98%	318,998	-	-	-
Bosnia and Herzegovina	$7,300,000,000	$1,900	$1,150,000,000	$2,800,000,000	6,553	90%	-	-	10%	8,414,623	-	50,000	75,000
Bulgaria	$49,230,000,000	$6,600	$5,300,000,000	$6,900,000,000	13,500	46%	-	-	53%	28,841,963	37,048	310,000	-
Croatia	$43,120,000,000	$8,800	$4,900,000,000	$10,700,000,000	4,962	-	42%	43%	15%	5,104	1,191,360	-	-
Czech Republic	$157,100,000,000	$15,300	$40,800,000,000	$43,200,000,000	39,843	85%	1%	1%	14%	63,466,671	283,097	-	-
Denmark	$155,300,000,000	$29,000	$56,300,000,000	$47,900,000,000	36,502	-	70%	29%	2%	-	16,701,163	-	-
Estonia	$15,520,000,000	$10,900	$3,400,000,000	$4,400,000,000	3,892	100%	-	-	-	-	-	-	-
Finland	$133,800,000,000	$26,200	$40,100,000,000	$31,800,000,000	11,933	15%	-	-	85%	-	-	-	-
France	$1,558,000,000,000	$25,700	$307,800,000,000	$303,700,000,000	175,306	2%	4%	1%	93%	3,616,981	1,446,228	12,000	-
Germany	$2,160,000,000,000	$26,600	$608,000,000,000	$487,300,000,000	181,697	47%	2%	13%	38%	204,685,080	3,044,206	5,000	-
Greece	$203,300,000,000	$19,000	$12,600,000,000	$31,400,000,000	12,988	92%	3%	1%	4%	64,503,999	166,807	583,000	1,975,000
Hungary	$134,000,000,000	$13,300	$31,400,000,000	$33,900,000,000	16,319	25%	19%	24%	32%	14,796,257	1,301,710	-	994,000
Iceland	$8,444,000,000	$25,000	$2,300,000,000	$2,100,000,000	1,638	-	-	-	100%	-	-	-	-
Ireland	$113,700,000,000	$30,500	$86,600,000,000	$48,600,000,000	3,232	47%	-	47%	6%	-	-	-	-
Italy	$1,455,000,000,000	$25,000	$259,200,000,000	$238,200,000,000	40,332	-	16%	54%	30%	47,666	4,144,278	-	-
Latvia	$20,990,000,000	$8,300	$2,300,000,000	$3,900,000,000	369	6%	-	-	94%	-	-	-	-
Lithuania	$30,080,000,000	$8,400	$5,400,000,000	$6,800,000,000	3,677	-	12%	-	87%	-	251,824	-	-
Luxembourg	$21,940,000,000	$44,000	$10,100,000,000	$13,250,000,000	113	-	-	-	100%	-	-	-	-
Macedonia	$10,570,000,000	$5,000	$1,100,000,000	$1,900,000,000	3,038	95%	-	-	5%	7,463,628	-	9,000	-
Moldova	$11,510,000,000	$2,500	$590,000,000	$980,000,000	7	-	-	-	100%	-	-	-	-
Netherlands	$437,800,000,000	$26,900	$243,300,000,000	$201,100,000,000	87,974	-	4%	94%	2%	-	1,437,293	-	-
Norway	$149,100,000,000	$31,800	$68,200,000,000	$37,300,000,000	324,396	-	72%	22%	5%	847,996	154,419,533	355,000	-
Poland	$373,200,000,000	$9,500	$32,400,000,000	$43,400,000,000	108,277	94%	1%	5%	-	164,737,813	645,072	-	-
Portugal	$195,200,000,000	$18,000	$25,900,000,000	$39,000,000,000	1,560	-	-	-	100%	-	-	6,000	-
Romania	$169,300,000,000	$7,400	$13,700,000,000	$16,700,000,000	37,598	19%	24%	46%	10%	27,392,191	6,038,110	24,000	-
Serbia and Montenegro	$23,150,000,000	$2,370	$2,400,000,000	$6,300,000,000	14,188	74%	8%	8%	10%	34,480,488	810,787	10,000	580,000
Slovakia	$67,340,000,000	$12,200	$12,900,000,000	$15,400,000,000	8,813	17%	1%	2%	79%	3,606,648	48,134	200,000	-
Slovenia	$37,060,000,000	$18,000	$10,300,000,000	$11,100,000,000	3,644	38%	-	-	62%	4,391,644	991	-	-
Spain	$850,700,000,000	$20,700	$122,200,000,000	$156,600,000,000	40,444	28%	2%	1%	68%	23,479,212	296,665	-	-
Sweden	$230,700,000,000	$25,400	$80,600,000,000	$68,600,000,000	31,413	1%	-	-	99%	-	-	12,114,000	-
Switzerland	$233,400,000,000	$31,700	$100,300,000,000	$94,400,000,000	14,710	-	-	-	100%	-	-	-	-
Ukraine	$218,000,000,000	$4,500	$18,100,000,000	$18,000,000,000	118,973	50%	5%	20%	25%	81,998,575	3,747,936	28,933,000	-
United Kingdom	$1,528,000,000,000	$25,300	$286,300,000,000	$330,100,000,000	397,906	7%	47%	38%	8%	32,758,497	119,820,635	1,000	-
Russia	$1,409,000,000,000	$9,300	$104,600,000,000	$60,700,000,000	1,412,286	10%	33%	52%	5%	253,376,954	324,436,632	48,300,000	3,983,000
ASIA													
Afghanistan	$19,000,000,000	$700	$1,200,000,000	$1,300,000,000	195	1%	-	79%	20%	1,000	-	-	-
Armenia	$12,130,000,000	$3,800	$525,000,000	$991,000,000	901	-	-	-	100%	-	-	-	-
Azerbaijan	$28,610,000,000	$3,500	$2,000,000,000	$1,800,000,000	27,748	-	72%	27%	1%	-	14,183,985	-	-
Bahrain	$9,910,000,000	$14,000	$5,800,000,000	$4,200,000,000	14,442	-	22%	78%	-	-	1,827,397	-	-
Bangladesh	$238,200,000,000	$1,700	$6,200,000,000	$8,500,000,000	11,713	-	-	99%	1%	-	120,476	-	-
Brunei	$6,500,000,000	$18,600	$3,000,000,000	$1,400,000,000	27,922	-	49%	51%	-	-	9,435,323	-	-
Cambodia	$20,420,000,000	$1,500	$1,380,000,000	$1,730,000,000	10	-	-	-	100%	-	-	-	-
China	$5,989,000,000,000	$4,400	$658,260,000,000	$618,930,000,000	1,023,314[5]	70%[5]	23%[5]	4%[5]	3%[5]	1,251,423,183	161,226,848	72,967,000	9,000,000
Cyprus	$9,400,000,000	$15,000	$1,030,000,000	$3,900,000,000	-	-	-	-	-	-	-	-	-
East Timor	$440,000,000	$500	$8,000,000	$237,000,000	-	-	-	-	-	-	-	-	-
Georgia	$16,050,000,000	$3,100	$515,000,000	$750,000,000	963	1%	16%	8%	75%	10,000	102,258	-	-
India	$2,664,000,000,000	$2,540	$44,500,000,000	$53,800,000,000	367,807	73%	14%	8%	4%	304,842,421	32,123,682	48,080,000	7,554,000
Indonesia	$714,200,000,000	$3,100	$52,300,000,000	$32,100,000,000	279,695	27%	45%	26%	2%	79,664,587	70,565,213	282,000	1,168,000
Iran	$458,300,000,000	$7,000	$24,800,000,000	$21,800,000,000	350,729	-	77%	23%	-	1,376,993	181,632,777	5,367,000	136,000
Iraq	$58,000,000,000	$2,400	$13,000,000,000	$7,800,000,000	186,519	-	97%	3%	-	-	124,281,583	-	-
Israel	$117,400,000,000	$19,000	$28,100,000,000	$30,800,000,000	334	94%	2%	4%	1%	-	5,957	-	-
Japan	$3,651,000,000,000	$28,000	$383,800,000,000	$292,100,000,000	142,731	2%	1%	2%	95%	3,286,983	351,650	1,000	-
Jordan	$22,630,000,000	$4,300	$2,500,000,000	$4,400,000,000	316	-	1%	97%	2%	-	1,986	-	-
Kazakhstan	$120,000,000,000	$6,300	$10,300,000,000	$9,600,000,000	113,390	40%	45%	14%	1%	70,311,969	30,508,827	7,467,000	3,668,000
Korea, North	$22,260,000,000	$1,000	$842,000,000	$1,314,000,000	65,932	96%	-	-	4%	94,174,845	-	3,000,000	-
Korea, South	$941,500,000,000	$19,400	$162,600,000,000	$148,400,000,000	43,892	6%	-	-	94%	4,054,646	-	175,000	-
Kuwait	$36,850,000,000	$15,000	$16,000,000,000	$7,300,000,000	161,322	-	92%	8%	-	-	98,844,823	-	-

CONTINENT/Country	GDP 2002 Total GDP[1]	GDP Per Capita[1]	Trade Value of Exports[1]	Value of Imports[1]	Commercial Energy Production Avg. 2000[2] Total (1,000 Metric Tons of Coal Equiv.)	Solid %	Liquid %	Gas %	Hydro & Nuclear %	Average Production 1999-2001 in Metric Tons Coal[3]	Petroleum[3]	Iron Ore[4]	Bauxite[4]
Kyrgyzstan	$13,880,000,000	$2,800	$488,000,000	$587,000,000	2,026	9%	5%	2%	83%	423,664	91,503	-	-
Laos	$10,400,000,000	$1,700	$345,000,000	$555,000,000	146	1%	-	-	99%	1,000	-	-	-
Lebanon	$17,610,000,000	$5,400	$1,000,000,000	$6,000,000,000	55	-	-	-	100%	-	-	-	-
Malaysia	$198,400,000,000	$9,300	$95,200,000,000	$76,800,000,000	110,069	-	41%	58%	1%	314,332	33,792,132	208,000	137,000
Mongolia	$5,060,000,000	$1,840	$501,000,000	$659,000,000	2,212	100%	-	-	-	5,099,640	-	-	-
Myanmar	$73,690,000,000	$1,660	$2,700,000,000	$2,500,000,000	9,297	3%	6%	88%	2%	358,331	587,374	-	-
Nepal	$37,320,000,000	$1,400	$720,000,000	$1,600,000,000	172	10%	-	-	90%	9,667	-	-	-
Oman	$22,400,000,000	$8,300	$10,600,000,000	$5,500,000,000	74,376	-	92%	8%	-	-	46,989,489	-	-
Pakistan	$295,300,000,000	$2,100	$9,800,000,000	$11,100,000,000	33,773	6%	12%	74%	7%	3,247,391	2,768,108	-	10,000
Philippines	$379,700,000,000	$4,200	$35,100,000,000	$33,500,000,000	16,244	6%	-	-	94%	1,306,993	173,128	-	-
Qatar	$15,910,000,000	$21,500	$10,900,000,000	$3,900,000,000	92,237	-	57%	43%	-	-	35,018,538	-	-
Saudi Arabia	$268,900,000,000	$10,500	$71,000,000,000	$39,500,000,000	736,996	-	91%	9%	-	-	401,559,222	-	-
Singapore	$112,400,000,000	$24,000	$127,000,000,000	$113,000,000,000	-	-	-	-	-	-	-	-	-
Sri Lanka	$73,700,000,000	$3,700	$4,600,000,000	$5,400,000,000	394	-	-	-	100%	-	-	-	-
Syria	$63,480,000,000	$3,500	$6,200,000,000	$4,900,000,000	47,898	-	83%	15%	2%	-	26,119,029	-	-
Taiwan	$406,000,000,000	$18,000	$130,000,000,000	$113,000,000,000	(6)	(6)	(6)	(6)	(6)	58,284	38,686	-	-
Tajikistan	$8,476,000,000	$1,250	$710,000,000	$830,000,000	1,790	-	1%	3%	95%	20,667	16,613	-	-
Thailand	$445,800,000,000	$6,900	$67,700,000,000	$58,100,000,000	44,127	25%	24%	50%	2%	18,551,756	5,080,720	20,000	-
Turkey	$489,700,000,000	$7,000	$35,100,000,000	$50,800,000,000	28,167	69%	14%	3%	14%	65,334,995	2,642,106	2,300,000	303,000
Turkmenistan	$31,340,000,000	$5,500	$2,970,000,000	$2,250,000,000	71,764	-	15%	85%	-	-	7,139,688	-	-
United Arab Emirates	$53,970,000,000	$22,000	$44,900,000,000	$30,800,000,000	199,656	-	83%	17%	-	-	112,737,023	-	-
Uzbekistan	$66,060,000,000	$2,500	$2,800,000,000	$2,500,000,000	85,806	1%	13%	85%	1%	2,736,319	4,419,300	-	-
Vietnam	$183,800,000,000	$2,250	$16,500,000,000	$16,800,000,000	39,300	30%	59%	5%	7%	9,688,950	15,926,911	-	-
Yemen	$15,070,000,000	$840	$3,400,000,000	$2,900,000,000	30,622	-	100%	-	-	-	21,304,264	-	-
AFRICA													
Algeria	$173,800,000,000	$5,300	$19,500,000,000	$10,600,000,000	222,648	-	47%	53%	-	24,000	61,651,110	757,000	-
Angola	$18,360,000,000	$1,600	$8,600,000,000	$4,100,000,000	53,315	-	98%	1%	-	-	36,961,745	-	-
Benin	$7,380,000,000	$1,070	$207,000,000	$479,000,000	69	-	100%	-	-	-	39,547	-	-
Botswana	$13,480,000,000	$9,500	$2,400,000,000	$1,900,000,000	(7)	(7)	(7)	(7)	(7)	956,767	-	-	-
Burkina Faso	$14,510,000,000	$1,080	$250,000,000	$525,000,000	15	-	-	-	100%	-	-	-	-
Burundi	$3,146,000,000	$600	$26,000,000	$135,000,000	21	29%	-	-	71%	-	-	-	-
Cameroon	$26,840,000,000	$1,700	$1,900,000,000	$1,700,000,000	10,722	-	96%	-	4%	1,000	4,326,440	-	-
Cape Verde	$600,000,000	$1,400	$30,000,000	$220,000,000	-	-	-	-	-	-	-	-	-
Central African Republic	$4,296,000,000	$1,300	$134,000,000	$102,000,000	10	-	-	-	100%	-	-	-	-
Chad	$9,297,000,000	$1,100	$197,000,000	$570,000,000	-	-	-	-	-	-	-	-	-
Comoros	$441,000,000	$720	$16,300,000	$39,800,000	-	-	-	-	-	-	-	-	-
Congo	$2,500,000,000	$900	$2,400,000,000	$73,000,000	19,097	-	99%	1%	-	-	13,651,000	-	-
Congo, Democratic Republic of the	$34,000,000,000	$610	$1,200,000,000	$890,000,000	2,630	4%	71%	-	25%	96,000	1,194,669	-	-
Cote d'Ivoire	$24,030,000,000	$1,500	$4,400,000,000	$2,500,000,000	4,439	-	50%	45%	5%	-	620,450	-	-
Djibouti	$619,000,000	$1,300	$70,000,000	$255,000,000	-	-	-	-	-	-	-	-	-
Egypt	$289,800,000,000	$3,900	$7,000,000,000	$15,200,000,000	86,315	-	65%	32%	2%	-	38,024,058	1,283,000	-
Equatorial Guinea	$1,270,000,000	$2,700	$2,500,000,000	$562,000,000	7,531	-	100%	-	-	-	7,461,521	-	-
Eritrea	$3,300,000,000	$740	$20,000,000	$500,000,000	-	-	-	-	-	-	-	-	-
Ethiopia	$48,530,000,000	$750	$433,000,000	$1,630,000,000	211	-	-	-	100%	-	-	-	-
Gabon	$8,354,000,000	$5,700	$2,600,000,000	$1,100,000,000	23,273	-	95%	5%	-	-	15,674,359	-	-
Gambia, The	$2,582,000,000	$1,800	$138,000,000	$225,000,000	-	-	-	-	-	-	-	-	-
Ghana	$41,250,000,000	$2,100	$2,200,000,000	$2,800,000,000	830	-	2%	-	98%	-	330,933	-	525,000
Guinea	$18,690,000,000	$2,000	$835,000,000	$670,000,000	25	-	-	-	100%	-	-	-	15,663,000
Guinea-Bissau	$901,400,000	$800	$71,000,000	$59,000,000	-	-	-	-	-	-	-	-	-
Kenya	$32,890,000,000	$1,020	$2,100,000,000	$3,000,000,000	642	-	-	-	100%	-	-	-	-
Lesotho	$5,106,000,000	$2,700	$422,000,000	$738,000,000	(7)	(7)	(7)	(7)	(7)	-	-	-	-
Liberia	$3,116,000,000	$1,100	$110,000,000	$165,000,000	24	-	-	-	100%	-	-	-	-
Libya	$33,360,000,000	$7,600	$11,800,000,000	$6,300,000,000	103,205	-	92%	8%	-	-	67,767,436	-	-
Madagascar	$12,590,000,000	$760	$700,000,000	$985,000,000	64	-	-	-	100%	-	-	-	-
Malawi	$6,811,000,000	$670	$435,000,000	$505,000,000	107	-	-	-	100%	-	-	-	-
Mali	$9,775,000,000	$860	$680,000,000	$630,000,000	29	-	-	-	100%	-	-	-	-
Mauritania	$4,891,000,000	$1,900	$355,000,000	$360,000,000	4	-	-	-	100%	-	-	7,492,000	-
Mauritius	$12,150,000,000	$11,000	$1,600,000,000	$1,800,000,000	12	-	-	-	100%	-	-	-	-
Morocco	$121,800,000,000	$3,900	$7,500,000,000	$10,400,000,000	201	14%	9%	33%	43%	61,000	15,223	4,000	-
Mozambique	$19,520,000,000	$1,000	$680,000,000	$1,180,000,000	874	2%	-	-	98%	18,667	-	-	8,000
Namibia	$13,150,000,000	$6,900	$1,210,000,000	$1,380,000,000	(7)	(7)	(7)	(7)	(7)	-	-	-	-
Niger	$8,713,000,000	$830	$293,000,000	$368,000,000	175	100%	-	-	-	151,666	-	-	-
Nigeria	$112,500,000,000	$875	$17,300,000,000	$13,600,000,000	172,641	-	90%	10%	-	61,000	108,397,478	-	-
Rwanda	$8,920,000,000	$1,200	$68,000,000	$253,000,000	20	-	-	-	100%	-	-	-	-
Sao Tome and Principe	$200,000,000	$1,200	$5,500,000	$24,800,000	1	-	-	-	100%	-	-	-	-
Senegal	$15,640,000,000	$1,500	$1,150,000,000	$1,460,000,000	1	-	-	100%	-	-	-	-	-
Sierra Leone	$2,826,000,000	$580	$35,000,000	$190,000,000	(7)	(7)	(7)	(7)	(7)	-	-	-	-
Somalia	$4,270,000,000	$550	$126,000,000	$343,000,000	-	-	-	-	-	-	-	-	-
South Africa	$427,700,000,000	$10,000	$31,800,000,000	$26,600,000,000	245,195[8]	92%[8]	5%[8]	1%[8]	2%[8]	224,286,505	1,277,485	20,751,000	-
Sudan	$52,900,000,000	$1,420	$1,800,000,000	$1,500,000,000	13,436	-	99%	-	1%	-	7,679,837	-	-
Swaziland	$5,542,000,000	$4,400	$820,000,000	$938,000,000	(7)	(7)	(7)	(7)	(7)	288,665	-	-	-
Tanzania	$20,420,000,000	$630	$863,000,000	$1,670,000,000	343	23%	-	-	77%	5,000	-	-	-
Togo	$7,594,000,000	$1,500	$449,000,000	$561,000,000	-	-	-	-	-	-	-	-	-
Tunisia	$67,130,000,000	$6,500	$6,800,000,000	$8,700,000,000	8,065	-	66%	34%	-	-	3,826,400	105,000	-
Uganda	$30,490,000,000	$1,260	$476,000,000	$1,140,000,000	193	-	-	-	100%	-	-	3,000	-
Zambia	$8,240,000,000	$890	$709,000,000	$1,123,000,000	1,117	15%	-	-	85%	192,358	-	-	-
Zimbabwe	$26,070,000,000	$2,400	$1,570,000,000	$1,739,000,000	4,801	92%	-	-	8%	4,508,643	-	237,000	-
OCEANIA													
Australia	$525,500,000,000	$27,000	$66,300,000,000	$68,000,000,000	331,923	71%	14%	14%	1%	307,176,075	31,728,994	104,014,000	51,834,000
Fiji	$4,822,000,000	$5,500	$442,000,000	$642,000,000	53	-	-	-	100%	-	-	-	-
Kiribati	$79,000,000	$840	$6,000,000	$44,000,000	-	-	-	-	-	-	-	-	-
Micronesia, Federated States of	$277,000,000	$2,000	$22,000,000	$149,000,000	-	-	-	-	-	-	-	-	-
New Zealand	$78,400,000,000	$20,200	$15,000,000,000	$12,500,000,000	19,812	14%	13%	40%	33%	3,452,315	1,839,394	660,000	-
Papua New Guinea	$10,860,000,000	$2,300	$1,800,000,000	$1,100,000,000	5,864	-	96%	2%	2%	-	3,874,601	-	-
Samoa	$1,000,000,000	$5,600	$15,500,000	$130,100,000	3	-	-	-	100%	-	-	-	-
Solomon Islands	$800,000,000	$1,700	$47,000,000	$82,000,000	-	-	-	-	-	-	-	-	-
Tonga	$236,000,000	$2,200	$8,900,000	$70,000,000	-	-	-	-	-	-	-	-	-
Vanuatu	$563,000,000	$2,900	$22,000,000	$93,000,000	-	-	-	-	-	-	-	-	-

This table presents data for most independent nations having an area greater than 200 square miles
- Zero, insignificant, or not available
(1) Source: United States Central Intelligence Agency World Factbook
(2) Source: United Nations Energy Statistics Yearbook
(3) Source: United States Energy Information Administration International Energy Annual
(4) Source: United States Geological Survey Minerals Yearbook
(5) Includes data for Taiwan
(6) Data for Taiwan is included with China
(7) Data for countries in the South Africa Customs Union are included with South Africa
(8) Includes data for countries in the South Africa Customs Union

WORLD ENVIRONMENT TABLE

CONTINENT/Country	Total Area Sq. Miles	Protected Area 2002[1,2] Sq. Miles	%	Mammal	Bird	Reptile	Amphib.	Fish	Invrt.	Forest Cover[4] Sq. Miles 2000	Percent Change 1990-2000
NORTH AMERICA											
Bahamas	5,382	-	-	5	4	6	0	15	1	3,251	-
Belize	8,867	3,999	45.1%	5	2	4	0	17	1	5,205	-20.9%
Canada	3,855,103	427,916	11.1%	16	8	2	1	25	11	944,294	-
Costa Rica	19,730	4,538	23.0%	13	13	7	1	13	9	7,598	-7.4%
Cuba	42,804	29,578	69.1%	11	18	7	0	23	3	9,066	13.4%
Dominica	290	-	-	1	3	4	0	11	0	178	-8.0%
Dominican Republic	18,730	9,721	51.9%	5	15	10	1	10	2	5,313	-
El Salvador	8,124	33	0.4%	2	0	4	0	5	1	467	-37.3%
Guatemala	42,042	8,408	20.0%	7	6	8	0	14	8	11,004	-15.9%
Haiti	10,714	43	0.4%	4	14	8	1	12	2	340	-44.3%
Honduras	43,277	2,770	6.4%	10	5	6	0	14	2	20,784	-9.9%
Jamaica	4,244	3,590	84.6%	5	12	8	4	12	5	1,255	-14.2%
Mexico	758,452	77,362	10.2%	72	40	18	4	106	41	213,148	-10.3%
Nicaragua	50,054	8,910	17.8%	6	5	7	0	17	2	12,656	-26.3%
Panama	29,157	6,327	21.7%	17	16	7	0	17	2	11,104	-15.3%
St. Lucia	238	-	-	2	5	6	0	10	0	35	-35.7%
Trinidad and Tobago	1,980	119	6.0%	1	1	5	0	15	0	1,000	-7.8%
United States	3,794,083	982,668	25.9%	39	56	27	25	155	557	872,563	1.7%
SOUTH AMERICA											
Argentina	1,073,519	70,852	6.6%	32	39	5	5	9	10	133,777	-7.6%
Bolivia	424,165	56,838	13.4%	25	28	2	1	0	1	204,897	-2.9%
Brazil	3,300,172	221,112	6.7%	74	113	22	6	33	34	2,100,028	-4.1%
Chile	291,930	55,175	18.9%	21	22	0	3	9	0	59,985	-1.3%
Colombia	439,737	44,853	10.2%	39	78	14	0	23	0	191,510	-3.7%
Ecuador	109,484	20,036	18.3%	34	62	10	0	11	48	40,761	-11.5%
Guyana	83,000	249	0.3%	13	2	6	0	13	1	65,170	-2.8%
Paraguay	157,048	5,497	3.5%	10	26	2	0	0	0	90,240	-5.0%
Peru	496,225	30,270	6.1%	46	76	6	1	8	2	251,796	-4.0%
Suriname	63,037	3,089	4.9%	12	1	6	0	12	0	54,491	-
Uruguay	67,574	203	0.3%	6	11	3	0	8	1	4,988	63.3%
Venezuela	352,145	224,669	63.8%	26	24	13	0	19	1	191,144	-4.2%
EUROPE											
Albania	11,100	422	3.8%	3	3	4	0	16	4	3,826	-7.3%
Austria	32,378	10,685	33.0%	7	3	0	0	7	44	15,004	2.0%
Belarus	80,155	5,050	6.3%	7	3	0	0	0	5	36,301	37.5%
Belgium	11,787	-	-	11	2	0	0	7	11	2,811	-1.8%
Bosnia and Herzegovina	19,767	99	0.5%	10	3	1	1	10	10	8,776	-
Bulgaria	42,855	1,928	4.5%	14	10	2	0	10	9	14,247	5.9%
Croatia	21,829	1,637	7.5%	9	4	1	1	26	11	6,884	1.1%
Czech Republic	30,450	4,902	16.1%	8	2	0	0	7	19	10,162	0.2%
Denmark	16,640	5,658	34.0%	5	1	0	0	7	11	1,757	2.2%
Estonia	17,462	2,061	11.8%	5	3	0	0	1	4	7,954	6.5%
Finland	130,559	12,142	9.3%	4	3	0	0	1	10	84,691	0.4%
France	208,482	27,728	13.3%	18	5	3	2	15	65	59,232	4.2%
Germany	137,847	43,973	31.9%	11	5	0	0	12	31	41,467	-
Greece	50,949	1,834	3.6%	13	7	6	1	26	11	13,896	9.1%
Hungary	35,919	2,514	7.0%	9	8	1	0	8	25	7,104	4.1%
Iceland	39,769	3,897	9.8%	7	0	0	0	8	0	120	24.0%
Ireland	27,133	461	1.7%	6	1	0	0	6	3	2,544	34.8%
Italy	116,342	9,191	7.9%	14	5	4	4	16	58	38,622	3.0%
Latvia	24,942	3,342	13.4%	5	3	0	0	3	8	11,286	4.5%
Lithuania	25,213	2,597	10.3%	6	4	0	0	3	5	7,699	2.5%
Luxembourg	999	-	-	3	1	0	0	0	4	-	-
Macedonia	9,928	705	7.1%	11	3	2	0	4	5	3,498	-
Moldova	13,070	183	1.4%	6	5	1	0	9	5	1,255	2.2%
Netherlands	16,164	2,295	14.2%	10	4	0	0	7	7	1,448	2.7%
Norway	125,050	8,503	6.8%	10	2	0	0	7	9	34,240	3.6%
Poland	120,728	14,970	12.4%	14	4	0	0	3	15	34,931	2.0%
Portugal	35,516	2,344	6.6%	17	7	0	1	19	82	14,154	18.4%
Romania	91,699	4,310	4.7%	17	8	2	0	10	22	24,896	2.3%
Serbia and Montenegro	39,449	1,302	3.3%	12	5	1	0	19	19	11,147	-0.5%
Slovakia	18,924	4,315	22.8%	9	4	1	0	8	19	8,405	9.0%
Slovenia	7,821	469	6.0%	9	1	0	1	15	42	4,274	2.0%
Spain	194,885	16,565	8.5%	24	7	7	3	23	63	55,483	6.4%
Sweden	173,732	15,810	9.1%	6	2	0	0	6	13	104,765	-
Switzerland	15,943	4,783	30.0%	5	2	0	0	4	30	4,629	3.7%
Ukraine	233,090	9,091	3.9%	16	8	2	0	11	14	37,004	3.3%
United Kingdom	93,788	19,602	20.9%	12	2	0	0	11	10	10,788	6.5%
Russia	6,592,849	514,242	7.8%	45	38	6	0	18	30	3,287,242	0.2%
ASIA											
Afghanistan	251,773	755	0.3%	13	11	1	1	0	1	5,216	-
Armenia	11,506	874	7.6%	11	4	5	0	1	7	1,355	13.6%
Azerbaijan	33,437	2,040	6.1%	13	8	5	0	5	6	4,224	13.5%
Bahrain	267	-	-	1	6	0	0	6	0	-	-
Bangladesh	55,598	445	0.8%	22	23	20	0	8	0	5,151	14.1%
Brunei	2,226	-	-	11	14	4	0	6	0	1,707	-2.2%
Cambodia	69,898	12,931	18.5%	24	19	10	0	11	0	36,043	-5.7%
China	3,690,045	287,824	7.8%	81	75	31	1	46	4	631,200	12.4%
Cyprus	3,572	-	-	3	3	3	0	6	0	664	44.5%
East Timor	5,743	-	-	0	6	0	0	2	0	1,958	-6.3%
Georgia	26,911	619	2.3%	13	3	7	1	6	10	11,537	-
India	1,222,510	63,571	5.2%	86	72	25	3	27	23	247,542	0.6%
Indonesia	735,310	151,474	20.6%	147	114	28	0	91	31	405,353	-11.1%
Iran	636,372	30,546	4.8%	22	13	8	2	14	3	28,182	-
Iraq	169,235	-	-	11	11	2	0	3	2	3,085	-
Israel	8,019	1,267	15.8%	15	12	4	0	10	10	510	61.0%
Japan	145,850	9,918	6.8%	37	35	11	10	27	45	92,957	0.1%
Jordan	34,495	1,173	3.4%	9	8	1	0	5	3	332	-
Kazakhstan	1,049,156	28,327	2.7%	17	15	2	1	7	4	46,904	24.5%
Korea, North	46,540	1,210	2.6%	13	19	0	0	5	1	31,699	-
Korea, South	38,328	2,645	6.9%	13	25	0	0	7	1	24,124	-0.8%

CONTINENT/Country	Total Area Sq. Miles	Protected Area 2002[1,2] Sq. Miles	%	Endangered Species 2003[3] Mammal	Bird	Reptile	Amphib.	Fish	Invrt.	Forest Cover[4] Sq. Miles 2000	Percent Change 1990-2000
Kuwait	6,880	103	1.5%	1	7	1	0	6	0	19	66.7%
Kyrgyzstan	77,182	2,779	3.6%	7	4	2	0	0	3	3,873	29.4%
Laos	91,429	11,429	12.5%	31	20	11	0	6	0	48,498	-4.0%
Lebanon	4,016	20	0.5%	6	7	1	0	8	1	139	-2.7%
Malaysia	127,320	7,257	5.7%	50	37	21	0	34	3	74,487	52.4%
Mongolia	604,829	69,555	11.5%	14	16	0	0	1	3	41,101	-5.3%
Myanmar	261,228	784	0.3%	39	35	20	0	7	2	132,892	-13.1%
Nepal	56,827	5,058	8.9%	29	25	6	0	0	1	15,058	-16.7%
Oman	119,499	16,730	14.0%	11	10	4	0	17	1	4	-
Pakistan	339,732	16,647	4.9%	17	17	9	0	14	0	9,116	-14.3%
Philippines	115,831	6,602	5.7%	50	67	8	23	48	19	22,351	-13.3%
Qatar	4,412	-	-	0	6	1	0	4	0	-	-
Saudi Arabia	830,000	317,890	38.3%	9	15	2	0	8	1	5,807	-
Singapore	264	13	4.9%	3	7	3	0	12	1	8	-
Sri Lanka	25,332	3,420	13.5%	22	14	8	0	22	2	7,490	-15.2%
Syria	71,498	-	-	4	8	3	0	8	3	1,780	-
Taiwan	13,901	-	-	12	21	8	0	23	0	-	-
Tajikistan	55,251	2,321	4.2%	9	7	1	0	3	2	1,544	5.3%
Thailand	198,115	27,538	13.9%	37	37	19	0	35	1	56,996	-7.1%
Turkey	302,541	4,841	1.6%	17	11	12	3	29	13	39,479	2.2%
Turkmenistan	188,457	7,915	4.2%	13	6	2	0	8	5	14,498	-
United Arab Emirates	32,278	-	-	4	8	1	0	6	0	1,239	32.1%
Uzbekistan	172,742	3,455	2.0%	9	9	2	0	4	1	7,602	2.4%
Vietnam	128,066	4,738	3.7%	42	37	24	1	22	0	37,911	5.5%
Yemen	203,850	-	-	6	12	2	0	10	2	1,734	-17.0%
AFRICA											
Algeria	919,595	45,980	5.0%	13	6	2	0	9	12	8,282	14.2%
Angola	481,354	31,769	6.6%	19	15	4	0	8	6	269,329	-1.7%
Benin	43,484	4,957	11.4%	9	2	1	0	7	0	10,232	-20.9%
Botswana	224,607	41,552	18.5%	7	7	0	0	0	0	47,981	-8.7%
Burkina Faso	105,869	12,175	11.5%	7	2	1	0	0	0	27,371	-2.1%
Burundi	10,745	612	5.7%	6	7	0	0	0	3	363	-61.0%
Cameroon	183,568	8,261	4.5%	38	15	1	1	34	4	92,116	-8.5%
Cape Verde	1,557	-	-	3	2	0	0	13	0	328	142.9%
Central African Republic	240,536	20,927	8.7%	14	3	1	0	0	0	88,444	-1.3%
Chad	495,755	45,114	9.1%	15	5	1	0	0	1	49,004	-6.0%
Comoros	863	-	-	2	9	2	0	3	4	31	-33.3%
Congo	132,047	6,602	5.0%	15	3	1	0	9	1	85,174	-0.8%
Congo, Democratic Republic of the	905,446	58,854	6.5%	40	28	2	0	9	45	522,037	-3.8%
Cote d'Ivoire	124,504	7,470	6.0%	19	12	2	1	10	1	27,479	-27.1%
Djibouti	8,958	-	-	5	5	0	0	9	0	23	-
Egypt	386,662	37,506	9.7%	13	7	6	0	13	1	278	38.5%
Equatorial Guinea	10,831	-	-	16	5	2	1	7	2	6,765	-5.7%
Eritrea	45,406	1,952	4.3%	12	7	6	0	8	0	6,120	-3.3%
Ethiopia	426,373	72,057	16.9%	35	16	1	0	0	4	17,734	-8.1%
Gabon	103,347	723	0.7%	14	5	1	0	11	1	84,271	-0.5%
Gambia, The	4,127	95	2.3%	3	2	1	0	10	0	1,857	10.3%
Ghana	92,098	5,157	5.6%	14	8	2	0	7	0	24,460	-15.9%
Guinea	94,926	664	0.7%	12	10	1	1	7	3	26,753	-4.8%
Guinea-Bissau	13,948	-	-	3	0	1	0	9	1	8,444	-9.0%
Kenya	224,961	17,997	8.0%	50	24	5	0	27	15	66,008	-5.2%
Lesotho	11,720	23	0.2%	6	7	0	0	1	1	54	-
Liberia	43,000	731	1.7%	16	11	2	0	7	2	13,440	-17.9%
Libya	679,362	679	0.1%	8	1	3	0	8	0	1,382	15.1%
Madagascar	226,658	9,746	4.3%	50	27	18	2	25	32	45,278	-9.1%
Malawi	45,747	5,124	11.2%	8	11	0	0	0	8	9,892	-21.6%
Mali	478,841	17,717	3.7%	13	4	1	0	1	0	50,911	-7.0%
Mauritania	397,956	6,765	1.7%	10	2	2	0	10	1	1,224	-23.6%
Mauritius	788	-	-	3	9	4	0	7	32	62	-5.9%
Morocco	172,414	1,207	0.7%	16	9	2	0	10	8	11,680	-0.4%
Mozambique	309,496	25,998	8.4%	15	16	5	0	19	7	118,151	-2.0%
Namibia	317,818	43,223	13.6%	14	11	3	1	11	1	31,043	-8.4%
Niger	489,192	37,668	7.7%	11	3	0	0	0	1	5,127	-31.7%
Nigeria	356,669	11,770	3.3%	27	9	2	0	11	1	52,189	-22.8%
Rwanda	10,169	630	6.2%	8	9	0	0	0	2	1,185	-32.8%
Sao Tome and Principe	372	-	-	3	9	1	0	6	2	104	-
Senegal	75,951	8,810	11.6%	12	4	6	0	17	0	23,958	-6.8%
Sierra Leone	27,699	582	2.1%	12	10	3	0	7	4	4,073	-25.5%
Somalia	246,201	1,970	0.8%	19	10	2	0	16	1	29,016	-9.3%
South Africa	470,693	25,888	5.5%	36	28	19	9	47	113	34,429	-0.9%
Sudan	967,500	50,310	5.2%	22	6	2	0	7	1	237,943	-13.5%
Swaziland	6,704	-	-	5	5	0	0	0	0	2,015	12.5%
Tanzania	364,900	108,740	29.8%	41	33	5	0	26	47	149,850	-2.3%
Togo	21,925	1,732	7.9%	9	0	2	0	7	0	1,969	-29.1%
Tunisia	63,170	190	0.3%	11	5	3	0	8	5	1,969	2.2%
Uganda	93,065	22,894	24.6%	20	13	0	0	27	10	16,178	-17.9%
Zambia	290,586	92,697	31.9%	11	11	0	0	0	6	120,641	-21.4%
Zimbabwe	150,873	18,256	12.1%	11	10	0	0	0	2	73,514	-14.4%
OCEANIA											
Australia	2,969,910	397,968	13.4%	63	35	38	35	74	282	596,678	-1.8%
Fiji	7,056	78	1.1%	5	13	6	1	8	2	3,147	-2.0%
Kiribati	313	-	-	0	4	1	0	4	1	108	-
Micronesia, Federated States of	271	-	-	6	5	2	0	6	4	58	-37.5%
New Zealand	104,454	30,918	29.6%	8	63	11	1	16	13	30,680	5.2%
Papua New Guinea	178,704	4,110	2.3%	58	32	9	0	31	12	118,151	-3.6%
Samoa	1,093	-	-	3	8	1	0	4	1	405	-19.2%
Solomon Islands	10,954	33	0.3%	20	23	4	0	4	6	9,792	-1.7%
Tonga	251	-	-	2	3	2	0	3	2	15	-
Vanuatu	4,707	-	-	5	8	2	0	4	0	1,726	1.4%

This table presents data for most independent nations having an area greater than 200 square miles

- Zero, insignificant, or not available

(1) Source: World Resources Institute, 2003. Earth Trends: The Environmental Information Portal. Available at http://earthtrends.wri.org. Washington D. C. World Resources Institute

(2) Source: United Nations Environment Programme - World Conservation Monitoring Centre (UNEP-WCMC); World Database on Protected Areas

(3) Source: International Union of Conservation of Nature and Natural Resources; IUCN 2003 Red List of Threatened Species <www.redlist.org>

(4) Source: United Nations Food and Agriculture Organization; Global Forest Resources Assessment 2000

WORLD COMPARISONS

General Information

Equatorial diameter of the earth, 7,926.38 miles.
Polar diameter of the earth, 7,899.80 miles.
Mean diameter of the earth, 7,917.52 miles.
Equatorial circumference of the earth, 24,901.46 miles.
Polar circumference of the earth, 24,855.34 miles.
Mean distance from the earth to the sun, 93,020,000 miles.
Mean distance from the earth to the moon, 238,857 miles.
Total area of the earth, 197,000,000 sq. miles.

Highest elevation on the earth's surface, Mt. Everest, Asia, 29,028 ft.
Lowest elevation on the earth's land surface, shores of the Dead Sea, Asia, 1,339 ft. below sea level.
Greatest known depth of the ocean, southwest of Guam, Pacific Ocean, 35,810 ft.
Total land area of the earth (incl. inland water and Antarctica), 57,900,000 sq. miles.

Area of Africa, 11,700,000 sq. miles.
Area of Antarctica, 5,400,000 sq. miles.
Area of Asia, 17,300,000 sq. miles.
Area of Europe, 3,800,000 sq. miles.
Area of North America, 9,500,000 sq. miles.
Area of Oceania (incl. Australia) 3,300,000 sq. miles.
Area of South America, 6,900,000 sq. miles.
Population of the earth (est. 1/1/04), 6,339,505,000.

Principal Islands and Their Areas

ISLAND	Area (Sq. Mi.)	ISLAND	Area (Sq. Mi.)	ISLAND	Area (Sq. Mi.)	ISLAND	Area (Sq. Mi.)	ISLAND	Area (Sq. Mi.)
Baffin I., Canada	195,928	Flores, Indonesia	5,502	Kyūshū, Japan	17,129	New Ireland, Papua New Guinea	3,475	Somerset I., Canada	9,570
Banks I., Canada	27,038	Great Britain, U.K.	88,795	Lyete, Philippines	2,785	North East Land, Norway	6,350	Southampton I., Canada	15,913
Borneo (Kalimantan), Asia	287,300	Greenland, N. America	840,000	Long Island, U.S.	1,377	North I., New Zealand	44,333	South I., New Zealand	57,708
Bougainville, Papua New Guinea	3,591	Guadalcanal, Solomon Is.	2,060	Luzon, Philippines	40,420	Novaya Zemlya, Russia	31,892	Spitsbergen, Norway	15,260
Cape Breton I., Canada	3,981	Hainan Dao, China	13,127	Madagascar, Africa	226,642	Palawan, Philippines	4,550	Sri Lanka, Asia	24,942
Celebes (Sulawesi), Indonesia	73,057	Hawaii, U.S.	4,028	Melville I., Canada	16,274	Panay, Philippines	4,446	Sumatra (Sumatera), Indonesia	182,860
Ceram (Seram), Indonesia	7,191	Hispaniola, N. America	29,300	Mindanao, Philippines	36,537	Prince of Wales I., Canada	12,872	Taiwan, Asia	13,900
Corsica, France	3,367	Hokkaidō, Japan	32,245	Mindoro, Philippines	3,759	Puerto Rico, N. America	3,514	Tasmania, Australia	26,178
Crete, Greece	3,189	Honshū, Japan	89,176	Negros, Philippines	4,907	Sakhalin, Russia	29,498	Tierra del Fuego, S. America	18,600
Cuba, N. America	42,780	Iceland, Europe	39,769	New Britain, Papua New Guinea	14,093	Samar, Philippines	5,050	Timor, Asia	5,743
Cyprus, Asia	3,572	Ireland, Europe	32,587	New Caledonia, Oceania	6,252	Sardinia, Italy	9,301	Vancouver I., Canada	12,079
Devon I., Canada	21,331	Jamaica, N. America	4,247	Newfoundland, Canada	42,031	Shikoku, Japan	7,258	Victoria I., Canada	83,897
Ellesmere I., Canada	75,767	Java (Jawa), Indonesia	51,038	New Guinea, Asia-Oceania	308,882	Sicily, Italy	9,926	Vrangelya (Wrangel), Russia	2,819
		Kodiak I., U.S.	3,670						

Principal Lakes, Oceans, Seas, and Their Areas

LAKE Country	Area (Sq. Mi.)	LAKE Country	Area (Sq. Mi.)	LAKE Country	Area (Sq. Mi.)	LAKE Country	Area (Sq. Mi.)	LAKE Country	Area (Sq. Mi.)
Arabian Sea	1,492,000	Black Sea, Europe-Asia	178,000	Hudson Bay, Canada	475,000	Michigan, L., U.S.	22,300	Southern Ocean	7,800,000
Aral Sea, Kazakhstan-Uzbekistan	13,000	Caribbean Sea, N.A.-S.A.	1,063,000	Huron, L., Canada-U.S.	23,000	Nicaragua, Lago de, Nicaragua	3,147	Superior, L., Canada-U.S.	31,700
Arctic Ocean	5,400,000	Caspian Sea, Asia-Europe	144,402	Indian Ocean	26,500,000	North Sea, Europe	222,000	Tanganyika. L., Africa	12,355
Athabasca, L., Canada	3,064	Chad, L., Cameroon-Chad-Nigeria	595	Japan, Sea of, Asia	389,000	Nyasa, L., Malawi-Mozambique-Tanzania	11,120	Titicaca, Lago, Bolivia-Peru	3,232
Atlantic Ocean	29,600,000	Erie, L., Canada-U.S.	9,910	Koko Nor (Qinghai Hu), China	1,722	Onezhskoye Ozero (L. Onega), Russia	3,819	Torrens, L., Australia	1,076
Balqash köli (L. Balkhash), Kazakhstan	7,027	Eyre, L., Australia	3,668	Ladozhskoye Ozero (L. Ladoga), Russia	7,002	Ontario, L., Canada-U.S.	7,340	Vänern (L.), Sweden	2,181
Baltic Sea, Europe	163,000	Gairdner, L., Australia	1,076	Manitoba, L., Canada	1,785	Pacific Ocean	60,100,000	Van Gölü (L.), Turkey	1,434
Baykal, Ozero (L. Baikal), Russia	12,162	Great Bear Lake, Canada	12,096	Mediterranean Sea, Europe-Africa-Asia	967,000	Red Sea, Africa-Asia	169,000	Victoria, L., Kenya-Tanzania-Uganda	26,564
Bering Sea, Asia-N.A.	876,000	Great Salt Lake, U.S.	1,700	Mexico, Gulf of, N. America	596,000	Rudolf, L., Ethiopia-Kenya	2,471	Winnipeg, L., Canada	9,416
		Great Slave Lake, Canada	11,030					Winnipegosis, L., Canada	2,075
								Yellow Sea, China-Korea	480,000

Principal Mountains and Their Heights

MOUNTAIN Country	Elev. (Ft.)	MOUNTAIN Country	Elev. (Ft.)	MOUNTAIN Country	Elev. (Ft.)	MOUNTAIN Country	Elev. (Ft.)	MOUNTAIN Country	Elev. (Ft.)
Aconcagua, Cerro, Argentina	22,831	Elgon, Mt., Kenya-Uganda	14,178	Kebnekaise, Sweden	6,926	Musala, Bulgaria	9,596	St. Elias, Mt., Alaska, U.S.-Canada	18,008
Annapurna, Nepal	26,504	Erciyeş, Dağı, Turkey	12,848	Kenya, Mt. (Kirinyaga), Kenya	17,058	Muztag, China	25,338	Sajama, Nevado, Bolivia	21,391
Aoraki, New Zealand	12,316	Etna, Mt., Italy	10,902	Kerinci, Gunung, Indonesia	12,467	Muztagata, China	24,757	Semeru, Gunung, Indonesia	12,060
Api, Nepal	23,399	Everest, Mt., China-Nepal	29,028	Kilimanjaro, Tanzania	19,340	Namjagbarwa Feng, China	25,446	Shām, Jabal ash, Oman	9,957
Apo, Philippines	9,692	Fairweather, Mt., Alaska-Canada	15,300	Kinabalu, Gunong, Malaysia	13,455	Nanda Devi, India	25,645	Shasta, Mt., California, U.S.	14,162
Ararat, Mt., Turkey	16,854	Folādī, Koh-e, Afghanistan	16,847	Klyuchevskaya, Russia	15,584	Nanga Parbat, Pakistan	26,660	Snowdon, United Kingdom	3,560
Barú, Volcán, Panama	11,401	Foraker, Mt., Alaska, U.S.	17,400	Kosciuszko, Mt., Australia	7,313	Narodnaya, Gora, Russia	6,217	Tahat, Algeria	9,541
Bangueta, Mt., Papua New Guinea	13,520	Fuji San, Japan	12,388	Koussi, Emi, Chad	11,204	Nevis, Ben, United Kingdom	4,406	Tajumulco, Guatemala	13,845
Belukha, Mt., Kazakhstan-Russia	14,783	Galdhøpiggen, Norway	8,100	Kula Kangri, Bhutan	24,784	Ojos del Salado, Nevado, Argentina-Chile	22,615	Taranaki, Mt., New Zealand	8,260
Bia, Phou, Laos	9,249	Gannett Pk., Wyoming, U.S.	13,804	La Selle, Massif de, Haiti	8,793	Ólimbos, Cyprus	6,401	Tirich Mīr, Pakistan	25,230
Blanc, Mont (Monte Bianco), France-Italy	15,771	Gasherbrum, China-Pakistan	26,470	Lassen Pk., California, U.S.	10,457	Ólympos, Greece	9,570	Tomanivi (Victoria), Fiji	4,341
Blanca Pk., Colorado, U.S.	14,345	Gerlachovský štít, Slovakia	8,711	Llullaillaco, Volcán, Argentina-Chile	22,110	Olympus, Mt., Washington, U.S.	7,965	Toubkal, Jebel, Morocco	13,665
Bolívar, Pico, Venezuela	16,427	Giluwe, Mt., Papua New Guinea	14,331	Logan, Mt., Canada	19,551	Orizaba, Pico de, Mexico	18,406	Triglav, Slovenia	9,396
Bonete, Cerro, Argentina	22,546	Gongga Shan, China	24,790	Longs Pk., Colorado, U.S.	14,255	Paektu San, North Korea-China	9,003	Trikora, Puncak, Indonesia	15,584
Borah Pk., Idaho, U.S.	12,662	Grand Teton, Wyoming, U.S.	13,770	Makalu, China-Nepal	27,825	Paricutín, Mexico	9,186	Tupungato, Cerro, Argentina-Chile	21,555
Boundary Pk., Nevada, U.S.	13,140	Grossglockner, Austria	12,457	Margherita Peak, Dem. Rep. of the Congo-Uganda	16,763	Parnassós, Greece	8,061	Turquino, Pico, Cuba	6,470
Cameroon Mtn., Cameroon	13,451	Hadūr Shu'ayb, Yemen	12,008	Markham, Mt., Antarctica	14,049	Pelée, Montagne, Martinique	4,583	Uluru (Ayers Rock), Australia	2,844
Carrauntoohil, Ireland	3,406	Haleakalā Crater, Hawaii, U.S.	10,023	Maromokotro, Madagascar	9,436	Pidurutalagala, Sri Lanka	8,281	Uncompahgre Pk, Colorado, U.S.	14,309
Chaltel, Cerro (Monte Fitzroy), Argentina-Chile	10,958	Hekla, Iceland	4,892	Massive, Mt., Colorado, U.S.	14,421	Pikes Pk., Colorado, U.S.	14,110	Vesuvio (Vesuvius), Italy	4,190
Chimborazo, Ecuador	20,702	Hood, Mt., Oregon, U.S.	11,239	Matterhorn, Italy-Switzerland	14,692	Pobedy, pik, China-Kyrgyzstan	24,406	Victoria, Mt., Papua New Guinea	13,238
Chirripó, Cerro, Costa Rica	12,530	Huascarán, Nevado, Peru	22,133	Mauna Kea, Hawaii, U.S.	13,796	Popocatépetl, Volcán, Mexico	17,930	Vinson Massif, Antarctica	16,066
Colima, Nevado de, Mexico	13,911	Huila, Nevado de, Colombia	18,865	Mauna Loa, Hawaii, U.S.	13,679	Pulog, Mt., Philippines	9,626	Waddington, Mt., Canada	13,163
Cotopaxi, Ecuador	19,347	Hvannadalshnúkur, Iceland	6,952	Mayon Volcano, Philippines	8,077	Rainier, Mt., Washington, U.S.	14,410	Washington, Mt., New Hampshire, U.S.	6,288
Cristóbal Colón, Pico, Colombia	19,029	Illampu, Nevado, Bolivia	21,066	McKinley, Mt., Alaska, U.S.	20,320	Ramm, Jabal, Jordan	5,755	Whitney, Mt., California, U.S.	14,494
Damāvand, Qolleh-ye, Iran	18,386	Illimani, Nevado, Bolivia	20,741	Meron, Hare, Israel	3,963	Ras Dashen Terara, Ethiopia	15,158	Wilhelm, Mt., Papua New Guinea	14,793
Dhawalāgiri, Nepal	26,810	Ismail Samani, pik, Tajikistan	24,590	Meru, Mt., Tanzania	14,978	Rinjani, Gunung, Indonesia	12,224	Wrangell, Mt., Alaska, U.S.	14,163
Duarte, Pico, Dominican Rep.	10,417	Iztaccíhuatl, Mexico	17,159	Misti, Volcán, Peru	19,101	Robson, Mt., Canada	12,972	Xixabangma Feng (Gosainthan), China	26,286
Dufourspitze (Monte Rosa), Italy-Switzerland	15,203	Jaya, Puncak, Indonesia	16,503	Mitchell, Mt., North Carolina, U.S.	6,684	Roraima, Mt., Brazil-Guyana-Venezuela	9,432	Yü Shan, Taiwan	13,114
Elbert, Mt., Colorado, U.S.	14,433	Jungfrau, Switzerland	13,642	Môco, Serra do, Angola	8,596	Ruapehu, Mt., New Zealand	9,177	Zugspitze, Austria-Germany	9,718
El'brus, Gora, Russia	18,510	K2 (Qogir Feng), China-Pakistan	28,250	Moldoveanu, Romania	8,346				
		Kāmet, China-India	25,447	Mulhacén, Spain	11,424				
		Kānchenjunga, India-Nepal	28,208						
		Kātrīna, Jabal, Egypt	8,668						

Principal Rivers and Their Lengths

RIVER Continent	Length (Mi.)	RIVER Continent	Length (Mi.)	RIVER Continent	Length (Mi.)	RIVER Continent	Length (Mi.)	RIVER Continent	Length (Mi.)
Albany, N. America	610	Don, Europe	1,162	Marañón, S. America	1,000	Pechora, Europe	1,125	Tagus, Europe	625
Aldan, Asia	1,412	Elbe, Europe	690	Mekong, Asia	2,796	Pecos, N. America	926	Tarim, Asia	1,328
Amazonas-Ucayali, S. America	4,000	Essequibo, S. America	603	Meuse, Europe	575	Pilcomayo, S. America	1,550	Tennessee, N. America	886
Amu Darya, Asia	1,578	Euphrates, Asia	1,510	Mississippi, N. America	2,340	Plata-Paraná, S. America	2,920	Tigris, Asia	1,180
Amur, Asia	1,752	Fraser, N. America	851	Mississippi-Missouri, N. America	3,710	Platte, N. America	990	Tisa, Europe	607
Araguaia, S. America	1,367	Ganges, Asia	1,864	Missouri, N. America	2,540	Purús, S. America	1,860	Tocantins, S. America	1,640
Arkansas, N. America	1,460	Gila, N. America	649	Murray-Darling, Australia	2,169	Red, N. America	1,290	Ucayali, S. America	1,220
Atchafalaya, N. America	1,420	Godāvari, Asia	932	Negro, S. America	1,305	Rhine, Europe	820	Ural, Asia	1,509
Athabasca, N. America	765	Huang (Yellow), Asia	2,902	Nelson, N. America	1,600	Rhône, Europe	503	Uruguay, S. America	1,025
Brahmaputra, Asia	1,770	Indigirka, Asia	1,072	Niger, Africa	2,585	Rio Grande, N. America	1,900	Verkhnyaya Tunguska (Angara), Asia	1,105
Brazos, N. America	1,280	Indus, Asia	1,118	Nile, Africa	4,132	Roosevelt, S. America	950	Vilyuy, Asia	1,647
Canadian, N. America	906	Irrawaddy, Asia	1,300	Ob', Asia	2,268	St. Lawrence, N. America	1,900	Volga, Europe	2,082
Churchill, N. America	1,000	Juruá, S. America	1,250	Oder, Europe	565	Salado, S. America	870	Volta, Africa	994
Colorado, N. America (U.S.-Mexico)	1,450	Kama, Europe	1,122	Ohio, N. America	1,310	Salween (Nu), Asia	1,750	Wisła (Vistula), Europe	630
Colorado, N. America (Texas)	862	Kasai, Africa	1,338	Oka, Europe	932	São Francisco, S. America	1,740	Xiang, Asia	930
Columbia, N. America	1,240	Kolyma, Asia	1,323	Orange, Africa	1,300	Saskatchewan-Bow, N. America	1,205	Xingú, S. America	1,230
Congo (Zaïre), Africa	2,715	Lena, Asia	2,734	Orinoco, S. America	1,703	Severnaya Dvina (Northern Dvina), Europe	462	Yangtze (Chang), Asia	3,915
Danube, Europe	1,777	Limpopo, Africa	1,100	Ottawa, N. America	790	Snake, N. America	1,040	Yellowstone, N. America	692
Darling, Australia	864	Loire, Europe	690	Paraguay, S. America	1,610	Sungari (Songhua), Asia	1,140	Yenisey, Asia	2,169
Dnieper (Dnipro), Europe	1,367	Mackenzie, N. America	2,635	Paraná, S. America	901	Syr Darya, Asia	1,370	Yukon, N. America	1,980
		Madeira, S. America	2,013	Peace, N. America	1,195			Zambezi, Africa	1,653
		Magdalena, S. America	951						

PRINCIPAL CITIES OF THE WORLD

Abidjan, Cote d'Ivoire 1,929,079
Abū Ẓaby (Abu Dhabi), United Arab Emirates 242,975
Accra, Ghana (1,390,000) 949,113
Addis Ababa, Ethiopia 2,424,000
Ahmadābād, India (4,519,278) 3,515,361
Aleppo (Ḩalab), Syria (1,640,000) 1,591,400
Alexandria (Al Iskandarīyah), Egypt (3,350,000) 3,339,076
Algiers (El Djazaïr), Algeria (2,547,983) 1,507,241
Al Jīzah (Giza), Egypt (*Al Qāhirah) 2,221,817
Almaty, Kazakhstan (1,190,000) 1,129,356
'Ammān, Jordan (1,500,000) 1,147,447
Amsterdam, Netherlands (1,121,303) 727,053
Ankara, Turkey (3,294,220) 2,984,099
Antananarivo, Madagascar 1,250,000
Antwerp (Antwerpen), Belgium (1,135,000) 453,030
Ashgabat (Ashkhabad), Turkmenistan 557,600
Asmera, Eritrea 358,100
Astana (Aqmola), Kazakhstan (319,324) 312,965
Asunción, Paraguay (700,000) 546,637
Athens (Athína), Greece (3,150,000) 772,072
Atlanta, Georgia, U.S. (4,112,198) 416,474
Auckland, New Zealand (1,074,510) 367,737
Baghdād, Iraq 3,841,268
Baku (Bakı), Azerbaijan (2,020,000) 1,792,300
Bamako, Mali 658,275
Bandung, Indonesia 5,919,400
Bangalore, India (5,686,844) 4,292,223
Banghāzī, Libya 800,000
Bangkok (Krung Thep), Thailand (7,060,000) 5,620,591
Bangui, Central African Republic 451,690
Barcelona, Spain (4,000,000) 1,496,266
Beijing, China (7,320,000) 6,690,000
Beirut (Bayrūt), Lebanon (1,675,000) 509,000
Belfast, N. Ireland, U.K. (730,000) 297,300
Belgrade (Beograd), Serbia and Montenegro 1,594,483
Belo Horizonte, Brazil (4,055,000) 1,366,301
Berlin, Germany (4,220,000) 3,386,667
Birmingham, England, U.K. (2,705,000) 1,020,589
Bishkek, Kyrgyzstan 753,400
Bogotá, Colombia 6,422,198
Bonn, Germany (600,000) 301,048
Boston, Massachusetts, U.S. (5,819,100) 589,141
Brasília, Brazil 1,947,133
Bratislava, Slovakia 451,395
Brazzaville, Congo 693,712
Brisbane, Australia (1,627,535) 888,449
Brussels (Bruxelles), Belgium (2,390,000) 133,845
Bucharest (București), Romania (2,300,000) 2,016,131
Budapest, Hungary (2,450,000) 1,825,153
Buenos Aires, Argentina (11,000,000) 2,960,976
Cairo (Al Qāhirah), Egypt (9,300,000) 6,800,992
Calgary, Alberta, Canada (951,395) 878,866
Cali, Colombia 2,128,920
Canberra, Australia (342,798) 311,518
Cape Town, South Africa (1,900,000) 854,616
Caracas, Venezuela (4,000,000) 1,822,465
Cardiff, Wales, U.K. (645,000) 315,040
Casablanca, Morocco (3,400,000) 3,022,000
Changchun, China 2,470,000
Chelyabinsk, Russia (1,320,000) 1,086,300
Chengdu, China 2,760,000
Chennai (Madras), India (6,424,624) 4,216,268
Chicago, Illinois, U.S. (9,157,540) 2,896,016
Chișinău (Kishinev), Moldova (746,500) 658,300
Chittagong, Bangladesh (2,342,662) 1,566,070
Chongqing, China 3,870,000
Cincinnati, Ohio, U.S. (1,979,202) 331,285
Cleveland, Ohio, U.S. (2,945,831) 478,403
Cologne (Köln), Germany (1,830,000) 962,507
Colombo, Sri Lanka (2,050,000) 615,000
Conakry, Guinea 950,000
Copenhagen (København), Denmark (2,030,000) 499,148
Córdoba, Argentina (1,260,000) 1,179,067
Cotonou, Benin 650,660
Curitiba, Brazil (2,595,000) 1,586,848
Dakar, Senegal (1,976,533) 879,703
Dalian, China 2,400,000
Dallas, Texas, U.S. (5,221,801) 1,188,580
Damascus (Dimashq), Syria (2,230,000) 1,549,932
Dar es Salaam, Tanzania 1,360,850
Delhi, India (12,791,458) 9,817,439
Denver, Colorado, U.S. (2,581,506) 554,636
Detroit, Michigan, U.S. (5,456,428) 951,270
Dhaka (Dacca), Bangladesh (6,537,308) 3,637,892
Djibouti, Djibouti 329,337
Dnipropetrovs'k, Ukraine (1,590,000) 1,108,682
Donets'k, Ukraine (2,090,000) 1,050,369
Douala, Cameroon 712,251
Dublin (Baile Átha Cliath), Ireland (1,175,000) 481,854
Durban, South Africa (1,740,000) 669,242
Dushanbe, Tajikistan (700,000) 528,600
Düsseldorf, Germany (1,200,000) 568,855
Edinburgh, Scotland, U.K. (640,000) 448,850
Edmonton, Alberta, Canada (937,845) 666,104
Eşfahān, Iran (1,525,000) 1,266,072
Essen, Germany (5,040,000) 599,515
Fortaleza, Brazil (2,780,000) 788,956
Frankfurt am Main, Germany (1,960,000) 643,821
Fukuoka, Japan (2,000,000) 1,341,489
Geneva (Génève), Switzerland (450,592) 172,598
Glasgow, Scotland, U.K. (1,870,000) 616,430
Goiânia, Brazil 1,075,761
Guadalajara, Mexico (3,669,021) 1,646,183
Guangzhou (Canton), China 3,750,000
Guatemala, Guatemala (1,500,000) 1,006,954
Guayaquil, Ecuador 2,117,553
Halifax, Nova Scotia, Canada (359,183) 119,300
Hamburg, Germany (2,460,000) 1,704,735
Hannover, Germany (1,015,000) 514,718
Hanoi, Vietnam (1,275,000) 1,073,760
Harare, Zimbabwe (1,470,000) 1,189,103
Harbin, China 3,120,000
Havana (La Habana), Cuba (2,285,000) 2,189,716
Helsinki, Finland (939,697) 548,720
Hiroshima, Japan (1,600,000) 1,126,282
Ho Chi Minh City (Saigon), Vietnam (3,300,000) 3,015,743
Hong Kong (Xianggang), China (4,770,000) 1,250,993
Honolulu, Hawaii, U.S. (876,156) 371,657
Houston, Texas, U.S. (4,669,571) 1,953,631
Hyderābād, India (5,533,640) 3,449,878
Ibadan, Nigeria 1,144,000
Islāmābād, Pakistan (*Rāwalpindi) 529,180
İstanbul, Turkey (8,506,026) 8,260,438
İzmir, Turkey (2,554,363) 2,081,556
Jaipur, India 2,324,319
Jakarta, Indonesia (10,200,000) 9,373,900
Jerusalem (Yerushalayim), Israel (685,000) 633,700
Jiddah, Saudi Arabia 1,450,000
Jinan, China 2,150,000
Johannesburg, South Africa (4,000,000) 752,349
Kābul, Afghanistan 1,424,400
Kampala, Uganda 773,463
Kānpur, India (2,690,486) 2,540,069
Kaohsiung, Taiwan (1,845,000) 1,468,586
Karāchi, Pakistan 9,339,023
Katowice, Poland (2,755,000) 343,158
Kharkiv, Ukraine (1,950,000) 1,494,235
Khartoum (Al Kharṭūm), Sudan (1,450,000) 947,483
Kiev (Kyïv), Ukraine (3,250,000) 2,589,541
Kingston, Jamaica (830,000) 516,500
Kinshasa, Dem. Rep. of the Congo 3,000,000
Kitakyūshū, Japan (1,550,000) 1,011,491
Kolkata (Calcutta), India (13,216,546) 4,580,544
Kuala Lumpur, Malaysia (2,500,000) 1,297,526
Kuwait (Al Kuwayt), Kuwait (1,126,000) 28,859
Lagos, Nigeria (3,800,000) 1,213,000
Lahore, Pakistan 5,143,495
La Paz, Bolivia (1,487,854) 792,611
Libreville, Gabon (418,616) 362,386
Lilongwe, Malawi 435,964
Lima, Peru (6,321,173) 340,422
Lisbon (Lisboa), Portugal (2,350,000) 563,210
Liverpool, England, U.K. (1,515,000) 467,995
Ljubljana, Slovenia 263,832
Lomé, Togo 450,000
London, England, U.K. (12,000,000) 7,074,265
Los Angeles, California, U.S. (16,373,645) 3,694,820
Luanda, Angola 1,459,900
Lucknow, India (2,266,933) 2,207,340
Lusaka, Zambia 1,269,848
Lyon, France (1,648,216) 445,452
Madrid, Spain (4,690,000) 2,882,860
Managua, Nicaragua 864,201
Manaus, Brazil 1,394,724
Manchester, England, U.K. (2,760,000) 430,818
Manila, Philippines (11,200,000) 1,654,761
Mannheim, Germany (1,525,000) 307,730
Maputo, Mozambique 966,837
Maracaibo, Venezuela 1,249,670
Marseille, France (1,516,340) 798,430
Mashhad, Iran 1,887,405
Mecca (Makkah), Saudi Arabia 630,000
Medan, Indonesia 1,988,200
Medellín, Colombia (2,290,000) 1,885,001
Melbourne, Australia (3,366,542) 67,784
Mexico City (Ciudad de México), Mexico (17,786,983) 8,605,239
Miami, Florida, U.S. (3,876,380) 362,470
Milan (Milano), Italy (3,790,000) 1,305,591
Milwaukee, Wisconsin, U.S. (1,689,572) 596,974
Minneapolis, Minnesota, U.S. (2,968,806) 382,618
Minsk, Belarus (1,680,567) 1,677,137
Mogadishu (Muqdisho), Somalia 600,000
Monrovia, Liberia 465,000
Monterrey, Mexico (3,236,604) 1,110,909
Montevideo, Uruguay (1,650,000) 1,303,182
Montréal, Quebec, Canada (3,426,350) 1,039,534
Moscow (Moskva), Russia (12,850,000) 8,389,700
Mumbai (Bombay), India (16,368,084) 11,914,398
Munich (München), Germany (1,930,000) 1,194,560
Nagoya, Japan (5,250,000) 2,171,378
Nāgpur, India (2,122,965) 2,051,320
Nairobi, Kenya 2,143,254
Nanjing, China 2,490,000
Naples (Napoli), Italy (3,150,000) 1,046,987
N'Djamena, Chad 546,572
Newcastle upon Tyne, England, U.K. (1,350,000) 282,338
New Delhi, India (*Delhi) 294,783
New York, New York, U.S. (21,199,865) 8,008,278
Niamey, Niger 392,165
Nizhniy Novgorod, Russia (1,950,000) 1,364,900
Nouakchott, Mauritania 393,325
Novosibirsk, Russia (1,505,000) 1,402,400
Nürnberg, Germany (1,065,000) 486,628
Odesa, Ukraine (1,150,000) 1,002,246
Omsk, Russia (1,190,000) 1,157,600
Ōsaka, Japan (17,050,000) 2,598,589
Oslo, Norway (773,498) 504,040
Ottawa, Ontario, Canada (1,063,664) 774,072
Ouagadougou, Burkina Faso 634,479
Palembang, Indonesia 1,415,500
Panamá, Panama (995,000) 415,964
Paris, France (11,174,743) 2,125,246
Patna, India (1,707,429) 1,376,950
Perm', Russia (1,110,000) 1,017,100
Perth, Australia (1,244,320) 10,195
Philadelphia, Pennsylvania, U.S. (6,188,463) 1,517,550
Phnom Penh (Phnum Pénh), Cambodia 570,155
Phoenix, Arizona, U.S. (3,251,876) 1,321,045
Port Moresby, Papua New Guinea 246,664
Port-au-Prince, Haiti (1,425,594) 990,558
Portland, Oregon, U.S. (2,265,223) 529,121
Porto, Portugal (1,230,000) 273,060
Porto Alegre, Brazil (3,375,000) 1,304,998
Prague (Praha), Czech Republic (1,328,000) 1,193,270
Pretoria, South Africa (1,100,000) 692,348
Pune, India (3,755,525) 2,540,069
Pusan, South Korea 3,814,325
P'yŏngyang, North Korea 2,741,260
Qingdao, China 2,300,000
Québec, Quebec, Canada (682,757) 169,076
Quezon City, Philippines (*Manila) 1,989,419
Quito, Ecuador 1,615,809
Rabat, Morocco (1,200,000) 717,000
Rangoon (Yangon), Myanmar (2,800,000) 2,705,039
Recife, Brazil (3,160,000) 1,421,993
Regina, Saskatchewan, Canada (192,800) 178,225
Reykjavík, Iceland (166,015) 107,684
Rīga, Latvia (1,000,000) 792,508
Rio de Janeiro, Brazil (10,465,000) 5,851,914
Riyadh (Ar Riyāḑ), Saudi Arabia 1,800,000
Rome (Roma), Italy (3,235,000) 2,649,765
Rosario, Argentina (1,190,000) 894,645
Rostov-na-Donu, Russia (1,160,000) 1,017,300
Rotterdam, Netherlands (1,089,979) 539,000
Sacramento, California, U.S. (1,796,857) 407,018
St. Louis, Missouri, U.S. (2,603,607) 348,189
St. Petersburg (Leningrad), Russia (6,000,000) 4,728,200
Salvador, Brazil (2,855,000) 2,439,823
Samara, Russia (1,450,000) 1,168,000
San Diego, California, U.S. (2,813,833) 1,223,400
San Francisco, California, U.S. (7,039,362) 776,733
San José, Costa Rica (996,194) 309,672
San Juan, Puerto Rico (1,967,627) 421,958
San Salvador, El Salvador (1,908,921) 473,372
Santiago, Chile 4,788,543
Santo Domingo, Dominican Republic 2,677,056
São Paulo, Brazil (17,380,000) 9,713,692
Sapporo, Japan (2,000,000) 1,822,300
Sarajevo, Bosnia and Herzegovina 367,703
Saratov, Russia (1,135,000) 881,000
Seattle, Washington, U.S. (3,554,760) 563,374
Seoul (Sŏul), South Korea (15,850,000) 10,231,217
Shanghai, China (11,010,000) 8,930,000
Shenyang (Mukden), China 4,050,000
Singapore, Singapore (4,400,000) 4,017,700
Skopje, Macedonia 440,577
Sofia (Sofiya), Bulgaria (1,189,794) 1,138,629
Stockholm, Sweden (1,643,366) 743,703
Stuttgart, Germany (2,020,000) 582,443
Surabaya, Indonesia 2,801,300
Sūrat, India (2,811,466) 2,433,787
Sydney, Australia (3,741,290) 11,115
T'aipei, Taiwan (6,200,000) 2,640,322
Tallinn, Estonia 403,981
Tashkent (Toshkent), Uzbekistan (2,325,000) 2,142,700
Tbilisi, Georgia (1,460,000) 1,279,000
Tegucigalpa, Honduras 576,661
Tehrān, Iran (8,800,000) 6,758,845
Tel Aviv-Yafo, Israel (1,890,000) 348,100
Tianjin (Tientsin), China 5,000,000
Tiranë, Albania 244,153
Tōkyō, Japan (30,300,000) 8,130,408
Toronto, Ontario, Canada (4,682,897) 2,481,494
Tripoli (Ṭarābulus), Libya 1,500,000
Tunis, Tunisia (1,300,000) 702,330
Turin (Torino), Italy (1,550,000) 921,485
Ufa, Russia (1,110,000) 1,088,900
Ulan Bator (Ulaanbaatar), Mongolia 672,882
Ürümqi, China 1,130,000
València, Spain (1,340,000) 739,014
Vancouver, British Columbia, Canada (1,986,965) 545,671
Viangchan (Vientiane), Laos 464,000
Vienna (Wien), Austria (1,950,000) 1,609,631
Vilnius, Lithuania 578,334
Volgograd (Stalingrad), Russia (1,358,000) 1,000,000
Warsaw (Warszawa), Poland (2,300,000) 1,615,369
Washington, D.C., U.S. (7,608,070) 572,059
Wellington, New Zealand (346,500) 167,400
Winnipeg, Manitoba, Canada (671,274) 619,544
Wuhan, China 3,870,000
Xi'an, China 2,410,000
Yekaterinburg, Russia (1,530,000) 1,272,900
Yerevan, Armenia (1,315,000) 1,249,202
Yokohama, Japan (*Tōkyō) 3,426,506
Zagreb, Croatia 867,865
Zürich, Switzerland (932,681) 337,553

Metropolitan area populations are shown in parentheses.
* City is located within the metropolitan area of another city; for example, Yokohama, Japan is located in the Tōkyō metropolitan area.

GLOSSARY OF FOREIGN GEOGRAPHICAL TERMS

Annam . . . Annamese
Arab . . . Arabic
Bantu . . . Bantu
Bur . . . Burmese
Camb . . . Cambodian
Celt . . . Celtic
Chn . . . Chinese
Czech . . . Czech
Dan . . . Danish
Du . . . Dutch
Fin . . . Finnish
Fr . . . French
Ger . . . German
Gr . . . Greek
Hung . . . Hungarian
Ice . . . Icelandic
India . . . India
Indian . . . American Indian
Indon . . . Indonesian
It . . . Italian
Jap . . . Japanese
Kor . . . Korean
Mal . . . Malayan
Mong . . . Mongolian
Nor . . . Norwegian
Per . . . Persian
Pol . . . Polish
Port . . . Portuguese
Rom . . . Romanian
Rus . . . Russian
Siam . . . Siamese
So. Slav . . . Southern Slavonic
Sp . . . Spanish
Swe . . . Swedish
Tib . . . Tibetan
Tur . . . Turkish
Yugo . . . Yugoslav

å, Nor., Swe . . . brook, river
aa, Dan., Nor . . . brook
aas, Dan., Nor . . . ridge
āb, Per . . . water, river
abad, India, Per . . . town, city
ada, Tur . . . island
adrar, Berber . . . mountain
air, Indon . . . stream
akrotírion, Gr . . . cape
älf, Swe . . . river
alp, Ger . . . mountain
altipiano, It . . . plateau
alto, Sp . . . height
archipel, Fr . . . archipelago
archipiélago, Sp . . . archipelago
arquipélago, Port . . . archipelago
arroyo, Sp . . . brook, stream
ås, Nor., Swe . . . ridge
austral, Sp . . . southern
baai, Du . . . bay
bab, Arab . . . gate, port
bach, Ger . . . brook, stream
backe, Swe . . . hill
bad, Ger . . . bath, spa
bahía, Sp . . . bay, gulf
bahr, Arab . . . river, sea, lake
baia, It . . . bay, gulf
baía, Port . . . bay
baie, Fr . . . bay, gulf
bajo, Sp . . . depression
bak, Indon . . . stream
bakke, Dan., Nor . . . hill
balkan, Tur . . . mountain range
bana, Jap . . . point, cape
banco, Sp . . . bank
bandar, Mal., Per. . . . town, port, harbor
bang, Siam . . . village
bassin, Fr . . . basin
batang, Indon., Mal . . . river
ben, Celt . . . mountain, summit
bender, Arab . . . harbor, port
bereg, Rus . . . coast, shore
berg, Du., Ger., Nor., Swe. . . . mountain, hill
bir, Arab . . . well
birkat, Arab . . . lake, pond, pool
bit, Arab . . . house
bjaerg, Dan., Nor . . . mountain
bocche, It . . . mouth
boğazı, Tur . . . strait
bois, Fr . . . forest, wood
boloto, Rus . . . marsh
bolsón, Sp. . . . flat-floored desert valley
boreal, Sp . . . northern
borg, Dan., Nor., Swe . . . castle, town
borgo, It . . . town, suburb
bosch, Du . . . forest, wood
bouche, Fr . . . river mouth
bourg, Fr . . . town, borough
bro, Dan., Nor., Swe . . . bridge
brücke, Ger . . . bridge
bucht, Ger . . . bay, bight
bugt, Dan., Nor., Swe . . . bay, gulf
bulu, Indon . . . mountain
burg, Du., Ger . . . castle, town
buri, Siam . . . town
burun, burnu, Tur . . . cape
by, Dan., Nor., Swe . . . village
caatinga, Port. (Brazil) . . . open brushland
cabezo, Sp . . . summit
cabo, Port., Sp . . . cape
campo, It., Port., Sp . . . plain, field
campos, Port. (Brazil) . . . plains
cañon, Sp . . . canyon
cap, Fr . . . cape
capo, It . . . cape
casa, It., Port., Sp . . . house
castello, It., Port . . . castle, fort
castillo, Sp . . . castle
càte, Fr . . . hill
çay, Tur . . . stream, river
cayo, Sp . . . rock, shoal, islet
cerro, Sp . . . mountain, hill
champ, Fr . . . field
chang, Chn . . . village, middle
château, Fr . . . castle
chen, Chn . . . market town
chiang, Chn . . . river
chott, Arab . . . salt lake
chou, Chn. . . . capital of district; island
chu, Tib . . . water, stream
cidade, Port . . . town, city
cima, Sp . . . summit, peak
città, It . . . town, city
ciudad, Sp . . . town, city
cochilha, Port . . . ridge
col, Fr . . . pass
colina, Sp . . . hill
cordillera, Sp . . . mountain chain
costa, It., Port., Sp . . . coast
côte, Fr . . . coast
cuchilla, Sp . . . mountain ridge
dağ, Tur . . . mountain(s)
dake, Jap . . . peak, summit
dal, Dan., Du., Nor., Swe . . . valley
dan, Kor . . . point, cape
danau, Indon . . . lake
dar, Arab . . . house, abode, country
darya, Per . . . river, sea
dasht, Per . . . plain, desert
deniz, Tur . . . sea
désert, Fr . . . desert
deserto, It . . . desert
desierto, Sp . . . desert
détroit, Fr . . . strait
dijk, Du . . . dam, dike
djebel, Arab . . . mountain
do, Kor . . . island
dorf, Ger . . . village
dorp, Du . . . village
duin, Du . . . dune
dzong, Tib. . . . fort, administrative capital
eau, Fr . . . water
ecuador, Sp . . . equator
eiland, Du . . . island
elv, Dan., Nor . . . river, stream
embalse, Sp . . . reservoir
erg, Arab . . . dune, sandy desert
est, Fr., It . . . east
estado, Sp . . . state
este, Port., Sp . . . east
estrecho, Sp . . . strait
étang, Fr . . . pond, lake
état, Fr . . . state
eyjar, Ice . . . islands
feld, Ger . . . field, plain
festung, Ger . . . fortress
fiume, It . . . river
fjäll, Swe . . . mountain
fjärd, Swe . . . bay, inlet
fjeld, Nor . . . mountain, hill
fjord, Dan., Nor . . . fiord, inlet
fjördur, Ice . . . fiord, inlet
fleuve, Fr . . . river
flod, Dan., Swe . . . river
flói, Ice . . . bay, marshland
fluss, Ger . . . river
foce, It . . . river mouth
fontein, Du . . . a spring
forêt, Fr . . . forest
fors, Swe . . . waterfall
forst, Ger . . . forest
fos, Dan., Nor . . . waterfall
fu, Chn . . . town, residence
fuente, Sp . . . spring, fountain
fuerte, Sp . . . fort
furt, Ger . . . ford
gang, Kor . . . stream, river
gangri, Tib . . . mountain
gat, Dan., Nor . . . channel
gàve, Fr . . . stream
gawa, Jap . . . river
gebergte, Du . . . mountain range
gebiet, Ger . . . district, territory
gebirge, Ger . . . mountains
ghat, India . . . pass, mountain range
gobi, Mong . . . desert
gol, Mong . . . river
göl, gölü, Tur . . . lake
golf, Du., Ger . . . gulf, bay
golfe, Fr . . . gulf, bay
golfo, It., Port., Sp . . . gulf, bay
gomba, gompa, Tib . . . monastery
gora, Rus., So. Slav . . . mountain
góra, Pol . . . mountain
gorod, Rus . . . town
grad, Rus., So. Slav . . . town
guba, Rus . . . bay, gulf
gundung, Indon . . . mountain
guntō, Jap . . . archipelago
gunung, Mal . . . mountain
haf, Swe . . . sea, ocean
hafen, Ger . . . port, harbor
haff, Ger . . . gulf, inland sea
hai, Chn . . . sea, lake
hama, Jap . . . beach, shore
hamada, Arab . . . rocky plateau
hamn, Swe . . . harbor
hāmūn, Per . . . swampy lake, plain
hantō, Jap . . . peninsula
hassi, Arab . . . well, spring
haus, Ger . . . house
haut, Fr . . . summit, top
hav, Dan., Nor . . . sea, ocean
havn, Dan., Nor . . . harbor, port
havre, Fr . . . harbor, port
háza, Hung . . . house, dwelling of
heim, Ger . . . hamlet, home
hem, Swe . . . hamlet, home
higashi, Jap . . . east
hisar, Tur . . . fortress
hissar, Arab . . . fort
ho, Chn . . . river
hoek, Du . . . cape
hof, Ger . . . court, farmhouse
höfn, Ice . . . harbor
hoku, Jap . . . north
holm, Dan., Nor., Swe . . . island
hora, Czech . . . mountain
horn, Ger . . . peak
hoved, Dan., Nor . . . cape
hsien, Chn . . . district, district capital
hu, Chn . . . lake
hügel, Ger . . . hill
huk, Dan., Swe . . . point
hus, Dan., Nor., Swe . . . house
île, Fr . . . island
ilha, Port . . . island
indsö, Dan., Nor . . . lake
insel, Ger . . . island
insjö, Swe . . . lake
irmak, irmagi, Tur . . . river
isla, Sp . . . island
isola, It . . . island
istmo, It., Sp . . . isthmus
järvi, jaur, Fin . . . lake
jebel, Arab . . . mountain
jima, Jap . . . island
jökel, Nor . . . glacier
joki, Fin . . . river
jökull, Ice . . . glacier
kaap, Du . . . cape
kai, Jap . . . bay, gulf, sea
kaikyō, Jap . . . channel, strait
kalat, Per . . . castle, fortress
kale, Tur . . . fort
kali, Mal . . . creek, river
kand, Per . . . village
kang, Chn . . . mountain ridge; village
kap, Dan., Ger . . . cape
kapp, Nor., Swe . . . cape
kasr, Arab . . . fort, castle
kawa, Jap . . . river
kefr, Arab . . . village
kei, Jap . . . creek, river
ken, Jap . . . prefecture
khor, Arab . . . bay, inlet
khrebet, Rus . . . mountain range
kiang, Chn . . . large river
king, Chn . . . capital city, town
kita, Jap . . . north
ko, Jap . . . lake
köbstad, Dan . . . market-town
kol, Mong . . . lake
kólpos, Gr . . . gulf
kong, Chn . . . river
kopf, Ger . . . head, summit, peak
köpstad, Swe . . . market-town
körfezi, Tur . . . gulf
kosa, Rus . . . spit
kou, Chn . . . river mouth
köy, Tur . . . village
kraal, Du. (Africa) . . . native village
ksar, Arab . . . fortified village
kuala, Mal . . . bay, river mouth
kuh, Per . . . mountain
kum, Tur . . . sand
kuppe, Ger . . . summit
küste, Ger . . . coast
kyo, Jap . . . town, capital
la, Tib . . . mountain pass
labuan, Mal . . . anchorage, port
lac, Fr . . . lake
lago, It., Port., Sp . . . lake
lagoa, Port . . . lake, marsh
laguna, It., Port., Sp . . . lagoon, lake
lahti, Fin . . . bay, gulf
län, Swe . . . county
landsby, Dan., Nor . . . village
liehtao, Chn . . . archipelago
liman, Tur . . . bay, port
ling, Chn . . . pass, ridge, mountain
llanos, Sp . . . plains
loch, Celt. (Scotland) . . . lake, bay
loma, Sp . . . long, low hill
lough, Celt. (Ireland) . . . lake, bay
machi, Jap . . . town
man, Kor . . . bay
mar, Port., Sp . . . sea
mare, It., Rom . . . sea
marisma, Sp . . . marsh, swamp
mark, Ger . . . boundary, limit
massif, Fr . . . block of mountains
mato, Port . . . forest, thicket
me, Siam . . . river
meer, Du., Ger . . . lake, sea
mer, Fr . . . sea
mesa, Sp . . . flat-topped mountain
meseta, Sp . . . plateau
mina, Port., Sp . . . mine
minami, Jap . . . south
minato, Jap . . . harbor, haven
misaki, Jap . . . cape, headland
mont, Fr . . . mount, mountain
montagna, It . . . mountain
montagne, Fr . . . mountain
montaña, Sp . . . mountain
monte, It., Port., Sp. . . . mount, mountain
more, Rus., So. Slav . . . sea
morro, Port., Sp . . . hill, bluff
mühle, Ger . . . mill
mund, Ger . . . mouth, opening
mündung, Ger . . . river mouth
mura, Jap . . . township
myit, Bur . . . river
mys, Rus . . . cape
nada, Jap . . . sea
nadi, India . . . river, creek
naes, Dan., Nor . . . cape
nafud, Arab . . . desert of sand dunes
nagar, India . . . town, city
nahr, Arab . . . river
nam, Siam . . . river, water
nan, Chn., Jap . . . south
näs, Nor., Swe . . . cape
nez, Fr . . . point, cape
nishi, nisi, Jap . . . west
njarga, Fin . . . peninsula
nong, Siam . . . marsh
noord, Du . . . north
nor, Mong . . . lake
nord, Dan., Fr., Ger., It., Nor., Swe . . . north
norte, Port., Sp . . . north
nos, Rus . . . cape
nyasa, Bantu . . . lake
ö, Dan., Nor., Swe . . . island
occidental, Sp . . . western
ocna, Rom . . . salt mine
odde, Dan., Nor . . . point, cape
oeste, Port., Sp . . . west
oka, Jap . . . hill
oost, Du . . . east
oriental, Sp . . . eastern
óros, Gr . . . mountain
ost, Ger., Swe . . . east
öster, Dan., Nor., Swe . . . eastern
ostrov, Rus . . . island
oued, Arab . . . river, stream
ouest, Fr . . . west
ozero, Rus . . . lake
pää, Fin . . . mountain
padang, Mal . . . plain, field
pampas, Sp. (Argentina) . . . grassy plains
pará, Indian (Brazil) . . . river
pas, Fr . . . channel, passage
paso, Sp . . . mountain pass, passage
passo, It., Port. mountain pass, passage, strait
patam, India . . . city, town
pei, Chn . . . north
pélagos, Gr . . . open sea
pegunungan, Indon . . . mountains
peña, Sp . . . rock
peresheyek, Rus . . . isthmus
pertuis, Fr . . . strait
peski, Rus . . . desert
pic, Fr . . . mountain peak
pico, Port., Sp . . . mountain peak
piedra, Sp . . . stone, rock
ping, Chn . . . plain, flat
planalto, Port . . . plateau
planina, Yugo . . . mountains
playa, Sp . . . shore, beach
pnom, Camb . . . mountain
pointe, Fr . . . point
polder, Du., Ger . . . reclaimed marsh
polje, So. Slav . . . plain, field
poluostrov, Rus . . . peninsula
pont, Fr . . . bridge
ponta, Port . . . point, headland
ponte, It., Port . . . bridge
pore, India . . . city, town
porthmós, Gr . . . strait
porto, It., Port . . . port, harbor
potamós, Gr . . . river
p'ov, Rus . . . peninsula
prado, Sp . . . field, meadow
presqu'île, Fr . . . peninsula
proliv, Rus . . . strait
pu, Chn . . . commercial village
pueblo, Sp . . . town, village
puerto, Sp . . . port, harbor
pulau, Indon . . . island
punkt, Ger . . . point
punt, Du . . . point
punta, It., Sp . . . point
pur, India . . . city, town
puy, Fr . . . peak
qal'a, qal'at, Arab . . . fort, village
qasr, Arab . . . fort, castle
rann, India . . . wasteland
ra's, Arab . . . cape, head
reka, Rus., So. Slav . . . river
represa, Port . . . reservoir
rettō, Jap . . . island chain
ría, Sp . . . estuary
ribeira, Port . . . stream
riberão, Port . . . river
rio, It., Port . . . stream, river
río, Sp . . . river
rivière, Fr . . . river
roca, Sp . . . rock
rt, Yugo . . . cape
rūd, Per . . . river
saari, Fin . . . island
sable, Fr . . . sand
sahara, Arab . . . desert, plain
saki, Jap . . . cape
sal, Sp . . . salt
salar, Sp . . . salt flat, salt lake
salto, Sp . . . waterfall
san, Jap., Kor . . . mountain, hill
sat, satul, Rom . . . village
schloss, Ger . . . castle
sebkha, Arab . . . salt marsh
see, Ger . . . lake, sea
şehir, Tur . . . town, city
selat, Indon . . . stream
selvas, Port. (Brazil) . . . tropical rain forests
seno, Sp . . . bay
serra, Port . . . mountain chain
serranía, Sp . . . mountain ridge
seto, Jap . . . strait
severnaya, Rus . . . northern
shahr, Per . . . town, city
shan, Chn . . . mountain, hill, island
shatt, Arab . . . river
shi, Jap . . . city
shima, Jap . . . island
shōtō, Jap . . . archipelago
si, Chn . . . west, western
sierra, Sp . . . mountain range
sjö, Nor., Swe . . . lake, sea
sö, Dan., Nor . . . lake, sea
söder, södra, Swe . . . south
song, Annam . . . river
sopka, Rus . . . peak, volcano
source, Fr . . . a spring
spitze, Ger . . . summit, point
staat, Ger . . . state
stad, Dan., Du., Nor., Swe. . . . city, town
stadt, Ger . . . city, town
stato, It . . . state
step', Rus . . . treeless plain, steppe
straat, Du . . . strait
strand, Dan., Du., Ger., Nor., Swe . . . shore, beach
stretto, It . . . strait
strom, Ger . . . river, stream
ström, Dan., Nor., Swe. . . . stream, river
stroom, Du . . . stream, river
su, suyu, Tur . . . water, river
sud, Fr., Sp . . . south
süd, Ger . . . south
suidō, Jap . . . channel
sul, Port . . . south
sund, Dan., Nor., Swe . . . sound
sungai, sungei, Indon., Mal . . . river
sur, Sp . . . south
syd, Dan., Nor., Swe . . . south
tafelland, Ger . . . plateau
take, Jap . . . peak, summit
tal, Ger . . . valley
tanjung, tanjong, Mal . . . cape
tao, Chn . . . island
târg, târgul, Rom . . . market, town
tell, Arab . . . hill
teluk, Indon . . . bay, gulf
terra, It . . . land
terre, Fr . . . earth, land
thal, Ger . . . valley
tierra, Sp . . . earth, land
tō, Jap . . . east; island
tonle, Camb . . . river, lake
top, Du . . . peak
torp, Swe . . . hamlet, cottage
tsangpo, Tib . . . river
tsi, Chn . . . village, borough
tso, Tib . . . lake
tsu, Jap . . . harbor, port
tundra, Rus . . . treeless arctic plains
tung, Chn . . . east
tuz, Tur . . . salt
udde, Swe . . . cape
ufer, Ger . . . shore, riverbank
ujung, Indon . . . point, cape
umi, Jap . . . sea, gulf
ura, Jap . . . bay, coast, creek
ust'ye, Rus . . . river mouth
valle, It., Port., Sp . . . valley
vallée, Fr . . . valley
valli, It . . . lake
vár, Hung . . . fortress
város, Hung . . . town
varoš, So. Slav . . . town
veld, Du . . . open plain, field
verkh, Rus . . . top, summit
ves, Czech . . . village
vest, Dan., Nor., Swe . . . west
vik, Swe . . . cove, bay
vila, Port . . . town
villa, Sp . . . town
villar, Sp . . . village, hamlet
ville, Fr . . . town, city
vostok, Rus . . . east
wad, wādī, Arab. . . . intermittent stream
wald, Ger . . . forest, woodland
wan, Chn., Jap . . . bay, gulf
weiler, Ger . . . hamlet, village
westersch, Du . . . western
wüste, Ger . . . desert
yama, Jap . . . mountain
yarimada, Tur . . . peninsula
yug, Rus . . . south
zaki, Jap . . . cape
zaliv, Rus . . . bay, gulf
zapad, Rus . . . west
zee, Du . . . sea
zemlya, Rus . . . land
zuid, Du . . . south

ABBREVIATIONS OF GEOGRAPHICAL NAMES AND TERMS

Afg.	Afghanistan
Afr.	Africa
Ak., U.S.	Alaska, U.S.
Al., U.S.	Alabama, U.S.
Alb.	Albania
Alg.	Algeria
Am. Sam.	American Samoa
And.	Andorra
Ang.	Angola
Ant.	Antarctica
Antig.	Antigua and Barbuda
aq.	Aqueduct
Ar., U.S.	Arkansas, U.S.
Arg.	Argentina
Arm.	Armenia
arpt.	Airport
Aus.	Austria
Austl.	Australia
Az., U.S.	Arizona, U.S.
Azer.	Azerbaijan
b.	Bay, Gulf, Inlet, Lagoon
Bah.	Bahamas
Bahr.	Bahrain
Barb.	Barbados
Bdi.	Burundi
Bel.	Belgium
Bela.	Belarus
Ber.	Bermuda
Bhu.	Bhutan
bk.	Undersea Bank
bldg.	Building
Blg.	Bulgaria
Bngl.	Bangladesh
Bol.	Bolivia
Bos.	Bosnia and Herzegovina
Bots.	Botswana
Braz.	Brazil
Bru.	Brunei
Br. Vir. Is.	British Virgin Islands
bt.	Bight
Burkina	Burkina Faso
c.	Cape, Point
Ca., U.S.	California, U.S.
Cam.	Cameroon
Camb.	Cambodia
can.	Canal
Can.	Canada
C.A.R.	Central African Republic
Cay. Is.	Cayman Islands
C. Iv.	Cote d'Ivoire
clf.	Cliff, Escarpment
co.	County, Parish
Co., U.S.	Colorado, U.S.
Col.	Colombia
Com.	Comoros
cont.	Continent
Cook Is.	Cook Islands
C.R.	Costa Rica
Cro.	Croatia
cst.	Coast, Beach
Ct., U.S.	Connecticut, U.S.
C.V.	Cape Verde
Cyp.	Cyprus
Czech Rep.	Czech Republic
d.	Delta
D.C., U.S.	District of Columbia, U.S.
De., U.S.	Delaware, U.S.
Den.	Denmark
dep.	Dependency, Colony
depr.	Depression
dept.	Department, District
des.	Desert
Dji.	Djibouti
Dom.	Dominica
Dom. Rep.	Dominican Republic
D.R.C.	Democratic Republic of the Congo
Ec.	Ecuador
educ.	Educational Facility
El Sal.	El Salvador
Eng., U.K.	England, U.K.
Eq. Gui.	Equatorial Guinea
Erit.	Eritrea
Est.	Estonia
est.	Estuary
Eth.	Ethiopia
E. Timor	East Timor
Eur.	Europe
Falk. Is.	Falkland Islands
Far. Is.	Faroe Islands
Fin.	Finland
fj.	Fjord
Fl., U.S.	Florida, U.S.
for.	Forest, Moor
Fr.	France
Fr. Gu.	French Guiana
Fr. Poly.	French Polynesia
Ga., U.S.	Georgia, U.S.
Gam.	The Gambia
Gaza	Gaza Strip
Geor.	Georgia
Ger.	Germany

Grc.	Greece
Gren.	Grenada
Grnld.	Greenland
Guad.	Guadeloupe
Guat.	Guatemala
Guern.	Guernsey
Gui.	Guinea
Gui.-B.	Guinea-Bissau
Guy.	Guyana
Hi., U.S.	Hawaii, U.S.
hist.	Historic Site, Ruins
hist. reg.	Historic Region
Hond.	Honduras
Hung.	Hungary
i.	Island
Ia., U.S.	Iowa, U.S.
ice	Ice Feature, Glacier
Ice.	Iceland
Id., U.S.	Idaho, U.S.
Il., U.S.	Illinois, U.S.
In., U.S.	Indiana, U.S.
Indon.	Indonesia
I. of Man	Isle of Man
I.R.	Indian Reservation
Ire.	Ireland
is.	Islands
Isr.	Israel
isth.	Isthmus
Jam.	Jamaica
Jord.	Jordan
Kaz.	Kazakhstan
Kir.	Kiribati
Kor., N.	Korea, North
Kor., S.	Korea, South
Ks., U.S.	Kansas, U.S.
Kuw.	Kuwait
Ky., U.S.	Kentucky, U.S.
Kyrg.	Kyrgyzstan
l.	Lake, Pond
La., U.S.	Louisiana, U.S.
Lat.	Latvia
Leb.	Lebanon
Leso.	Lesotho
Lib.	Liberia
Liech.	Liechtenstein
Lith.	Lithuania
Lux.	Luxembourg
Ma., U.S.	Massachusetts, U.S.
Mac.	Macedonia
Madag.	Madagascar
Malay.	Malaysia
Mald.	Maldives
Marsh. Is.	Marshall Islands
Mart.	Martinique
Maur.	Mauritania
May.	Mayotte
Md., U.S.	Maryland, U.S.
Me., U.S.	Maine, U.S.
Mex.	Mexico
Mi., U.S.	Michigan, U.S.
Micron.	Micronesia, Federated States of
Mn., U.S.	Minnesota, U.S.
Mo., U.S.	Missouri, U.S.
Mol.	Moldova
Mong.	Mongolia
Monts.	Montserrat
Mor.	Morocco
Moz.	Mozambique
Ms., U.S.	Mississippi, U.S.
Mt., U.S.	Montana, U.S.
mth.	River Mouth or Channel
mtn.	Mountain
mts.	Mountains
Mwi.	Malawi
Mya.	Myanmar
N.A.	North America
N.C., U.S.	North Carolina, U.S.
N. Cal.	New Caledonia
N.D., U.S.	North Dakota, U.S.
Ne., U.S.	Nebraska, U.S.
neigh.	Neighborhood
Neth.	Netherlands
Neth. Ant.	Netherlands Antilles
N.H., U.S.	New Hampshire, U.S.
Nic.	Nicaragua
Nig.	Nigeria
N. Ire., U.K.	Northern Ireland, U.K.
N.J., U.S.	New Jersey, U.S.
N.M., U.S.	New Mexico, U.S.
N. Mar. Is.	Northern Mariana Islands
Nmb.	Namibia
Nor.	Norway
Nv., U.S.	Nevada, U.S.
N.Y., U.S.	New York, U.S.
N.Z.	New Zealand
o.	Ocean
Oc.	Oceania
Oh., U.S.	Ohio, U.S.

Ok., U.S.	Oklahoma, U.S.
Or., U.S.	Oregon, U.S.
p.	Pass
Pa., U.S.	Pennsylvania, U.S.
Pak.	Pakistan
Pan.	Panama
Pap. N. Gui.	Papua New Guinea
Para.	Paraguay
pen.	Peninsula
Phil.	Philippines
Pit.	Pitcairn
pl.	Plain, Flat
plat.	Plateau, Highland
Pol.	Poland
Port.	Portugal
P.R.	Puerto Rico
prov.	Province, Region
pt. of i.	Point of Interest
r.	River, Creek
Reu.	Reunion
rec.	Recreational Site, Park
reg.	Physical Region
rel.	Religious Institution
res.	Reservoir
rf.	Reef, Shoal
R.I., U.S.	Rhode Island, U.S.
Rom.	Romania
Rw.	Rwanda
S.A.	South America
S. Afr.	South Africa
Sau. Ar.	Saudi Arabia
S.C., U.S.	South Carolina, U.S.
sci.	Scientific Station
Scot., U.K.	Scotland, U.K.
S.D., U.S.	South Dakota, U.S.
sea feat.	Undersea Feature
Sen.	Senegal
Serb.	Serbia and Montenegro
Sey.	Seychelles
S. Geor.	South Georgia
Sing.	Singapore
S.L.	Sierra Leone
Slvk.	Slovakia
Slvn.	Slovenia
S. Mar.	San Marino
Sol. Is.	Solomon Islands
Som.	Somalia
Sp. N. Afr.	Spanish North Africa
Sri L	Sri Lanka
St. Hel.	St. Helena
St. K./N.	St. Kitts and Nevis
St. Luc.	St. Lucia
St. P./M.	St. Pierre and Miquelon
strt.	Strait, Channel, Sound
S. Tom./P.	Sao Tome and Principe
St. Vin.	St. Vincent and the Grenadines
Sur.	Suriname
Sval.	Svalbard
sw.	Swamp, Marsh
Swaz.	Swaziland
Swe.	Sweden
Switz.	Switzerland
Tai.	Taiwan
Taj.	Tajikistan
Tan.	Tanzania
T./C. Is.	Turks and Caicos Islands
ter.	Territory
Thai.	Thailand
Tn., U.S.	Tennessee, U.S.
trans.	Transportation Facility
Trin.	Trinidad and Tobago
Tun.	Tunisia
Tur.	Turkey
Turkmen.	Turkmenistan
Tx., U.S.	Texas, U.S.
U.A.E.	United Arab Emirates
Ug.	Uganda
U.K.	United Kingdom
Ukr.	Ukraine
Ur.	Uruguay
U.S.	United States
Ut., U.S.	Utah, U.S.
Uzb.	Uzbekistan
Va., U.S.	Virginia, U.S.
val.	Valley, Watercourse
Ven.	Venezuela
Viet.	Vietnam
V.I.U.S.	Virgin Islands (U.S.)
vol.	Volcano
Vt., U.S.	Vermont, U.S.
Wa., U.S.	Washington, U.S.
W.B.	West Bank
Wi., U.S.	Wisconsin, U.S.
W. Sah.	Western Sahara
wtfl.	Waterfall
W.V., U.S.	West Virginia, U.S.
Wy., U.S.	Wyoming, U.S.
Zam.	Zambia
Zimb.	Zimbabwe

PRONUNCIATION OF GEOGRAPHICAL NAMES

Key to the Sound Values of Letters and Symbols Used in the Index to Indicate Pronunciation

ă-ăt; băttle
ã-fĭnăl; appeăl
ā-rāte; elāte
å-senåte; inanimåte
ä-ärm; cälm
å-åsk; båth
à-sofà; màrine (short neutral or indeterminate sound)
â-fâre; prepâre
ch-choose; church
dh-as th in other; either
ē-bē; ēve
ĕ-ĕvent; crêate
ĕ-bĕt; ĕnd
ĕ-recĕnt (short neutral or indeterminate sound)
ẽ-cratẽr; cindẽr
g-gō; gāme
gh-guttural g
ī-bĭt; wĭll
ĭ-(short neutral or indeterminate sound)
ī-rīde; bīte
κ-gutteral k as ch in German ich
ng-sing
ŋ-baŋk; liŋger
ɴ-indicates nasalized
ŏ-nŏd; ŏdd
ŏ-cŏmmit; cŏnnect
ō-ōld; bōld
ô-ôbey; hôtel
ô-ôrder; nôrth
oi-boil
o͞o-fo͞od; ro͞ot
ȯ-as oo in foot; wood
ou-out; thou
s-soft; so; sane
sh-dish; finish
th-thin; thick
ū-pūre; cūre
ů-ůnite; ůsůrp
û-ûrn; fûr
ŭ-stŭd; ŭp
ŭ-circŭs; sŭbmit
ü-as in French tu
zh-as z in azure
'-indeterminate vowel sound

In many cases the spelling of foreign geographical names does not even remotely indicate the pronunciation to an American, i.e., Słupsk in Poland is pronounced swȯpsk; Jujuy in Argentina is pronounced ho͞oȯhwē', La Spezia in Italy is lä-spē'zyä.

This condition is hardly surprising, however, when we consider that in our own language Worcester, Massachusetts, is pronounced wȯs'tẽr; Sioux City, Iowa, so͞o sĭ'tĭ; Schuylkill Haven, Pennsylvania, sko͞ol'kĭl hā-vĕn; Poughkeepsie, New York, pŏ-kĭp'sĕ.

The indication of pronunciation of geographic names presents several peculiar problems:

1. Many foreign tongues use sounds that are not present in the English language and which an American cannot normally articulate. Thus, though the nearest English equivalent sound has been indicated, only approximate results are possible.

2. There are several dialects in each foreign tongue which cause variation in the local pronunciation of names. This also occurs in identical names in the various divisions of a great language group, as the Slavic or the Latin.

3. Within the United States there are marked differences in pronunciation, not only of local geographic names, but also of common words, indicating that the sound and tone values for letters as well as the placing of the emphasis vary considerably from one part of the country to another.

4. A number of different letters and diacritical combinations could be used to indicate essentially the same or approximate pronunciations.

Some variation in pronunciation other than that indicated in this index may be encountered, but such a difference does not necessarily indicate that either is in error, and in many cases it is a matter of individual choice as to which is preferred. In fact, an exact indication of pronunciation of many foreign names using English letters and diacritical marks is extremely difficult and sometimes impossible.

PRONOUNCING INDEX

This universal index includes in a single alphabetical list approximately 6,500 names of features that appear on the reference maps. Each name is followed by a page reference and geographical coordinates.

Abbreviation and Capitalization Abbreviations of names on the maps have been standardized as much as possible. Names that are abbreviated on the maps are generally spelled out in full in the index. Periods are used after all abbreviations regardless of local practice. The abbreviation "St." is used only for "Saint". "Sankt" and other forms of this term are spelled out.

Most initial letters of names are capitalized, except for a few Dutch names, such as "s-Gravenhage". Capitalization of noninitial words in a name generally follows local practice.

Alphabetization Names are alphabetized in the order of the letters of the English alphabet. Spanish *ll* and *ch*, for example, are not treated as direct letters. Furthermore, diacritical marks are disregarded in alphabetization — German or Scandinavian *ä* or *ö* are treated as *a* or *o*.

The names of physical features may appear inverted, since they are always alphabetized under the proper, not the generic, part of the name, thus: "Gibraltar, Strait of". Otherwise every entry, whether consisting of one word or more, is alphabetized as a single continuous entity. "Lakeland", for example, appears after "La Crosse" and before "La Salle". Names beginning with articles (Le Harve, Den Helder, Al Manāmah, Ad Dawhah) are not inverted.

In the case of identical names, towns are listed first, then political divisions, then physical features.

Generic Terms Except for cities, the names of all features are followed by terms that represent broad classes of features, for example, Mississippi, r. or Alabama, state. A list of all abbreviations used in the index is on page 111.

Country names and the names of features that extend beyond the boundaries of one county are followed by the name of the continent in which each is located. Country designations follow the names of all other places in the index. The locations of places in the United States and the United Kingdom are further defined by abbreviations that include the state or political division in which each is located.

Pronunciations Pronunciations are included for most names listed. An explanation of the pronunciation system used appears on page 111.

Page References and Geographical Coordinates The geographical coordinates and page references are found in the last columns of each entry.

If a page contains several maps or insets, a lowercase letter identifies the specific map or inset.

Latitude and longitude coordinates for point features, such as cities and mountain peaks, indicate the location of the symbols. For extensive areal features, such as countries or mountain ranges, or linear features, such as canals and rivers, locations are given for the position of the type as it appears on the map.

PLACE (Pronunciation)	PAGE	LAT.	LONG.
A			
Aalborg, Den. (ôl′bôr)	50	57°02′N	9°55′E
Ābādān, Iran (ä-bŭ-dän′)	68	30°15′N	48°30′E
Abakan, Russia (ū-bá-kän′)	57	53°43′N	91°28′E
'Abbāsah, Tur'at al, can., Egypt	87d	30°45′N	32°15′E
'Abd al Kūrī, i., Yemen (äbd-ĕl-kó′rĕ)	87a	12°12′N	51°00′E
Aberdeen, Scot., U.K. (ăb-ĕr-dēn′)	50	57°10′N	2°05′W
Aberdeen, S.D., U.S. (ăb-ĕr-dēn′)	38	45°28′N	98°29′W
Aberdeen, Wa., U.S. (ăb-ĕr-dēn′)	38	47°00′N	123°48′W
Abidjan, C. Iv. (ä-bĕd-zhän′)	85	5°19′N	4°02′W
Abilene, Tx., U.S.	38	32°25′N	99°45′W
Abnūb, Egypt (ab-nōōb′)	87b	27°18′N	31°11′E
Åbo see Turku, Fin.	50	60°28′N	22°12′E
Abohar, India	72	30°12′N	74°13′E
Abra, r., Phil. (ä′brä)	75a	17°16′N	120°38′E
Abū Arīsh, Sau. Ar. (ä-bōō à-rēsh′)	68	16°48′N	43°00′E
Abu Dhabi see Abū Ẓaby, U.A.E.	68	24°15′N	54°28′E
Abuja, Nig.	85	9°12′N	7°11′E
Abū Kamāl, Syria	68	34°45′N	40°46′E
Abū Qīr, Egypt (ä′bōō kēr′)	87b	31°18′N	30°06′E
Abū Qurūn, Ra's, mtn., Egypt	67a	30°22′N	33°32′E
Abu Road, India (ä′bōō)	69	24°38′N	72°45′E
Abū Tīj, Egypt	87b	27°03′N	31°19′E
Abū Ẓaby, U.A.E.	68	24°15′N	54°28′E
Abū Zanīmah, Egypt	67a	29°03′N	33°08′E
Abyy, Russia	57	68°24′N	134°00′E
Acapulco, Mex. (ä-kä-pōōl′kō)	42	16°49′N	99°57′W
Accra, Ghana (ä′krä)	85	5°33′N	0°13′W
Acklins, i., Bah. (ăk′lĭns)	43	22°30′N	73°55′W
Aconcagua, Cerro, mtn., Arg. (ä-kŏn-kä′gwä)	46	32°38′S	70°00′W
Açores (Azores), is., Port.	86	37°44′N	29°25′W
A Coruña, Spain	50	43°20′N	8°20′W
Adak, Ak., U.S. (ä-däk′)	40a	56°50′N	176°48′W
Adak, i., Ak., U.S. (ä-däk′)	40a	51°40′N	176°28′W
Adak Strait, strt., Ak., U.S. (ä-däk′)	40a	51°42′N	177°16′W
Adana, Tur. (ä-dä′nä)	68	37°05′N	35°20′E
Adapazarı, Tur. (ä-dä-pä-zä′rĕ)	55	40°45′N	30°20′E
Ad Dahnā, des., Sau. Ar.	68	26°05′N	47°15′E
Ad Dammām, Sau. Ar.	68	26°27′N	49°59′E
Ad Dāmūr, Leb.	67a	33°44′N	35°27′E
Ad Dawhah, Qatar	68	25°02′N	51°28′E
Ad Dilam, Sau. Ar.	68	23°47′N	47°03′E
Ad Dilinjāt, Egypt	87b	30°48′N	30°32′E
Addis Ababa, Eth.	85	9°00′N	38°44′E
Adelaide, Austl. (ăd′ĕ-lād)	80	34°46′S	139°08′E
Adelaide Island, i., Ant. (ăd′ĕ-lād)	82	67°15′S	68°40′W
Aden ('Adan), Yemen (ä′dĕn)	68	12°48′N	45°00′E
Aden, Gulf of, b.	68	11°45′N	45°45′E
Adi, Pulau, i., Indon. (ä′dē)	75	4°25′S	133°52′E
Adige, r., Italy (ä′dē-jā)	54	46°38′N	10°43′E
Adigrat, Eth.	71	14°17′N	39°28′E
Adilābād, India (ŭ-dīl-ä-bäd′)	72	19°47′N	78°30′E
Adirondack Mountains, mts., N.Y., U.S. (ăd-ĭ-rŏn′dăk)	39	43°45′N	74°40′W
Admiralty, i., Ak., U.S. (ăd′mī-rál-tĕ)	40	57°50′N	133°50′W
Admiralty Island National Monument, rec., Ak., U.S. (ăd′mī-rál-tĕ)	40	57°50′N	137°30′W
Admiralty Islands, is., Pap. N. Gui. (ăd′mī-rál-tĕ)	75	1°40′S	146°45′E
Ādoni, India	73	15°42′N	77°18′E
Adrianople	50	41°41′N	26°35′E
Adriatic Sea, sea, Eur.	52	43°30′N	14°27′E
Aegean Sea, sea, (ē-jē′án)	52	39°04′N	24°56′E
Afghanistan, nation, Asia (ăf-găn-ĭ-stăn′)	68	33°00′N	63°00′E
Afgooye, Som. (äf-gô′ē)	87a	2°08′N	45°08′E
Afognak, i., Ak., U.S. (ä-fŏg-năk′)	40	58°28′N	151°35′W
Africa, cont.	86	10°00′N	22°00′E
'Afula, Isr. (ä-fō′lä)	67a	32°36′N	35°17′E
Afyon, Tur. (ä-fè-ōn)	68	38°45′N	30°20′E
Agartala, India	72	23°53′N	91°22′E
Agāshi, India	73b	19°28′N	72°46′E
Agattu, i., Ak., U.S. (ä′gä-tōō)	40a	52°14′N	173°40′E
Aginskoye, Russia (ä-hĭn′skô-yĕ)	57	51°15′N	113°15′E
Ágios Efstrátios, i., Grc.	55	39°30′N	24°58′E
Agno, Phil. (äg′nō)	75a	16°07′N	119°49′E
Agno, r., Phil.	75a	15°42′N	120°28′E
Agra, India (ä′grä)	69	27°18′N	78°00′E
Agrínio, Grc.	55	38°38′N	21°06′E
Aguadilla, P.R. (ä-gwä-dēl′yä)	43b	18°27′N	67°10′W
Aguascalientes, Mex. (ä′gwäs-käl-yĕn′tās)	42	21°52′N	102°17′W
Aguilas, Spain (ä-gē′läs)	54	37°26′N	1°35′W
Agulhas, Cape, c., S. Afr. (ä-gōōl′yäs)	85	34°47′S	20°00′E
Agusan, r., Phil. (ä-gōō′sän)	75	8°12′N	126°07′E
Ahar, Iran	71	38°28′N	47°04′E
Ahmadābād, India (ŭ-mĕd-ä-bäd′)	69	23°04′N	72°38′E
Ahmadnagar, India (ä′mŭd-nŭ-gŭr)	69	19°09′N	74°45′E
Ahvāz, Iran	68	31°15′N	48°54′E
Aïn Témouchent, Alg. (ä′ĕntĕ-mōō-shän′)	54	35°20′N	1°23′W
Airhitam, Selat, strt., Indon.	67b	0°58′N	102°38′E
Aitape, Pap. N. Gui. (ä-ē-tä′pä)	75	3°00′S	142°10′E
Aitutaki, i., Cook Is. (ī-tōō-tä′kē)	89	19°00′S	162°00′W
Aiud, Rom. (ä′ē-ôd)	55	46°19′N	23°40′E
Ajaccio, Fr. (ä-yät′chō)	50	41°55′N	8°42′E
Ajana, Austl. (äj-än′ĕr)	80	28°00′S	114°45′E
Ajaria, state, Geor.	58	41°40′N	42°00′E
Ajmah, Jabal al, mts., Egypt	67a	29°12′N	34°03′E
Ajman, U.A.E.	68	25°15′N	54°30′E
Ajmer, India (ŭj-mēr′)	69	26°26′N	74°42′E
Akhḍar, Al Jabal al, mts., Oman	68	23°30′N	56°43′W
Akhisar, Tur. (äk-hĭs-sär′)	55	38°58′N	27°58′E
Akiak, Ak., U.S. (äk′yák)	40	61°00′N	161°02′W
Akita, Japan (ä′kĕ-tä)	77	39°40′N	140°12′E
'Akko, Isr.	67a	32°56′N	35°05′E
Akola, India (ä-kō′lä)	69	20°47′N	77°00′E
Akron, Oh., U.S. (ăk′rŭn)	39	41°05′N	81°30′W
Aksaray, Tur. (äk-sä-rī′)	55	38°30′N	34°05′E
Akşehir, Tur. (äk′shä-hēr)	55	38°20′N	31°20′E
Akşehir Gölü, l., Tur. (äk′shä-hēr)	68	38°40′N	31°30′E
Aksha, Russia (äk′shà)	57	50°28′N	113°00′E
Aksu, China (ä-kü-sōō)	76	41°29′N	80°15′E
Akutan, i., Ak., U.S. (ä-kōō-tän′)	40a	53°58′N	169°54′W
Alabama, state, U.S. (ăl-á-băm′á)	39	32°50′N	87°30′W
Alabat, i., Phil. (ä-lä-bät′)	75a	14°14′N	122°05′E
Alacant, Spain	54	38°20′N	0°30′W
Al Aflaj, des., Sau. Ar.	68	24°00′N	44°47′E
Alajuela, Lago, l., Pan. (ä-lä-hwa′lä)	42a	9°15′N	79°34′W
Alaköl, l., Kaz.	59	45°45′N	81°13′E
Al 'Amārah, Iraq	71	31°50′N	47°09′E
Alameda, Ca., U.S. (ăl-á-mā′dá)	38	37°46′N	122°15′W
Alaminos, Phil. (ä-lä-mē′nôs)	75a	16°09′N	119°58′E
Al 'Amirīyah, Egypt	55	31°01′N	29°52′E
Alanya, Tur.	55	36°40′N	32°10′E
Alapayevsk, Russia (ä-lä-pä′yĕfsk)	56	57°50′N	61°35′E
Al 'Aqabah, Jord.	68	29°32′N	35°00′E
Al 'Arīsh, Egypt (äl-a-rēsh′)	67a	31°08′N	33°48′E
Alaska, state, U.S.	40	64°00′N	150°00′W
Alaska, Gulf of, b., Ak., U.S. (à-lăs′ká)	40	57°42′N	147°40′W
Alaska Highway, Ak., U.S. (à-lăs′ká)	40	63°00′N	142°00′W
Alaska Peninsula, pen., Ak., U.S. (à-lăs′ká)	40	55°50′N	162°10′W
Alaska Range, mts., Ak., U.S. (à-lăs′ká)	40	62°00′N	152°18′W
Alatyr', Russia (ä′lä-tür)	56	54°55′N	46°30′E
Alazani, r., Asia	58	41°05′N	46°40′E
Albacete, Spain (äl-bä-thä′tä)	54	39°00′N	1°49′W
Alba Iulia, Rom. (äl-bä yōō′lyä)	55	46°05′N	23°32′E
Albania, nation, Eur. (ăl-bā-nī-á)	50	41°45′N	20°00′E
Albany, Austl. (ôl′bá-nī)	80	35°00′S	118°00′E
Albany, Ga., U.S. (ôl′bá-nī)	39	31°35′N	84°10′W
Albany, N.Y., U.S. (ôl′bá-nī)	39	42°40′N	73°50′W
Albany, Or., U.S. (ôl′bá-nī)	38	44°38′N	123°06′W
Al Başrah, Iraq	68	30°35′N	47°59′E
Al Batrūn, Leb. (äl-bä-trōōn′)	67a	34°16′N	35°39′E
Albert, l., Afr. (ăl′bĕrt) (äl-bär′)	85	1°50′N	30°40′E
Albert Edward, Mount, mtn., Pap. N. Gui. (ăl′bĕrt ĕd′wĕrd)	75	8°25′S	147°25′E
Alboran, Isla del, i., Spain (ē′s-lä-dĕl-äl-bō-rä′n)	52	35°58′N	3°02′W
Albuquerque, N.M., U.S. (äl-bŭ-kûr′kĕ)	38	35°05′N	106°40′W
Albury, Austl. (ôl′bĕr-è)	81	36°00′S	147°00′E
Alcañiz, Spain (äl-kän-yēth′)	54	41°03′N	0°08′W

ăt; fināl; rāte; senâte; ärm; ásk; sofá; fâre; ch-choose; dh-as th in other; bē; ĕvent; bĕt; recĕnt; cratĕr; g-gō; gh-guttural g; bīt; ī-short neutral; rīde; κ-guttural k as ch in German ich;

114 Alcázar de San Juan – As Salt PRONOUNCING INDEX

PLACE (Pronunciation)	PAGE	LAT.	LONG.
Alcázar de San Juan, Spain (äl-kä′thär dā sän hwän′)	54	39°22′N	3°12′W
Alcoi, Spain	54	38°42′N	0°30′W
Aldan, Russia	57	58°46′N	125°19′E
Aldan, r., Russia	57	63°00′N	134°00′E
Aldanskaya, Russia	57	61°52′N	135°29′E
Aleksandrovsk, Russia (ä-lyĕk-sän′drôfsk)	57	51°02′N	142°21′E
Aleppo, Syria (*ȧ*-lĕp-ō)	68	36°10′N	37°18′E
Alessandria, Italy (ä-lĕs-sän′drē-ä)	54	44°53′N	8°35′E
Aleutian Islands, is., Ak., U.S. (*ȧ*-lu′sh*ă*n)	40b	52°40′N	177°30′W
Aleutian Trench, deep,	40a	50°40′N	177°10′E
Alevina, Mys, c., Russia	57	58°49′N	151°44′E
Alexander Archipelago, is., Ak., U.S. (ăl-ĕg-zăn′dẽr)	40	57°05′N	138°10′W
Alexander Island, i., Ant.	82	71°00′S	71°00′W
Alexandra, S. Afr. (ȧl-ex-än′dr*ȧ*)	87c	26°07′S	28°07′E
Alexandria, Austl. (ăl-ĕg-zăn′drĭ-*ȧ*)	80	19°00′S	136°56′E
Alexandria, Egypt (ăl-ĕg-zăn′drĭ-*ȧ*)	70	31°12′N	29°58′E
Alexandria, La., U.S. (ăl-ĕg-zăn′drĭ-*ȧ*)	39	31°18′N	92°28′W
Alexandria, Va., U.S. (ăl-ĕg-zăn′drĭ-*ȧ*)	39	38°50′N	77°05′W
Alexandroúpoli, Grc.	55	40°41′N	25°51′E
Al Fashn, Egypt	87b	28°47′N	30°53′E
Al Firdān, Egypt (äl-fer-dän′)	87b	30°43′N	32°20′E
Algeria, nation, Afr. (ăl-gē′rĭ-*ȧ*)	85	28°45′N	1°00′E
Al Ghayḑah, Yemen	71	16°12′N	52°15′E
Alghero, Italy (äl-gâ′rō)	54	40°32′N	8°22′E
Algiers, Alg. (ăl-jērs)	85	36°51′N	2°56′E
Al Ḩammān, Egypt	55	30°46′N	29°42′E
Al Ḩawrah, Yemen	71	13°49′N	47°37′E
Al Ḩawţah, Yemen	68	15°58′N	48°26′E
Al Ḩijāz, reg., Sau. Ar.	68	23°45′N	39°08′E
Al Hirmil, Leb.	67a	34°23′N	36°22′E
Al Ḩudaydah, Yemen	68	14°43′N	43°03′E
Al Hufūf, Sau. Ar.	68	25°15′N	49°43′E
Al Ḩulwān, Egypt (äl-hĕl′wän)	87b	29°51′N	31°20′E
Aliákmonas, r., Grc.	55	40°26′N	22°17′E
Äli Bayramlı, Azer.	58	39°56′N	48°56′E
Alice Springs, Austl. (ăl′ĭs)	80	23°38′S	133°56′E
Alīgarh, India (ä-lē-gŭr′)	69	27°58′N	78°08′E
Al Iskandarīyah	87b	31°12′N	29°58′E
Al Jafr, Qa‘al, pl., Jord.	67a	30°15′N	36°24′E
Al Jawārah, Oman	71	18°55′N	57°17′E
Al Jawf, Sau. Ar.	68	29°45′N	39°30′E
Al Jīzah, Egypt	87b	30°01′N	31°12′E
Al Jubayl, Sau. Ar.	68	27°01′N	49°40′E
Al Junaynah, Sudan	70	13°27′N	22°27′E
Al Kāb, Egypt	87d	30°56′N	32°19′E
Al Karak, Jord. (kĕ-räk′)	67a	31°11′N	35°42′E
Al Karnak, Egypt (kär′nak)	87b	25°42′N	32°43′E
Al Khābūrah, Oman	68	23°45′N	57°30′E
Al Khalīl, W.B.	67a	31°31′N	35°07′E
Al Khārijah, Egypt	70	25°26′N	30°33′E
Al Khurmah, Sau. Ar.	68	21°37′N	41°44′E
Al Kiswah, Syria	67a	33°31′N	36°13′E
Al Kuntillah, Egypt	67a	29°59′N	34°42′E
Al Kūt, Iraq	71	32°30′N	45°49′E
Al Kuwayt, Kuw. (äl-ko͞o-wit)	68	29°04′N	47°59′E
Al Lādhiqīyah, Syria	68	35°32′N	35°51′E
Allāhābād, India (ŭl-ŭ-hä-bäd′)	69	25°32′N	81°53′E
Allariz, Spain (äl-yä-rēth′)	54	42°10′N	7°48′W
Allaykha, Russia (ä-lī′kȧ)	57	70°32′N	148°53′E
Allentown, Pa., U.S. (ălĕn-toun)	39	40°35′N	75°30′W
Alleppey, India (ȧ-lĕp′ē)	73	9°33′N	76°22′E
Alliance, Ne., U.S. (*ȧ*-lī′*ă*ns)	38	42°06′N	102°53′W
Al Lidām, Sau. Ar.	68	20°45′N	44°12′E
Al Līth, Sau. Ar.	71	20°09′N	40°16′E
Al Luḩayyah, Yemen	68	15°58′N	42°48′E
Alma, S. Afr.	87c	24°30′S	28°05′E
Alma-Ata *see* Almaty, Kaz.	59	43°19′N	77°08′E
Al Madīnah, Sau. Ar.	68	24°26′N	39°42′E
Al Mafraq, Jord.	67a	32°21′N	36°13′E
Al Maḩallah al Kubrā, Egypt	87b	30°58′N	31°10′E
Al Manāmah, Bahr.	68	26°01′N	50°33′E
Al Manshāh, Egypt	87b	26°31′N	31°46′E
Al Manzilah, Egypt (män′za-la)	87b	31°09′N	32°05′E
Al Marāghah, Egypt	87b	26°41′N	31°35′E
Al Maşīrah, i., Oman	68	20°43′N	58°58′E
Almaty (Alma-Ata), Kaz.	59	43°19′N	77°08′E
Almaty, val., Sau. Ar.	67a	29°16′N	35°12′E
Al Mawşil, Iraq	68	36°00′N	42°53′E
Al Mazār, Jord.	67a	31°04′N	35°41′E
Al Mazra‘ah, Jord.	67a	31°17′N	35°33′E
Almería, Spain (äl-mā-rē′ä)	50	36°52′N	2°28′W
Almora, India	69	29°20′N	79°40′E
Al Mubarraz, Sau. Ar.	68	22°31′N	46°27′E
Al Mudawwarah, Jord.	67a	29°20′N	36°01′E
Al Mukhā (Mocha), Yemen	68	13°11′N	43°20′E
Alor, Pulau, i., Indon. (ä′lôr)	75	8°07′S	125°00′E
Alor Gajah, Malay.	67b	2°23′N	102°13′E
Alor Setar, Malay. (ä′lôr stär)	74	6°10′N	100°16′E
Alpena, Mi., U.S. (ăl-pē′n*ȧ*)	39	45°05′N	83°30′W
Alps, mts., Eur. (ălps)	52	46°18′N	8°42′E
Al Qanţarah, Egypt	87d	30°51′N	32°20′E
Al Qaşr, Egypt	70	25°42′N	28°53′E
Al Qaţīf, Sau. Ar.	68	26°30′N	50°00′E
Al Qayşūmah, Sau. Ar.	68	28°15′N	46°20′E
Al Qunaytirah, Syria	67a	33°09′N	35°49′E
Al Qunfudhah, Sau. Ar.	68	19°08′N	41°05′E
Al Quşaymah, Egypt	67a	30°40′N	34°23′E
Al Qusayr, Syria	67a	34°32′N	36°33′E
Altai Mountains, mts., Asia (äl′tī′)	76	49°11′N	87°15′E
Altamura, Italy (äl-tä-mo͞o′rä)	55	40°40′N	16°35′E
Altay, China (äl-tā)	76	47°52′N	86°50′E
Alton, Il., U.S. (ôl′tŭn)	39	38°53′N	90°11′W
Altoona, Pa., U.S. (ăl-to͞o′n*ȧ*)	39	40°25′N	78°25′W

PLACE (Pronunciation)	PAGE	LAT.	LONG.
Altun Shan, mts., China (äl-to̊n shän)	76	36°58′N	85°09′E
Al ‘Ubaylah, Sau. Ar.	71	21°59′N	50°57′E
Al Wajh, Sau. Ar.	68	26°15′N	36°32′E
Alwar, India (ŭl′wŭr)	69	27°39′N	76°39′E
Al Wāsiţah, Egypt	87b	29°21′N	31°15′E
Amalner, India	72	21°07′N	75°06′E
Amami, i., Japan	77	28°10′N	129°55′E
Amarillo, Tx., U.S. (ăm-*ȧ*-rĭl′ō)	38	35°14′N	101°49′W
Amaro, Mount, mtn., Italy (ä-mä′rō)	54	42°07′N	14°07′E
Amasya, Tur. (ä-mä′sē-ȧ)	55	40°40′N	35°50′E
Amatignak, i., Ak., U.S. (ä-mȧ′tē-näk)	40a	51°12′N	178°30′W
Amazon (Amazonas) (Solimões), r., S.A.	46	2°03′S	53°18′W
Ambāla, India (ŭm-bä′lŭ)	69	30°31′N	76°48′E
Ambarchik, Russia (*ŭ*m-bär′chĭk)	57	69°39′N	162°18′E
Ambarnāth, India	73b	19°12′N	73°10′E
Ambil Island, i., Phil. (äm′bēl)	75a	13°51′N	120°25′E
Ambon, Indon.	75	3°45′S	128°17′E
Ambon, Pulau, i., Indon.	75	4°50′S	128°45′E
Amchitka, i., Ak., U.S. (ăm-chĭt′kȧ)	40a	51°25′N	178°10′E
Amchitka Passage, strt., Ak., U.S. (ăm-chĭt′kȧ)	40a	51°30′N	179°36′W
Ameca, Mex. (ä-mĕ′kä)	42	20°34′N	104°02′W
American Highland, plat., Ant.	82	72°00′S	79°00′E
American Samoa, dep., Oc.	2	14°20′S	170°00′W
Americus, Ga., U.S. (*ȧ*-mĕr′ĭ-k*ŭ*s)	39	32°04′N	84°15′W
Amga, Russia (ŭm-gä′)	57	61°08′N	132°09′E
Amirante Islands, is., Sey.	5	6°02′S	52°30′E
Amlia, i., Ak., U.S. (ȧ′mlēä)	40a	52°00′N	173°28′W
‘Ammān, Jord. (äm′män)	68	31°57′N	35°57′E
Amorgós, i., Grc. (ä-môr′gōs)	55	36°47′N	25°47′E
Amoy *see* Xiamen, China	77	24°30′N	118°10′E
Amrāvati, India	69	20°58′N	77°47′E
Amritsar, India (ŭm-rĭt′s*ŭ*r)	69	31°43′N	74°52′E
Amsterdam, Neth. (äm-stẽr-däm′)	50	52°21′N	4°52′E
Amsterdam, Île, i., Afr.	82	37°52′S	77°32′E
Amu Darya, r., Asia (ä-mo̊-dä′rēä)	56	38°30′N	64°00′E
Amukta Passage, strt., Ak., U.S. (ä-mo͞ok′tȧ)	40a	52°30′N	172°00′W
Amundsen Sea, sea, Ant. (ä′mŭn-s*ĕ*n-sē′)	82	72°00′S	110°00′W
Amur, r., Asia	57	49°00′N	136°00′E
Amuyao, Mount, mtn., Phil. (ä-mo͞o-yä′ō)	75a	17°04′N	121°09′E
Amyun, Leb.	67a	34°18′N	35°48′E
Anaconda, Mt., U.S. (ăn-*ȧ*-kŏn′d*ȧ*)	38	46°07′N	112°55′W
Anadyr‘, Russia (ŭ-n*ȧ*-dĭr′)	57	64°47′N	177°01′E
Anadyrskiy Zaliv, b., Russia	56	64°10′N	178°00′W
Ānai Mudi, mtn., India	73	10°10′N	77°00′E
Anambas, Kepulauan, is., Indon. (ä-näm-bäs)	74	2°41′N	106°38′E
Anchorage, Ak., U.S. (ăŋ′kẽr-åj)	40	61°12′N	149°48′W
Ancón, Pan. (äŋ-kōn′)	42a	8°55′N	79°32′W
Ancona, Italy (än-kō′nä)	50	43°37′N	13°32′E
Andaman Islands, is., India (ăn-dȧ-măn′)	74	11°38′N	92°17′E
Andaman Sea, sea, Asia	74	12°44′N	95°45′E
Anderson, S.C., U.S. (ăn′dẽr-s*ŭ*n)	39	34°30′N	82°40′W
Andes Mountains, mts., S.A. (ăn′dēz) (än′dās)	47	13°00′S	75°00′W
Andheri, neigh., India	73b	19°08′N	72°50′E
Andhra Pradesh, state, India	69	16°00′N	79°00′E
Andikýthira, i., Grc.	55	35°50′N	23°20′E
Andizhan, Uzb. (än-dē-zhän′)	59	40°45′N	72°22′E
Andong, Kor., S. (än′dŭng′)	77	36°31′N	128°42′E
Andorra, nation, Eur. (än-dôr′rä)	50	42°30′N	2°00′E
Andria, Italy (än′drē-ä)	55	41°17′N	15°55′E
Ándros, i., Grc. (ȧn′drŏs)	55	37°59′N	24°55′E
Andros Island, i., Bah. (ăn′drŏs)	43	24°30′N	78°00′W
Aneto, Pico de, mtn., Spain (pē′kō-dĕ-ä-nĕ′tō)	52	42°35′N	0°38′E
Ang‘angxi, China (äŋ-äŋ-shyē)	77	47°05′N	123°58′E
Angarsk, Russia	57	52°48′N	104°15′E
Ángel de la Guarda, i., Mex. (ȧ′n-hĕl-dĕ-lä-gwä′r-dä)	42	29°30′N	113°00′W
Angeles, Phil. (än′hå-lās)	75a	15°09′N	120°35′E
Angkor, hist., Camb. (äng′kôr)	74	13°52′N	103°50′E
Angmagssalik, Grnld. (ȧŋ-mȧ′sȧ-lĭk)	37	65°40′N	37°40′W
Angola, nation, Afr. (ăŋ-gō′lȧ)	85	14°15′S	16°00′E
Angora *see* Ankara, Tur.	68	39°55′N	32°50′E
Anguilla, dep., N.A.	43	18°15′N	62°54′W
Anhui, prov., China (än-hwā)	77	31°30′N	117°15′E
Aniak, Ak., U.S. (ä-nyȧ′k)	40	61°32′N	159°35′W
Aniakchak National Monument, rec., Ak., U.S.	40	56°50′N	157°50′W
Ankang, China (än-ĸäŋ)	76	32°38′N	109°10′E
Ankara, Tur. (än′k*ȧ*-r*ȧ*)	68	39°55′N	32°50′E
An Nafūd, des., Sau. Ar.	68	28°30′N	40°30′E
An Najaf, Iraq (än nä-jäf′)	68	32°00′N	44°25′E
An Nakhl, Egypt	67a	29°55′N	33°45′E
Annamese Cordillera, mts., Asia	74	17°34′N	105°38′E
Annapolis, Md., U.S. (ă-năp′ō-lĭs)	39	39°00′N	76°25′W
Ann Arbor, Mi., U.S. (ăn är′bẽr)	39	42°15′N	83°45′W
An Nāşirīyah, Iraq	68	31°08′N	46°15′E
An Nhon, Viet.	74	13°55′N	109°00′E
Anniston, Al., U.S. (ăn′ĭs-t*ŭ*n)	39	33°39′N	85°47′W
Annobón, i., Eq. Gui.	86	2°00′S	3°30′E
Anpu, China (än-po͞o)	76	21°28′N	110°00′E
Anshun, China (än-sho͞on′)	76	26°12′N	105°50′E
Antalya, Tur. (än-tä′lē-ä) (ä-dä′lē-ä)	55	37°00′N	30°50′E
Antalya Körfezi, b., Tur.	55	36°40′N	31°20′E
Antananarivo, Madag.	85	18°51′S	47°40′E
Antarctica, cont.	82	80°15′S	127°00′E
Antarctic Peninsula, pen., Ant.	82	70°00′S	65°00′W
Antequera, Spain (än-tĕ-kĕ′rä)	54	37°01′N	4°34′W
Antigua, Guat. (än-tē′gwä)	42	14°32′N	90°43′W
Antigua and Barbuda, nation, N.A.	43	17°15′N	61°15′W

PLACE (Pronunciation)	PAGE	LAT.	LONG.
Antofagasta, Chile (än-tō-fä-gäs′tä)	46	23°32′S	70°21′W
Antón, Pan. (än-tōn′)	43	8°24′N	80°15′W
Antwerp, Bel.	50	51°13′N	4°24′E
Antwerpen *see* Antwerp, Bel.	50	51°13′N	4°24′E
Anūpgarh, India (ŭ-no̊p′gŭr)	72	29°22′N	73°20′E
Anuradhapura, Sri L. (ŭ-no͞o′rä-dŭ-po͞o′r*ŭ*)	73	8°24′N	80°25′E
Anxi, China (än-shyē)	76	40°36′N	95°49′E
Anyang, China (än′yäng)	77	36°05′N	114°22′E
Anzhero-Sudzhensk, Russia (än′zhâ-rō-so̊d′zhĕnsk)	56	56°08′N	86°08′E
Aoba, i., Vanuatu	78f	15°25′S	167°50′E
Aomori, Japan (äō-mō′rē)	77	40°45′N	140°52′E
Aoraki (Cook, Mount), mtn., N.Z.	81a	43°27′S	170°13′E
Aparri, Phil. (ä-pär′rē)	74	18°15′N	121°40′E
Apennines *see* Appennino, mts., Italy	52	43°48′N	11°06′E
Apia, Samoa	78a	13°50′S	171°44′W
Apo, Mount, mtn., Phil. (ä′pō)	75	6°56′N	125°05′E
Appalachian Mountains, mts., N.A. (ăp-*ȧ*-lăch′ĭ-*ȧ*n)	36	37°20′N	82°00′W
Appennino, mts., Italy (äp-pĕn-nē′nō)	52	43°48′N	11°06′E
Appleton, Wi., U.S.	39	44°14′N	88°27′W
Apsheronsk, Russia	58	44°28′N	39°44′E
Aqaba, Gulf of, b., (ä′kȧ-bȧ)	68	28°30′N	34°40′E
Aqabah, Wādī al, r., Egypt	67a	29°48′N	34°05′E
Aqmola *see* Astana, Kaz.	59	51°10′N	71°43′E
Aqtaū, Kaz.	59	43°35′N	51°05′E
Aqtöbe, Kaz.	59	50°20′N	57°00′E
‘Arabah, Wādī, val., Egypt	87b	29°02′N	32°10′E
Arabian Sea, sea, (ȧ-rā′bĭ-*ă*n)	66	16°00′N	65°15′E
‘Arad, Isr.	67a	31°20′N	35°15′E
Arad, Rom. (ŏ′rŏd)	55	46°10′N	21°18′E
Arafura Sea, sea, (ä-rä-fo͞o′rä)	75	8°40′S	130°00′E
Aragats, Gora, mtn., Arm.	58	40°32′N	44°14′E
Araj, oasis, Egypt (ä-räj′)	55	29°05′N	26°51′E
Arāk, Iran	68	34°08′N	49°57′E
Arakan Yoma, mts., Mya. (ŭ-rŭ-kŭn′yō′mȧ)	69	19°51′N	94°13′E
Aral, Kaz.	59	46°47′N	62°00′E
Aral Sea, sea, Asia	56	45°17′N	60°02′E
Aranjuez, Spain (ä-rän-hwāth′)	54	40°02′N	3°24′W
Arapkir, Tur. (ä-räp-kēr′)	55	39°00′N	38°10′E
Ararat, Austl. (ăr′*ȧ*răt)	81	37°17′S	142°56′E
Ararat, Mount, mtn., Tur.	68	39°50′N	44°20′E
Aras, r., Asia (ȧ-räs)	68	39°15′N	47°10′E
Aravalli Range, mts., India (ä-rä′vŭ-lē)	69	24°15′N	72°40′E
Arayat, Phil. (ä-rä′yät)	75a	15°10′N	120°44′E
Arbīl, Iraq	68	36°10′N	44°00′E
Arctic Ocean, ocean	93	85°00′N	170°00′E
Ardabīl, Iran	68	38°15′N	48°00′E
Ardmore, Ok., U.S. (ärd′mōr)	38	34°10′N	97°08′W
Arecibo, P.R. (ä-rå-sē′bō)	43b	18°28′N	66°45′W
Arequipa, Peru (ä-rå-kē′pä)	46	16°27′S	71°30′W
Arezzo, Italy (ä-rĕt′sō)	54	43°28′N	11°54′E
Argentina, nation, S.A. (är-jĕn-tē′n*ȧ*)	46	35°30′S	67°00′W
Argun‘, r., Asia (ȧr-go͞on′)	57	50°00′N	119°00′E
Aringay, Phil. (ä-rĭŋ-gä′ē)	75a	16°25′N	120°20′E
‘Arīsh, Wādī al, r., Egypt (ȧ-rēsh′)	67a	30°36′N	34°07′E
Arizona, state, U.S. (ăr-ĭ-zō′n*ȧ*)	38	34°00′N	113°00′W
Arkansas, state, U.S. (är′kăn-sô) (är-kăn′s*ȧ*s)	39	34°50′N	93°40′W
Arkhangelsk (Archangel), Russia (ȧr-ĸän′gĕlsk)	56	64°30′N	40°25′E
Arkonam, India (ār-kō-nām′)	73	13°05′N	79°43′E
Arlington, S. Afr.	87c	28°02′S	27°52′E
Arltunga, Austl. (ärl-to̊ŋ′gȧ)	80	23°19′S	134°45′E
Armant, Egypt (är-mänt′)	87b	25°37′N	32°32′E
Armavir, Russia (ȧr-mȧ-vĭr′)	56	45°00′N	41°00′E
Armenia, nation, Asia	56	41°00′N	44°39′E
Armidale, Austl. (är′mĭ-dāl)	81	30°27′S	151°50′E
Arno, r., Italy (ä′r-nō)	54	43°30′N	11°00′E
Aroroy, Phil. (ä-rō-rō′ē)	75a	12°30′N	123°24′E
Ar Ramādī, Iraq	68	33°26′N	43°19′E
Ar Rawḑah, Egypt	87b	27°47′N	30°52′E
Ar Rub‘ al Khālī, des., Asia	68	20°00′N	51°00′E
Ar Ruţbah, Iraq	71	33°02′N	40°17′E
Arsen‘yev, Russia	57	44°13′N	133°32′E
Árta, Grc. (är′tā)	55	39°08′N	21°02′E
Artëm, Russia (*ȧ*r-tyôm′)	57	43°28′N	132°29′E
Aru, Kepulauan, is., Indon.	75	6°20′S	133°00′E
Aruba, dep., N.A. (ä-ro͞o′bȧ)	43	12°29′N	70°00′W
Arunachal Pradesh, state, India	69	27°35′N	92°56′E
Arziw, Alg.	54	35°50′N	0°20′W
Asahikawa, Japan	77	43°50′N	142°09′E
Asansol, India	69	23°45′N	86°58′E
Ascension, i., St. Hel. (ȧ-sĕn′sh*ŭ*n)	86	8°00′S	13°00′W
Ascent, S. Afr. (ăs-ĕnt′)	87c	27°14′S	29°06′E
Ashdod, Isr.	67a	31°46′N	34°39′E
Asheville, N.C., U.S. (ăsh′vĭl)	39	35°35′N	82°35′W
Ashgabat, Turkmen.	59	37°57′N	58°23′E
Ashland, Wi., U.S.	39	46°34′N	90°55′W
Ashmūn, Egypt (ȧsh-mo͞on′)	87b	30°19′N	30°57′E
Ashqelon, Isr. (äsh′k*ĕ*-lōn)	67a	31°40′N	34°36′E
Ash Shallūfah, Egypt (shäl′lo̊-fä)	87b	30°09′N	32°33′E
Ash Shaqrā‘, Sau. Ar.	68	25°10′N	45°08′E
Ash Shāriqah, U.A.E.	71	25°22′N	55°23′E
Ash Shawbak, Jord.	67a	30°31′N	35°35′E
Ash Shiḩr, Yemen	68	14°45′N	49°32′E
Ashtabula, Oh., U.S. (ăsh-t*ȧ*-bū′l*ȧ*)	39	41°55′N	80°50′W
Asia, cont.	66	50°00′N	100°00′E
Asia Minor, reg., Tur. (ā′zh*ȧ*)	53	38°18′N	31°18′E
Asīr, reg., Sau. Ar. (ä-sēr′)	68	19°30′N	42°00′E
Asmara *see* Asmera, Erit.	85	15°17′N	38°56′E
Asmera, Erit. (äs-mä′rä)	85	15°17′N	38°56′E
Aş Şaff, Egypt	87b	29°33′N	31°23′E
As Salt, Jord.	67a	32°02′N	35°44′E

ăt; fin*ă*l; rāte; senåte; ärm; ȧsk; sof*ȧ*; fâre; ch-choose; dh-as th in other; bē; ėvent; bĕt; recĕnt; cratẽr; g-gō; gh-guttural g; bĭt; *ĭ*-short neutral; rīde; ĸ-guttural k as ch in German ich;

PRONOUNCING INDEX Assam – Bellpat 115

PLACE (Pronunciation)	PAGE	LAT.	LONG.
Assam, state, India (ăs-săm′)	69	26°00′N	91°00′E
As Samāwah, Iraq	71	31°18′N	45°17′E
As Sinbillāwayn, Egypt	87b	30°53′N	31°37′E
Assisi, Italy	54	43°04′N	12°37′E
As Sulaymānīyah, Iraq	68	35°47′N	45°23′E
As Sulaymānīyah, Sau. Ar.	71	24°09′N	46°19′E
As Suwaydā', Syria	68	32°41′N	36°41′E
Astana (Aqmola), Kaz.	59	51°10′N	71°43′E
Asti, Italy (äs′tē)	54	44°54′N	8°12′E
Astoria, Or., U.S. (ăs-tō′rĭ-ȧ)	38	46°11′N	123°51′W
Astrakhan', Russia (äs-trȧ-kän′)	56	46°15′N	48°00′E
Astypalaia, i., Grc.	55	36°31′N	26°19′E
Asunción, Para. (ä-sōōn-syōn′)	46	25°25′S	57°30′W
Aswān, Egypt (ä-swän′)	70	24°05′N	32°57′E
Atacama, Desierto de, des., Chile (dĕ-syĕ′r-tô-dĕ-ä-tä-kä′mä)	47	23°50′S	69°00′W
Atā 'itah, Jabal al, mtn., Jord.	67a	30°48′N	35°19′E
'Atāqah, Jabal, mts., Egypt	87d	29°59′N	32°20′E
Atauro, Ilha de, i., E. Timor (dĕ-ä-tä′ōō-rô)	75	8°20′S	126°15′E
Atbasar, Kaz. (ät′bä-sär′)	59	51°42′N	68°28′E
Atchison, Ks., U.S. (ăch′ĭ-sŭn)	39	39°33′N	95°08′W
Athens, Ga., U.S.	39	33°55′N	83°24′W
Athens (Athína), Grc.	50	38°00′N	23°38′E
Athína see Athens, Grc.	50	38°00′N	23°38′E
Ath Thamad, Egypt	67a	29°41′N	34°17′E
Atimonan, Phil. (ä-tē-mō′nän)	75a	13°59′N	121°56′E
Atka, Ak., U.S. (ăt′kȧ)	40a	52°18′N	174°18′W
Atlanta, Ga., U.S. (ăt-lăn′tȧ)	39	33°45′N	84°23′W
Atlantic City, N.J., U.S.	39	39°20′N	74°30′W
Atlantic Ocean, o.	4	5°00′S	25°00′W
Atlas Mountains, mts., Afr. (ăt′lăs)	85	31°22′N	4°57′W
Atrak, r., Asia	68	37°45′N	56°30′E
Aṭ Ṭafīlah, Jord. (tä-fē′la)	67a	30°50′N	35°36′E
Aṭ Ṭā'if, Sau. Ar.	68	21°03′N	41°00′E
Aṭ Ṭūr, Egypt	55	28°09′N	33°47′E
Aṭ Ṭurayf, Sau. Ar.	68	31°32′N	38°30′E
Atyraū, Kaz.	59	47°10′N	51°50′E
Auburn, Me., U.S.	39	44°04′N	70°24′W
Auckland, N.Z. (ôk′lănd)	81a	36°53′S	174°45′E
Auckland Islands, is., N.Z.	3	50°30′S	166°30′E
Audo Range, mts., Eth.	87a	6°58′N	41°18′E
Augusta, Ga., U.S.	39	33°26′N	82°00′W
Augusta, Me., U.S.	39	44°19′N	69°42′W
Auki, Sol. Is.	78e	8°46′S	160°42′E
Aur, i., Malay.	67b	2°27′N	104°51′E
Aurangābād, India (ou-rŭŋ-gä-bäd′)	69	19°56′N	75°19′E
Aurès, Massif de l', mts., Alg.	54	35°16′N	5°53′E
Aurora, Il., U.S. (ô-rō′rȧ)	39	41°45′N	88°18′W
Austin, Tx., U.S.	38	30°15′N	97°42′W
Australia, nation, Oc.	80	25°00′S	135°00′E
Australian Capital Territory, ter., Austl. (ôs-trā′lĭ-ăn)	81	35°30′S	148°40′E
Austria, nation, Eur. (ôs′trĭ-ȧ)	50	47°15′N	11°53′E
Autlán, Mex. (ä-ōōt-län′)	42	19°47′N	104°24′W
Aveiro, Port. (ä-vā′rō)	54	40°38′N	8°38′W
Avilés, Spain (ä-vē-lās′)	54	43°33′N	5°55′W
Ayachi, Arin', mtn., Mor.	54	32°29′N	4°57′W
Ayaköz, Kaz.	59	48°00′N	80°12′E
Ayamonte, Spain (ä-yä-mō′n-tĕ)	54	37°14′N	7°28′W
Ayan, Russia (ȧ-yän′)	57	56°26′N	138°18′E
Aydın, Tur. (āĭy-dĕn)	68	37°40′N	27°40′E
Ayer Hitam, Malay.	67b	1°55′N	103°11′E
Ayers Rock see Uluru, mtn., Austl.	80	25°23′S	131°05′E
Ayon, i., Russia (ī-ôn′)	57	69°50′N	168°40′E
Ayvalık, Tur. (āĭy-wä-lĭk)	55	39°19′N	26°40′E
Azerbaijan, nation, Asia	56	40°30′N	47°30′E
Azores see Açores, is., Port.	86	37°44′N	29°25′W
Azov, Sea of, sea, Eur.	56	46°00′N	36°20′E
Azuero, Península de, pen., Pan.	43	7°30′N	80°34′W
Aẓ Ẓahrān (Dhahran), Sau. Ar.	68	26°13′N	50°00′E
Az Zarqā', Jord.	67a	32°03′N	36°07′E
B			
Baadheere (Bardera), Som.	87a	2°13′N	42°24′E
Baao, Phil. (bä′ō)	75a	13°27′N	123°22′E
Babar, Pulau, i., Indon. (bä′bär)	75	7°50′S	129°15′E
Bab-el-Mandeb see Mandeb, Bab-el-, strt.,	68	13°17′N	42°49′E
Babelthuap, i., Palau	78b	7°30′N	134°36′E
Bābol, Iran	68	36°30′N	52°48′E
Babuyan Islands, is., Phil. (bä-bōō-yän′)	74	19°30′N	122°38′E
Babylon, hist., Iraq	68	32°15′N	45°23′E
Bacan, Pulau, i., Indon.	75	0°30′S	127°00′E
Bacău, Rom.	55	46°34′N	27°00′E
Bachu, China (bä-chōō)	76	39°50′N	78°23′E
Back Bay, India (băk)	73b	18°55′N	72°45′E
Bac Lieu, Viet.	74	9°45′N	105°50′E
Baco, Mount, mtn., Phil. (bä′kō)	75a	12°50′N	121°11′E
Bacolod, Phil. (bä-kō′lôd)	75	10°42′N	123°03′E
Badajoz, Spain (bȧ-dhä-hōth′)	54	38°52′N	6°56′W
Badanah, Sau. Ar.	68	30°49′N	40°45′E
Badīn, Pak.	72	24°47′N	69°51′E
Badlands, reg., S.D., U.S.	34	43°43′N	102°36′W
Badlāpur, India	73b	19°12′N	73°12′E
Badulla, Sri L.	73	6°55′N	81°07′E
Baena, Spain (bä-ā′nä)	54	37°38′N	4°20′W
Baffin Bay, b., N.A. (băf′ĭn)	37	72°00′N	65°00′W
Baffin Island, i., Can.	37	67°20′N	71°00′W
Bāfq, Iran (bäfk)	68	31°48′N	55°23′E
Bafra, Tur. (bäf′rä)	55	41°30′N	35°50′E
Bagabag, Phil. (bä-gä-bäg′)	75a	16°38′N	121°16′E
Bāgalkot, India	73	16°14′N	75°40′E
Bagdad see Baghdād, Iraq	68	33°14′N	44°22′E
Baghdād, Iraq (bȧgh-dȧd′) (băg′dăd)	68	33°14′N	44°22′E
Bago, Mya.	74	17°17′N	96°29′E
Baguio, Phil. (bä-gē-ō′)	74	16°24′N	120°36′E
Bahamas, nation, N.A. (bȧ-hā′mȧs)	43	26°15′N	76°00′W
Bahau, Malay.	67b	2°48′N	102°25′E
Bahāwalpur, Pak. (bŭ-hä′wŭl-pōōr)	69	29°29′N	71°41′E
Bahía, Islas de la, i., Hond. (ē′s-läs-dĕ-lä-bä-ē′ä)	42	16°15′N	86°30′W
Bahrain, nation, Asia (bä-rān′)	68	26°15′N	51°17′E
Baḥrīyah, oasis, Egypt (bȧ-hȧ-rē′yä)	55	28°34′N	29°01′E
Baia Mare, Rom. (bä′yä mä′rā)	55	47°40′N	23°35′E
Baidyabāti, India	72a	22°47′N	88°21′E
Baikal, Lake see Baykal, Ozero, l., Russia	57	53°00′N	109°28′E
Baird Mountains, mts., Ak., U.S.	40	67°35′N	160°10′W
Bairnsdale, Austl. (bârnz′dāl)	81	37°50′S	147°39′E
Baja California, state, Mex. (bä-hä)	42	30°15′N	117°25′W
Baja California, pen., Mex.	37	28°00′N	113°30′W
Baja California Sur, state, Mex.	42	26°00′N	113°30′W
Baker, Or., U.S.	38	44°46′N	117°52′W
Baker, i., Oc.	2	1°00′N	176°00′W
Bakersfield, Ca., U.S. (bā′kẽrz-fēld)	38	35°23′N	119°00′W
Bakhtarān, Iran	68	34°01′N	47°00′E
Bakhtegan, Daryācheh-ye, l., Iran	68	29°29′N	54°31′E
Baku (Bakı), Azer. (bȧ-kōō′)	56	40°28′N	49°45′E
Balabac Island, i., Phil. (bä′lä-bäk)	74	8°00′N	116°28′E
Balabac Strait, strt., Asia	74	7°23′N	116°30′E
Ba'labakk, Leb.	67a	34°00′N	36°13′E
Balakhta, Russia (bȧ′lȧk-tȧ′)	57	55°22′N	91°43′E
Balanga, Phil. (bä-läŋ′gä)	75a	14°41′N	120°31′E
Balasore, India (bä-lȧ-sōr′)	69	21°38′N	86°59′E
Balaton Lake, l., Hung. (bô′lô-tôn)	55	46°47′N	17°55′E
Balayan, Phil. (bä-lä-yän′)	75a	13°56′N	120°44′E
Balayan Bay, b., Phil.	75a	13°46′N	120°46′E
Balboa Mountain, mtn., Pan.	42a	9°05′N	79°44′W
Balearic Islands see Balears, Illes, is., Spain	52	39°25′N	1°28′E
Balears, Illes, is., Spain	52	39°25′N	1°28′E
Baler, Phil. (bä-lar′)	75a	15°46′N	121°33′E
Baler Bay, b., Phil.	75a	15°51′N	121°40′E
Balesin, i., Phil.	75a	14°28′N	122°10′E
Balfour, S. Afr. (băl′fŏŏr)	87c	26°41′S	28°37′E
Bali, i., Indon. (bä′lē)	74	8°00′S	115°22′E
Balikpapan, Indon. (bä′lēk-pä′pän)	74	1°13′S	116°52′E
Balintang Channel, strt., Phil. (bä-lĭn-täng′)	74	19°50′N	121°08′E
Balkan Mountains see Stara Planina, mts., Blg.	52	42°50′N	24°45′E
Balkh, Afg. (bälk)	69	36°48′N	66°50′E
Balkhash, Lake see Balqash köli, l., Kaz.	59	45°58′N	72°15′E
Ballarat, Austl. (băl′ȧ-răt)	81	37°37′S	144°00′E
Balleny Islands, is., Ant. (băl′ĕ nē)	82	67°00′S	164°00′E
Bālotra, India	72	25°56′N	72°12′E
Balqash, Kaz.	59	46°58′N	75°00′E
Balqash köli, l., Kaz.	59	45°58′N	72°15′E
Balsas, r., Mex.	42	18°00′N	101°00′W
Baltic Sea, sea, Eur. (bôl′tĭk)	52	55°20′N	16°50′E
Baltīm, Egypt (bȧl-tēm′)	87b	31°33′N	31°04′E
Baltimore, Md., U.S. (bôl′tĭ-môr)	39	39°20′N	76°38′W
Baluchistān, hist. reg., Asia (bȧ-lò-chĭ-stän′)	69	27°30′N	65°30′E
Bamako, Mali (bä-mä-kō′)	85	12°39′N	8°00′W
Bambang, Phil. (bäm-bäng′)	75a	16°24′N	121°08′E
Bampūr, Iran (bŭm-pōōr′)	68	27°15′N	60°22′E
Banahao, Mount, mtn., Phil. (bä-nä-hä′ô)	75a	14°04′N	121°45′E
Banās, r., India (bän-äs′)	69	25°20′N	75°20′E
Bānda, India (bän′dä)	69	25°36′N	80°21′E
Banda, Kepulauan, is., Indon.	75	4°40′S	129°56′E
Banda, Laut (Banda Sea), sea, Indon.	75	6°05′S	127°28′E
Banda Aceh, Indon.	74	5°10′N	95°10′E
Bandar Beheshtī, Iran	68	25°18′N	60°45′E
Bandar-e 'Abbās, Iran (bän-där′ ȧb-bäs′)	68	27°04′N	56°22′E
Bandar-e Būshehr, Iran	68	28°48′N	50°53′E
Bandar-e Lengeh, Iran	68	26°44′N	54°47′E
Bandar-e Torkeman, Iran	68	37°05′N	54°08′E
Bandar Lampung, Indon.	74	5°16′S	105°06′E
Bandar Maharani, Malay. (bän-där′ mä-hä-rä′nē)	67b	2°02′N	102°34′E
Bandar Seri Begawan, Bru.	74	5°00′N	114°59′E
Bandirma, Tur. (bän-dĭr′mȧ)	55	40°25′N	27°50′E
Bāndra, India	73b	19°04′N	72°49′E
Bandung, Indon.	74	7°00′S	107°22′E
Bangalore, India (băŋ′gȧ-lôr)	69	13°03′N	77°39′E
Bangeta, Mount, mtn., Pap. N. Gui.	75	6°20′S	147°00′E
Banggai, Kepulauan, is., Indon. (bäng-gī′)	75	1°05′S	123°45′E
Banggi, Pulau, i., Malay.	74	7°12′N	117°10′E
Banghāzī, Libya	85	32°07′N	20°04′E
Bangka, i., Indon. (bäŋ′kȧ)	74	2°24′S	106°55′E
Bangkalan, Indon. (bäng-kȧ-län′)	74	6°07′S	112°50′E
Bangkok, Thai.	74	13°50′N	100°29′E
Bangladesh, nation, Asia	69	24°15′N	90°00′E
Bangong Co, l., Asia (bäŋ-gôŋ tswo)	72	33°40′N	79°30′E
Bangor, Me., U.S. (băn′gẽr)	39	44°47′N	68°47′W
Bangued, Phil. (bän-gād′)	75a	17°36′N	120°38′E
Bangui, C.A.R. (bän-gē′)	85	4°22′N	18°35′E
Bani, Phil. (bä′nē)	75a	16°11′N	119°51′E
Banī Mazār, Egypt	70	28°29′N	30°48′E
Banja Luka, Bos. (bän-yä-lōō′kä)	55	44°45′N	17°11′E
Banjarmasin, Indon. (bän-jẽr-mä′sēn)	74	3°18′S	114°32′E
Banjul, Gam.	85	13°28′N	16°39′W
Banks Island, i., Can.	37	73°00′N	123°00′W
Bannu, Pak.	72	33°03′N	70°39′E
Banton, i., Phil. (bän-tōn′)	75a	12°54′N	121°55′E
Banyak, Kepulauan, is., Indon.	74	2°08′N	97°15′E
Banyuwangi, Indon. (bän-jò-wäŋ′gē)	74	8°15′S	114°15′E
Baoding, China (bou-dĭŋ)	77	38°52′N	115°31′E
Baoshan, China (bou-shän)	76	25°14′N	99°03′E
Baotou, China (bou-tō)	77	40°28′N	110°10′E
Baraawe, Som.	87a	1°20′N	44°00′E
Baranagar, India	72	22°38′N	88°25′E
Baranof, i., Ak., U.S. (bä-rä′nôf)	40	56°48′N	136°08′W
Baranpauh, Indon.	67b	0°40′N	103°28′E
Bārāsat, India	72a	22°42′N	88°29′E
Barbados, nation, N.A. (bär-bā′dōz)	43	13°30′N	59°00′W
Barbuda, i., Antig. (bär-bōō′dȧ)	43	17°45′N	61°15′W
Barcaldine, Austl. (bär′kôl-dĭn)	81	23°33′S	145°17′E
Barcelona, Spain (bär-thȧ-lō′nä)	50	41°25′N	2°08′E
Bardawīl, Sabkhat al, b., Egypt	67a	31°20′N	33°24′E
Bareilly, India	69	28°21′N	79°25′E
Barents Sea, sea, Eur. (bä′rĕnts)	56	72°14′N	37°28′E
Barguzin, Russia (bär′gōō-zĭn)	57	53°44′N	109°28′E
Bari, Italy (bä′rē)	50	41°08′N	16°53′E
Barisan, Pegunungan, mts., Indon. (bä-rē-sän′)	74	2°38′S	101°45′E
Barito, r., Indon. (bä-rē′tō)	74	2°10′S	114°38′E
Barkol, China (bär-kůl)	76	43°43′N	92°50′E
Bârlad, Rom.	55	46°15′N	27°43′E
Barletta, Italy (bär-lĕt′tä)	55	41°19′N	16°20′E
Barnaul, Russia (bär-nä-ōl′)	56	53°18′N	83°23′E
Baroda, India (bär-rō′dä)	69	22°21′N	73°12′E
Barrackpore, India	72a	22°46′N	88°21′E
Barreiro, Port. (bär-rĕ′ē-rò)	54	38°39′N	9°05′W
Barrow, Point, c., Ak., U.S.	40	71°20′N	156°00′W
Barrow Creek, Austl.	80	21°23′S	133°55′E
Bartın, Tur. (bär′tĭn)	55	41°35′N	32°12′E
Bashi Channel, strt., Asia (bäsh′ē)	77	21°20′N	120°22′E
Basilan Island, i., Phil.	74	6°37′N	122°07′E
Basra see Al Baṣrah, Iraq	68	30°35′N	47°59′E
Basse Terre, Guad. (bȧs′ târ′)	43	16°00′N	61°43′W
Batāla, India	72	31°54′N	75°18′E
Batam, i., Indon. (bä-täm′)	67b	1°03′N	104°00′E
Batang, China (bä-täŋ)	76	30°08′N	99°00′E
Batangas, Phil. (bä-tän′gäs)	74	13°45′N	121°04′E
Batan Islands, is., Phil. (bä-tän′)	74	20°58′N	122°20′E
Bătdâmbâng, Camb. (bȧt-tȧm-bäng′)	74	13°14′N	103°15′E
Bathurst, Austl. (băth′ŭrst)	81	33°28′S	149°30′E
Baton Rouge, La., U.S. (băt′ŭn rōōzh′)	39	30°28′N	91°10′W
Batticaloa, Sri L.	73	7°40′N	81°10′E
Battle Creek, Mi., U.S. (băt′'l krēk′)	39	42°20′N	85°15′W
Batu, Kepulauan, is., Indon. (bä′tōō)	74	0°10′S	98°00′E
Batumi, Geor. (bŭ-tōō′mē)	56	41°40′N	41°30′E
Batu Pahat, Malay.	74	1°51′N	102°56′E
Batupanjang, Indon.	67b	1°42′N	101°35′E
Bauang, Phil. (bä′wäng)	75a	16°31′N	120°19′E
Bāuria, India	72a	22°29′N	88°08′E
Bawean, Pulau, i., Indon. (bȧ′vē-än)	74	5°50′S	112°40′E
Bay, Laguna de, l., Phil. (lä-gōō′nä dä bä′ē)	75a	14°24′N	121°13′E
Bay al Kabīr, Wādī, val., Libya	54	29°52′N	14°28′E
Bayambang, Phil. (bä-yäm-bäng′)	75a	15°50′N	120°26′E
Bayamón, P.R.	43b	18°27′N	66°13′W
Bayanaūyl, Kaz.	59	50°43′N	75°37′E
Bay City, Mi., U.S. (bā)	39	43°35′N	83°55′W
Baydhabo (Baidoa), Som.	87a	3°19′N	44°20′E
Baydrag, r., Mong.	76	46°09′N	98°52′E
Baykal, Ozero (Lake Baikal), l., Russia	57	53°00′N	109°28′E
Baykal'skiy Khrebet, mts., Russia	57	53°30′N	107°30′E
Baykit, Russia (bī-kēt′)	57	61°43′N	96°39′E
Bayombong, Phil. (bä-yōm-bōng′)	75a	16°28′N	121°09′E
Bayonne, Fr. (bȧ-yōn′)	50	43°28′N	1°30′W
Bayqongyr, Kaz.	59	47°46′N	66°11′E
Bayt Lahm, W.B. (bĕth′lē-hĕm)	67a	31°42′N	35°13′E
Baza, Spain (bä′thä)	54	37°29′N	2°46′W
Beatrice, Ne., U.S. (bē′ȧ-trĭs)	38	40°16′N	96°45′W
Beaufort Sea, sea, N.A.	40	70°30′N	138°40′W
Beaumont, Tx., U.S.	39	30°05′N	94°06′W
Becharof, l., Ak., U.S. (bĕk-ȧ-rôf)	40	57°58′N	156°58′W
Be'er Sheva', Isr. (bēr-shē′bȧ)	67a	31°15′N	34°48′E
Be'er Sheva', r., Isr.	67a	31°23′N	34°30′E
Beestekraal, S. Afr.	87c	25°22′S	27°34′E
Bega, Austl. (bā′gȧ)	81	36°50′S	149°49′E
Behala, India	72a	22°31′N	88°19′E
Behbehān, Iran	71	30°35′N	50°14′E
Beihai, China (bā-hī)	76	21°30′N	109°10′E
Beijing, China	77	39°55′N	116°23′E
Beirut, Leb. (bā-rōōt′)	68	33°53′N	35°30′E
Beja, Port. (bā′zhä)	54	38°03′N	7°53′W
Béja, Tun.	54	36°52′N	9°20′E
Bejestān, Iran	68	34°30′N	58°22′E
Békéscsaba, Hung. (bā′kāsh-chô′bô)	55	46°39′N	21°06′E
Belarus, nation, Eur.	56	53°30′N	25°33′E
Belau see Palau, nation, Oc.	3	7°15′N	134°30′E
Belawan, Indon. (bā-lä′wän)	74	3°43′N	98°43′E
Belfast, N. Ire., U.K.	50	54°36′N	5°45′W
Belgaum, India	69	15°57′N	74°32′E
Belgium, nation, Eur. (bĕl′jĭ-ŭm)	50	51°00′N	2°52′E
Belgrade (Beograd), Serb.	50	44°48′N	20°32′E
Belitung, i., Indon.	74	3°30′S	107°30′E
Belize, nation, N.A.	42	17°00′N	88°40′W
Belize City, Belize (bĕ-lēz′)	42	17°31′N	88°10′W
Bellary, India (bĕl-lä′rē)	69	15°15′N	76°56′E
Bellingham, Wa., U.S.	38	48°46′N	122°29′W
Bellingshausen Sea, sea, Ant. (bĕl′ĭngz houz′n)	82	72°00′S	80°30′W
Bellpat, Pak.	72	29°08′N	68°00′E

116 Belmopan – Cabugao PRONOUNCING INDEX

PLACE (Pronunciation)	PAGE	LAT.	LONG.
Belmopan, Belize	42	17°15'N	88°47'W
Belogorsk, Russia	57	51°09'N	128°32'E
Belo Horizonte, Braz. (bĕ'lôre-sô'n-tĕ)	46	19°54'S	43°56'W
Beloit, Wi., U.S.	39	42°31'N	89°04'W
Belukha, Mount, mtn., Asia	56	49°47'N	86°23'E
Belyy, i., Russia	56	73°19'N	72°00'E
Benalla, Austl. (bĕn-ăl'à)	81	36°30'S	146°00'E
Benares *see* Vārānasi, India	69	25°25'N	83°00'E
Benavente, Spain (bā-nä-vĕn'tä)	54	42°01'N	5°43'W
Bend, Or., U.S. (bĕnd)	38	44°04'N	121°17'W
Bendeleben, Mount, mtn., Ak., U.S. (bĕn-dĕl-bĕn)	40	65°18'N	163°45'W
Bender Beyla, Som.	87a	9°40'N	50°45'E
Bendigo, Austl. (bĕn'dĭ-gō)	81	36°39'S	144°20'E
Benevento, Italy (bā-nā-vĕn'tō)	54	41°08'N	14°46'E
Bengal, Bay of, b., Asia (bĕn-gôl')	66	17°30'N	87°00'E
Bengbu, China (bŭŋ-bōō)	77	32°52'N	117°22'E
Benghazi *see* Banghāzī, Libya	85	32°07'N	20°04'E
Bengkalis, Indon. (bĕng-kä'lĭs)	74	1°29'N	102°06'E
Bengkulu, Indon.	74	3°46'S	102°18'E
Benin, nation, Afr.	85	8°00'N	2°00'E
Benut, r., Malay.	67b	1°43'N	103°20'E
Beograd *see* Belgrade, Serb.	50	44°48'N	20°32'E
Berakit, Tanjung, c., Indon.	67b	1°16'N	104°44'E
Berau, Teluk, b., Indon.	75	2°22'S	131°40'E
Berbera, Som. (bûr'bûr-*à*)	87a	10°25'N	45°05'E
Berdychiv, Ukr.	56	49°53'N	28°32'E
Berëzovo, Russia (bĭr-yô'zĕ-vû)	56	64°10'N	65°10'E
Bergama, Tur. (bĕr'gä-mä)	68	39°08'N	27°09'E
Bergamo, Italy (bĕr'gä-mō)	54	45°43'N	9°41'E
Bergen, Nor.	50	60°24'N	5°20'E
Berhampur, India	69	19°19'N	84°48'E
Bering Sea, sea, (bē'rĭng)	88	58°00'N	175°00'W
Bering Strait, strt.,	38a	64°50'N	169°50'W
Berkeley, Ca., U.S. (bûrk'lĭ)	38	37°52'N	122°17'W
Berlin, Ger. (bĕr-lēn')	50	52°31'N	13°28'E
Bermuda, dep., N.A.	43	32°20'N	65°45'W
Bern, Switz. (bĕrn)	50	46°55'N	7°25'E
Berriyyane, Alg.	54	32°50'N	3°49'E
Besar, Gunong, mtn., Malay.	67b	2°31'N	103°09'E
Beslan, Russia	58	43°12'N	44°33'E
Bethal, S. Afr. (bĕth'*ă*l)	87c	26°27'S	29°28'E
Bethlehem *see* Bayt Lahm, W.B.	67a	31°42'N	35°13'E
Bet She'an, Isr.	67a	32°30'N	35°30'E
Bettles Field, Ak., U.S. (bĕt'tŭls)	40	66°58'N	151°48'W
Betwa, r., India (bĕt'wȧ)	69	25°00'N	78°00'E
Beypazari, Tur. (bā-pȧ-zä'rĭ)	55	40°10'N	31°40'E
Bhadreswar, India	72a	22°49'N	88°22'E
Bhāgalpur, India (bä'g*ŭ*l-pôr)	69	25°15'N	86°59'E
Bhamo, Mya. (bŭ-mō')	69	24°00'N	96°15'E
Bhāngar, India	72a	22°30'N	88°36'E
Bharatpur, India (bĕrt'pôr)	69	27°21'N	77°33'E
Bhatinda, India (b*ŭ*-tĭn-dȧ)	69	30°19'N	74°56'E
Bhātpāra, India	69	22°52'N	88°24'E
Bhaunagar, India (bäv-nŭg'*ŭ*r)	69	21°45'N	72°58'E
Bhayandar, India	73b	19°20'N	72°50'E
Bhilai, India	72	21°14'N	81°23'E
Bhīma, r., India (bē'mȧ)	69	18°00'N	74°45'E
Bhiwandi, India	73b	19°18'N	73°03'E
Bhiwāni, India	72	28°53'N	76°08'E
Bhopāl, India (bō-päl)	69	23°20'N	77°25'E
Bhubaneswar, India (bō-bŭ-năsh'vûr)	69	20°21'N	85°53'E
Bhuj, India (bōōj)	69	23°22'N	69°39'E
Bhutan, nation, Asia (bōō-tän')	69	27°15'N	90°30'E
Biak, i., Indon. (bē'äk)	75	1°00'S	136°00'E
Bialystok, Pol. (byä-wĭs'tŏk)	50	53°08'N	23°12'E
Bien Hoa, Viet.	74	10°59'N	106°49'E
Big Delta, Ak., U.S. (bĭg dĕl'*tà*)	40	64°08'N	145°48'W
Bihār, state, India (bē-här')	69	25°30'N	87°00'E
Bijāpur, India	73	16°53'N	75°42'E
Bīkaner, India (bĭ-kă'nûr)	69	28°07'N	73°19'E
Bilāspur, India (bē-läs'pōōr)	69	22°08'N	82°12'E
Bilauktaung, mts., Asia	74	14°40'N	98°50'E
Bilbao, Spain (bĭl-bä'ō)	50	43°12'N	2°48'W
Bilbays, Egypt	87b	30°26'N	31°37'E
Bilecik, Tur. (bē-lĕd-zhĕk')	55	40°10'N	29°58'E
Billings, Mt., U.S. (bĭl'ĭngz)	38	45°47'N	108°29'W
Biloxi, Ms., U.S. (bĭ-lŏk'sĭ)	39	30°24'N	88°50'W
Bilqās Qism Awwal, Egypt	87b	31°14'N	31°25'E
Binalonan, Phil. (bē-nä-lō'nän)	75a	16°03'N	120°35'E
Binghamton, N.Y., U.S. (bĭng'*ă*m-t*ŭ*n)	39	42°05'N	75°55'W
Binjai, Indon.	74	3°59'N	108°00'E
Bintan, i., Indon. (bĭn'tän)	67b	1°09'N	104°43'E
Bintulu, Malay. (bēn'tōō-lōō)	74	3°07'N	113°06'E
Birātnagar, Nepal (bĭ-rät'nŭ-gŭr)	72	26°35'N	87°18'E
Birdsville, Austl. (bûrdz'vĭl)	80	25°50'S	139°31'E
Birdum, Austl. (bûrd'*ŭ*m)	80	15°45'S	133°25'E
Birecik, Tur. (bē-rĕd-zhĕk')	55	37°10'N	37°50'E
Bīrjand, Iran (bēr'jänd)	68	33°07'N	59°16'E
Birmingham, Eng., U.K.	50	52°29'N	1°53'W
Birmingham, Al., U.S. (bûr'mĭng-häm)	39	33°31'N	86°49'W
Birobidzhan, Russia (bē'rō-bē-jän')	57	48°42'N	133°28'E
Birsk, Russia (bĭrsk)	56	55°25'N	55°30'E
Bi'r Za'farānah, Egypt	67a	29°07'N	32°38'E
Bisbee, Az., U.S. (bĭz'bē)	38	31°30'N	109°55'W
Biscay, Bay of, b., Eur. (bĭs'kā')	52	45°19'N	3°51'W
Bishkek, Kyrg.	59	42°49'N	74°42'E
Bismarck, N.D., U.S. (bĭz'märk)	38	46°48'N	100°46'W
Bismarck Archipelago, is., Pap. N. Gui.	75	3°15'S	150°45'E
Bismarck Range, mts., Pap. N. Gui.	75	5°15'S	144°15'E
Bissau, Gui.-B. (bē-sa'ōō)	85	11°51'N	15°35'W
Bistrita, Rom. (bĭs-trĭt-sȧ)	55	47°09'N	24°29'E
Bitlis, Tur. (bĭt-lēs')	68	38°30'N	42°00'E
Biysk, Russia (bēsk)	56	52°32'N	85°28'E
Black, r., Asia	74	21°00'N	103°30'E
Blackall, Austl. (blăk'*ŭ*l)	81	24°23'S	145°37'E
Blackburn Mount, mtn., Ak., U.S.	40	61°50'N	143°12'W
Black Hills, mts., U.S.	38	44°08'N	103°47'W
Black Sea, sea,	53	43°01'N	32°16'E
Blagoveshchensk, Russia (blä'gō-vyĕsh'chĕnsk)	57	50°16'N	127°47'E
Blanc, Cap, c., Afr.	85	20°39'N	18°08'W
Blanc, Mont, mtn., Eur. (môɴ bläɴ)	52	45°50'N	6°53'E
Bloomington, Il., U.S.	39	40°30'N	89°00'W
Bluefields, Nic. (blōō'fēldz)	43	12°03'N	83°45'W
Blue Nile, r., Afr.	85	12°30'N	34°00'E
Blue Ridge, mtn., U.S. (blōō rĭj)	39	35°30'N	82°50'W
Blumut, Gunong, mtn., Malay.	67b	2°03'N	103°34'E
Boac, Phil.	75a	13°26'N	121°50'E
Bodaybo, Russia (bō-dī'bō)	57	57°12'N	114°46'E
Bodensee, l., Eur. (bō'dĕn zā)	52	47°48'N	9°22'E
Bogor, Indon.	74	6°45'S	106°45'E
Bogotá, Col.	46	4°36'N	74°05'W
Bogotol, Russia (bŏ'gō-tŏl)	57	56°15'N	89°45'E
Bo Hai, b., China	77	38°30'N	120°00'E
Bohemian Forest, mts., Eur. (bō-hē'mĭ-*ă*n)	52	49°35'N	12°27'E
Bohol, i., Phil. (bō-hôl')	75	9°28'N	124°35'E
Boise, Id., U.S. (boi'zē)	38	43°38'N	116°12'W
Bojnūrd, Iran	68	37°29'N	57°13'E
Bolan, mtn., Pak. (bō-län')	72	30°13'N	67°09'E
Bolan Pass, p., Pak.	69	29°50'N	67°10'E
Boli, China (bwo-lē)	77	45°40'N	130°38'E
Bolinao, Phil. (bō-lē-nä'ō)	75a	16°24'N	119°53'E
Bolivia, nation, S.A. (bō-lĭv'ĭ-ȧ)	46	17°00'S	64°00'W
Bologna, Italy (bō-lōn'yä)	50	44°30'N	11°18'E
Bol'shoy Begichëv, i., Russia	57	74°30'N	114°40'E
Bolu, Tur. (bō'lŏ)	55	40°45'N	31°45'E
Bolvadin, Tur. (bŏl-vä-dēn')	55	38°50'N	30°50'E
Bolzano, Italy (bŏl-tsä'nō)	54	46°31'N	11°22'E
Bombala, Austl. (bŭm-bä'lȧ)	81	36°55'S	149°07'E
Bombay *see* Mumbai, India	69	18°58'N	72°50'E
Bombay Harbour, b., India	73b	18°55'N	72°52'E
Bon, Cap, c., Tun. (bôɴ)	54	37°04'N	11°13'E
Bonaire, i., Neth. Ant. (bō-nâr')	43	12°10'N	68°15'W
Bondoc Peninsula, pen., Phil. (bōn-dŏk')	75a	13°24'N	122°30'E
Bone, Teluk, b., Indon.	74	4°09'S	121°00'E
Bonifacio, Strait of, strt., Eur.	54	41°14'N	9°02'E
Bonin Islands, is., Japan (bō'nĭn)	89	26°30'N	141°00'E
Bonn, Ger. (bōn)	50	50°44'N	7°06'E
Bonthain, Indon. (bōn-tīn')	74	5°30'S	119°52'E
Bontoc, Phil. (bŏn-tŏk')	75a	17°10'N	121°01'E
Boons, S. Afr.	87c	25°59'S	27°15'E
Boorama, Som.	87a	10°05'N	43°08'E
Boosaaso, Som.	87a	11°19'N	49°10'E
Boothia Peninsula, pen., Can.	37	73°30'N	95°00'W
Borāzjān, Iran (bō-räz-jän')	68	29°13'N	51°13'E
Bordeaux, Fr. (bôr-dō')	50	44°50'N	0°37'W
Bordj-bou-Arréridj, Alg. (bōrj-bōō-ȧ-rā-rēj')	54	36°03'N	4°48'E
Borisoglebsk, Russia (bŏ-rē sō-glyĕpsk')	56	51°20'N	42°00'E
Borivli, India	73b	19°15'N	72°48'E
Borneo, i., Asia	74	0°25'N	112°39'E
Bornholm, i., Den. (bôrn-hôlm)	52	55°16'N	15°15'E
Borovichi, Russia (bŏ-rō-vē'chē)	56	58°22'N	33°56'E
Borraan, Som.	87a	10°38'N	48°30'E
Borriana, Spain	54	39°53'N	0°05'W
Borroloola, Austl. (bŏr-rō-lōō'*lȧ*)	80	16°15'S	136°19'E
Borūjerd, Iran	68	33°45'N	48°53'E
Borzya, Russia (bôrz'yȧ)	57	50°37'N	116°53'E
Boshan, China (bwo-shan)	77	36°32'N	117°51'E
Bosnia and Herzegovina, nation, Eur.	50	44°15'N	17°30'E
Bosporus *see* İstanbul Boğazı, strt., Tur.	68	41°10'N	29°10'E
Bosten Hu, l., China (bwo-stŭn hōō)	76	42°06'N	88°01'E
Boston, Ma., U.S.	39	42°15'N	71°07'W
Bothaville, S. Afr. (bō'tȧ-vĭl)	87c	27°24'S	26°38'E
Bothnia, Gulf of, b., Eur. (bŏth'nĭ-ȧ)	52	63°40'N	21°30'E
Botswana, nation, Afr. (bŏtswănă)	85	22°10'S	23°13'E
Bougainville, i., Pap. N. Gui.	78e	6°00'S	155°00'E
Bougainville Trench, deep, (bōō-gāɴ-vēl')	89	7°00'S	152°00'E
Bouïra, Alg. (boo-ē'rȧ)	54	36°25'N	3°55'E
Boulder, Co., U.S.	38	40°02'N	105°19'W
Boulder City, Nv., U.S.	38	35°57'N	114°50'W
Bounty Islands, is., N.Z.	5	47°42'S	179°05'E
Bourail, N. Cal.	78f	21°34'S	165°30'E
Bourke, Austl. (bŭrk)	81	30°10'S	146°00'E
Bou Saâda, Alg. (bōō-sä'dä)	54	35°13'N	4°17'E
Bouvetøya, i., Ant.	3	55°00'S	3°00'E
Bowen, Austl. (bō'ĕn)	81	20°02'S	148°14'E
Bowling Green, Ky., U.S. (bōlĭng grēn)	39	37°00'N	86°26'W
Bozeman, Mt., U.S. (bōz'm*ă*n)	38	45°41'N	111°00'W
Braga, Port. (brä'gä)	54	41°20'N	8°25'W
Brahmaputra, r., Asia (bräˈmȧ-pōō'trȧ)	69	26°45'N	92°45'E
Brāhui, mts., Pak.	69	28°32'N	66°15'E
Brăila, Rom. (brē'ēlȧ)	50	45°15'N	27°58'E
Brandfort, S. Afr. (brăn'd-fôrt)	87c	28°42'S	26°29'E
Brasília, Braz. (brä-sē'lvä)	46	15°49'S	47°39'W
Braşov, Rom.	55	45°39'N	25°35'E
Bratislava, Slvk. (brä'tĭs-lä-vä)	50	48°09'N	17°07'E
Bratsk, Russia (brätsk)	57	56°10'N	102°04'E
Bratskoye Vodokhranilishche, res., Russia	57	56°10'N	102°05'E
Brawley, Ca., U.S. (brô'lĭ)	38	32°59'N	115°32'W
Brazil, nation, S.A.	46	9°00'S	53°00'W
Brazilian Highlands, mts., Braz. (brȧ zĭl y*ă*n hī-lăndz)	47	14°00'S	48°00'W
Brazzaville, Congo (brȧ-zȧ-vēl')	85	4°16'S	15°17'E
Bremen, Ger. (brā-m*ĕ*n)	50	53°05'N	8°50'E
Brescia, Italy (brā'shä)	54	45°33'N	10°15'E
Brest, Bela.	56	52°06'N	23°43'E
Brest, Fr. (brĕst)	50	48°24'N	4°30'W
Brewarrina, Austl. (brōō-ĕr-rē'n*ȧ*)	81	29°54'S	146°50'E
Bridgeport, Ct., U.S.	39	41°12'N	73°12'W
Bridgetown, Barb. (brĭj' toun)	43	13°08'N	59°37'W
Brindisi, Italy (brēn'dē-zē)	50	40°38'N	17°57'E
Brisbane, Austl. (brĭz'bān)	81	27°30'S	153°10'E
Bristol, Tn., U.S.	39	36°35'N	82°10'W
Bristol, Va., U.S.	39	36°36'N	82°00'W
Bristol Bay, b., Ak., U.S.	40	58°05'N	158°54'W
British Indian Ocean Territory, dep., Afr.	2	7°00'S	72°00'E
British Isles, is., Eur.	52	54°00'N	4°00'W
Brits, S. Afr.	87c	25°39'S	27°47'E
Brno, Czech Rep. (b'r'nô)	50	49°18'N	16°37'E
Broach, India	72	21°47'N	72°58'E
Broken Hill, Austl. (brōk'*ĕ*n)	81	31°55'S	141°35'E
Bronkhorstspruit, S. Afr.	87c	25°50'S	28°48'E
Brooks Range, mts., Ak., U.S. (brŏŏks)	40	68°20'N	159°00'W
Broome, Austl. (brōōm)	80	18°00'S	122°15'E
Brownsville, Tx., U.S.	38	25°55'N	97°30'W
Brownwood, Tx., U.S. (broun'wŏŏd)	38	31°44'N	98°58'W
Brunei, nation, Asia (brō-nī')	74	4°52'N	113°38'E
Brunswick, Ga., U.S. (brŭnz'wĭk)	39	31°08'N	81°30'W
Brussels, Bel.	50	50°51'N	4°21'E
Bruxelles *see* Brussels, Bel.	50	50°51'N	4°21'E
Bryansk, Russia	56	53°15'N	34°22'E
Buala, Sol. Is.	78e	8°08'S	159°35'E
Buatan, Indon.	67b	0°45'N	101°49'E
Bucharest, Rom.	50	44°23'N	26°10'E
Buckingham, can., India (bŭk'ĭng-*ă*m)	73	15°18'N	79°50'E
Bucureşti *see* Bucharest, Rom.	50	44°23'N	26°10'E
Budapest, Hung. (bōō'dȧ-pĕsht')	50	47°30'N	19°05'E
Budge Budge, India	72a	22°28'N	88°08'E
Budyonnovsk, Russia	58	44°46'N	44°09'E
Buenos Aires, Arg. (bwā'nōs ī'rās)	46	34°20'S	58°30'W
Buffalo, N.Y., U.S.	39	42°54'N	78°51'W
Bugul'ma, Russia (bō-gōōl'mȧ)	56	54°40'N	52°40'E
Buguruslan, Russia (bō-gō-rŏs-lȧn')	56	53°30'N	52°32'E
Buhi, Phil. (bōō'ē)	75a	13°26'N	123°31'E
Buir Nur, l., Asia (bōō-ēr nōōr)	77	47°50'N	117°00'E
Buka Island, i., Pap. N. Gui.	78e	5°15'S	154°35'E
Bukhara, Uzb. (bō-kä'rä)	59	39°31'N	64°22'E
Bukitbatu, Indon.	67b	1°25'N	101°58'E
Bukittinggi, Indon.	74	0°25'S	100°28'E
Bula, Indon. (bōō'lä)	75	3°00'S	130°30'E
Bulalacao, Phil. (bōō-lȧ-lä'kä-ō)	75a	12°30'N	121°20'E
Buldir, i., Ak., U.S. (bŭl dĭr)	40a	52°22'N	175°50'E
Bulgaria, nation, Eur. (bŏl-gā'rĭ-ȧ)	50	42°12'N	24°13'E
Bultfontein, S. Afr. (bŏlt'fŏn-tān')	87c	28°18'S	26°10'E
Bulun, Russia (bōō-lōōn')	57	70°48'N	127°27'E
Buna, Pap. N. Gui. (bōō'nȧ)	75	8°58'S	148°38'E
Bunbury, Austl. (bŭn'b*ŭ*rĭ)	80	33°25'S	115°45'E
Bundaberg, Austl. (bŭn'd*ȧ*-bûrg)	81	24°45'S	152°18'E
Bunguran Utara, Kepulauan, is., Indon.	74	3°22'N	108°00'E
Buor Khaya, Mys, c., Russia	57	71°47'N	133°22'E
Buraydah, Sau. Ar.	68	26°23'N	44°14'E
Burco, Som.	87a	9°20'N	45°45'E
Burdur, Tur. (bōōr-dōr')	55	37°50'N	30°15'E
Burdwān, India (bŏd-wän')	69	23°29'N	87°53'E
Bureinskiy, Khrebet, mts., Russia	57	51°15'N	133°30'E
Bureya, Russia (bŏrā'*ȧ*)	57	49°55'N	130°00'E
Burgas, Blg. (bŏr-gäs')	55	42°29'N	27°30'E
Burgas, Gulf of, b., Blg.	55	42°30'N	27°40'E
Burgos, Phil.	75a	16°03'N	119°52'E
Burgos, Spain (bōō'r-gōs)	54	42°20'N	3°44'W
Burhānpur, India (bŏr'hȧn-pōōr)	69	21°26'N	76°08'E
Burias Island, i., Phil. (bōō'rē-äs)	75a	12°56'N	122°56'E
Burias Pass, strt., Phil. (bōō'rē-äs)	75a	13°04'N	123°11'E
Burketown, Austl. (bûrk'toun)	80	17°50'S	139°30'E
Burkina Faso, nation, Afr.	85	13°00'N	2°00'W
Burlington, Ia., U.S.	39	40°48'N	91°05'W
Burlington, Vt., U.S.	39	44°30'N	73°15'W
Burma *see* Myanmar, nation, Asia	64	21°00'N	95°15'E
Burnie, Austl. (bûr'nē)	81	41°15'S	146°05'E
Bursa, Tur. (bōōr'sȧ)	68	40°10'N	28°10'E
Būr Sūdān, Sudan (sōō-dȧn')	85	19°30'N	37°10'E
Buru, i., Indon.	75	3°30'S	126°30'E
Burullus, l., Egypt	87b	31°20'N	30°58'E
Burundi, nation, Afr.	85	3°00'S	29°30'E
Būsh, Egypt (bōōsh)	87b	29°13'N	31°08'E
Busselton, Austl. (bûs'l-t*ŭ*n)	80	33°40'S	115°30'E
Busuanga, i., Phil. (bōō-swäŋ'gä)	75a	12°20'N	119°43'E
Butte, Mt., U.S. (būt)	38	46°00'N	112°31'W
Butuan, Phil. (bōō-tōō'än)	75	8°40'N	125°33'E
Buuhoodle, Som.	87a	8°15'N	46°20'E
Buulo Berde, Som.	87a	3°53'N	45°30'E
Buy, Russia (bwē)	56	58°30'N	41°48'E
Büyükmenderes, r., Tur.	68	37°50'N	28°20'E
Buzuluk, Russia (bō-zō-lōk')	56	52°50'N	52°10'E
Byblos *see* Jubayl, Leb.	67a	34°07'N	35°38'E
Byelorussia *see* Belarus, nation, Eur.	56	53°30'N	25°33'E

C

PLACE (Pronunciation)	PAGE	LAT.	LONG.
Cabagan, Phil. (kä-bä-gän')	75a	17°27'N	121°50'E
Cabalete, i., Phil. (kä-bä-lā'tȧ)	75a	14°19'N	122°00'E
Cabanatuan, Phil. (kä-bä-nä-twän')	75a	15°30'N	120°56'E
Cabarruyan, i., Phil. (kä-bä-rōō'yän)	75a	16°21'N	120°10'E
Cabra, i., Phil.	75a	13°55'N	119°55'E
Cabugao, Phil. (kä-bōō'gä-ō)	75a	17°48'N	120°28'E

ăt; finȧl; rāte; senȧte; ärm; ȧsk; sofȧ; fâre; ch-choose; dh-as th in other; bē; ėvent; bĕt; recĕnt; cratẽr; g-gō; gh-guttural g; bĭt; ĭ-short neutral; rīde; ᴋ-guttural k as ch in German ich;

PRONOUNCING INDEX Cáceres – Ciudad Rodrigo 117

PLACE (Pronunciation)	PAGE	LAT.	LONG.
Cáceres, Spain (kä′thå-räs)	54	39°28′N	6°20′W
Cadale, Som.	87a	2°45′N	46°15′E
Cádiz, Spain (kà′dēz)	50	36°34′N	6°20′W
Cádiz, Golfo de, b., Spain (gôl-fô-dĕ-kà′dēz)	54	36°50′N	7°00′W
Cagayan, Phil. (kä-gä-yän′)	75	8°13′N	124°30′E
Cagayan, r., Phil.	74	16°45′N	121°55′E
Cagayan Islands, is., Phil.	74	9°40′N	120°30′E
Cagayan Sulu, i., Phil. (kä-gä-yän sōō′lōō)	74	7°00′N	118°30′E
Cagliari, Italy (käl′yä-rē)	50	39°16′N	9°08′E
Cagliari, Golfo di, b., Italy (gôl-fô-dē-käl′yä-rē)	54	39°08′N	9°12′E
Caguas, P.R. (kä′gwäs)	43b	18°12′N	66°01′W
Caicos Islands, is., T./C. Is.	43	21°45′N	71°50′W
Caiman Point, c., Phil. (kī′män)	75a	15°56′N	119°33′E
Caimito, r., Pan. (kä-ē-mē′tô)	42a	8°50′N	79°45′W
Cairns, Austl. (kârnz)	81	17°02′S	145°49′E
Cairo, Egypt	70	30°00′N	31°17′E
Cairo, Il., U.S.	39	36°59′N	89°11′W
Calaguas Islands, is., Phil. (kä-läg′wäs)	75a	14°30′N	123°06′E
Calahorra, Spain (kä-lä-ôr′rä)	54	42°18′N	1°58′W
Calais, Fr. (kà-lĕ′)	50	50°56′N	1°51′E
Calais, Me., U.S.	39	45°11′N	67°15′W
Calamba, Phil. (kä-läm′bä)	75a	14°12′N	121°10′E
Calamian Group, is., Phil. (kä-lä-myän′)	74	12°14′N	118°38′E
Calapan, Phil. (kä-lä-pän′)	75a	13°25′N	121°11′E
Călăraşi, Rom. (kŭ-lŭ-rȧsh′ĭ)	55	44°09′N	27°20′E
Calatayud, Spain (kä-lä-tä-yōōdh′)	54	41°23′N	1°37′W
Calauag Bay, b., Phil.	75a	14°07′N	122°10′E
Calavite, Cape, c., Phil. (kä-lä-vē′tä)	75a	13°29′N	120°00′E
Calcutta *see* Kolkata, India (kăl-kŭt′ȧ)	69	22°32′N	88°22′E
Calexico, Ca., U.S. (kȧ-lĕk′sĭ-kō)	38	32°41′N	115°30′W
Calgary, Can. (kăl′gȧ-rĭ)	36	51°03′N	114°05′W
California, state, U.S.	38	38°10′N	121°20′W
California, Golfo de, b., Mex. (gôl-fô-dĕ-kä-lê-fôr-nyä)	42	30°30′N	113°45′W
Calimere, Point, c., India	73	10°20′N	80°20′E
Caltagirone, Italy (käl-tä-jē-rō′nā)	54	37°14′N	14°32′E
Caltanissetta, Italy (käl-tä-nê-sĕt′tä)	54	37°30′N	14°02′E
Caluula, Som.	87a	11°53′N	50°40′E
Camagüey, Cuba (kä-mä-gwā′)	43	21°25′N	78°00′W
Ca Mau, Mui, c., Viet.	74	8°36′N	104°43′E
Cambay, India (kăm-bā′)	72	22°22′N	72°39′E
Cambodia, nation, Asia	74	12°15′N	104°00′E
Camden, N.J., U.S.	39	39°56′N	75°06′W
Cameroon, nation, Afr.	85	5°48′N	11°00′E
Cameroon Mountain, mtn., Cam.	85	4°12′N	9°11′E
Camiling, Phil. (kä-mê-lĭng′)	75a	15°42′N	120°24′E
Camooweal, Austl.	80	20°00′S	138°13′E
Campbell, is., N.Z.	3	52°30′S	169°00′E
Campbellpore, Pak.	72	33°49′N	72°24′E
Campeche, Mex. (käm-pā′chä)	42	19°51′N	90°32′W
Campeche, state, Mex.	42	18°55′N	90°20′W
Campeche, Bahía de, b., Mex. (bä-ē′ä-dĕ-käm-pā′chā)	42	19°30′N	93°40′W
Câmpulung, Rom.	55	45°15′N	25°03′E
Canada, nation, N.A. (kăn′ȧ-dȧ)	36	50°00′N	100°00′W
Çanakkale, Tur. (chä-näk-kä′lĕ)	55	40°10′N	26°26′E
Çanakkale Boğazi (Dardanelles), strt., Tur.	55	40°05′N	25°50′E
Cananea, Mex. (kä-nä-nĕ′ä)	42	31°00′N	110°20′W
Canarias, Islas (Canary Is.), is., Spain (ē′s-läs-kä-nä′ryäs)	86	29°15′N	16°30′W
Canary Islands *see* Canarias, Islas, is., Spain	86	29°15′N	16°30′W
Canaveral, Cape, c., Fl., U.S.	39	28°30′N	80°23′W
Canberra, Austl. (kăn′bĕr-ȧ)	81	35°21′S	149°10′E
Candelaria, Phil. (kän-då-lä′rê-ä)	75a	15°39′N	119°55′E
Candia *see* Irákleio, Grc.	50	35°20′N	25°10′E
Candle, Ak., U.S. (kăn′d′l)	40	65°00′N	162°04′W
Candon, Phil. (kän-dōn′)	75a	17°13′N	120°26′E
Cantabrica, Cordillera, mts., Spain	52	43°05′N	6°05′W
Canton *see* Guangzhou, China	77	23°07′N	113°15′E
Canton, Oh., U.S.	39	40°50′N	81°25′W
Capalonga, Phil. (kä-pä-lôŋ′gä)	75a	14°20′N	122°30′E
Cape Breton, i., Can. (kāp brĕt′ŭn)	36	45°48′N	59°50′W
Cape Girardeau, Mo., U.S. (jē-rär-dō′)	39	37°17′N	89°32′W
Cape Krusenstern National Monument, rec., Ak., U.S.	40	67°30′N	163°40′W
Cape Romanzof, Ak., U.S. (rō′ män zôf)	40	61°50′N	165°45′W
Cape Town, S. Afr. (kāp toun)	85	33°48′S	18°28′E
Cape Verde, nation, Afr.	2	15°48′N	26°02′W
Cape York Peninsula, pen., Austl. (kāp yôrk)	81	12°30′S	142°35′E
Cap-Haïtien, Haiti (kȧp ȧ-ē-syăn′)	43	19°45′N	72°15′W
Capraia, i., Italy (kä-prä′yä)	54	43°02′N	9°51′E
Capua, Italy (kä′pwä)	54	41°07′N	14°14′E
Caracas, Ven. (kä-rä′käs)	46	10°30′N	66°58′W
Carballiño, Spain	54	42°26′N	8°04′W
Carbonara, Cape, c., Italy (kär-bō-nä′rä)	54	39°08′N	9°33′E
Cárdenas, Cuba (kär′då-näs)	43	23°00′N	81°10′W
Caribbean Sea, sea (kăr-ĭ-bē′ăn)	43	14°30′N	75°30′W
Carletonville, S. Afr.	87c	26°20′S	27°23′E
Carlisle, Eng., U.K. (kär-līl′)	50	54°54′N	3°03′W
Carnarvon, Austl. (kär-när′vŭn)	80	24°45′S	113°45′E
Caroline Islands, is., Oc.	5	8°00′N	140°00′E
Carpathians, mts., Eur. (kär-pā′thĭ-ăn)	52	49°23′N	20°14′E
Carpaţii Meridionali (Transylvanian Alps), mts., Rom.	52	45°30′N	23°30′E
Carpentaria, Gulf of, b., Austl. (kär-pĕn-târ′ĭȧ)	80	14°45′S	138°50′E
Carrara, Italy (kä-rä′rä)	54	44°05′N	10°05′E
Carsamba, Tur. (chär-shäm′bä)	55	41°05′N	36°40′E
Carson City, Nv., U.S.	38	39°10′N	119°45′W
Cartagena, Spain (kär-tä-κē′nä)	50	37°46′N	1°00′W
Cartago, C.R.	43	9°52′N	83°56′W
Casablanca, Mor.	85	33°32′N	7°41′W
Cascade Range, mts., N.A.	36	42°50′N	122°20′W
Casiguran, Phil. (käs-sē-gōō′rän)	75a	16°15′N	122°10′E
Casiguran Sound, strt., Phil.	75a	16°02′N	121°51′E
Casper, Wy., U.S. (kăs′pẽr)	38	42°51′N	106°18′W
Caspian Depression, depr., (kăs′pĭ-ăn)	56	47°40′N	52°35′E
Caspian Sea, sea	56	40°00′N	52°00′E
Cassino, Italy (käs-sē′nō)	54	41°30′N	13°50′E
Castelló de la Plana, Spain	54	39°59′N	0°05′W
Castelo Branco, Port. (käs-tā′lò brän′kò)	54	39°48′N	7°37′W
Castro-Urdiales, Spain	54	43°23′N	3°11′W
Catanaun, Phil. (kä-tä-nä′wän)	75a	13°36′N	122°20′E
Catanduanes Island, i., Phil. (kä-tän-dwä′nĕs)	75	13°55′N	125°00′E
Catania, Italy (kä-tä′nyä)	50	37°30′N	15°09′E
Catanzaro, Italy (kä-tän-dzä′rō)	55	38°53′N	16°34′E
Catbalogan, Phil. (kät-bä-lō′gän)	75	11°45′N	124°52′E
Catoche, Cabo, c., Mex. (kä-tō′chĕ)	42	21°30′N	87°15′W
Cauayan, Phil. (kou-ä′yän)	75a	16°56′N	121°46′E
Caucasus, mts.,	56	43°20′N	42°00′E
Cauvery, r., India	69	12°00′N	77°00′E
Cavite, Phil. (kä-vē′tä)	75a	14°30′N	120°54′E
Cayenne, Fr. Gu. (kā-ĕn′)	46	4°56′N	52°18′W
Cayey, P.R.	43b	18°05′N	66°12′W
Cayman Islands, dep., N.A.	43	19°30′N	80°30′W
Cebu, Phil. (sā-bōō′)	75	10°22′N	123°49′E
Cedar Rapids, Ia., U.S.	39	42°00′N	91°43′W
Cedros, i., Mex.	42	28°10′N	115°10′W
Ceduna, Austl. (sĕ-dò′nȧ)	80	32°15′S	133°55′E
Ceel Buur, Som.	87a	4°35′N	46°40′E
Celaya, Mex. (sā-lä′yä)	42	20°33′N	100°49′W
Celebes (Sulawesi), i., Indon.	74	2°15′S	120°30′E
Celebes Sea, sea, Asia	74	3°45′N	121°52′E
Cenderawasih, Teluk, b., Indon.	75	2°20′S	135°30′E
Central, Cordillera, mts., Phil. (kôr-dēl-yĕ′rä-sĕn′träl)	75a	17°05′N	120°55′E
Central African Republic, nation, Afr.	85	7°50′N	21°00′E
Central America, reg., N.A. (ä-mĕr′ĭ-kȧ)	42	10°45′N	87°15′W
Ceram (Seram), i., Indon.	75	2°45′S	129°30′E
Cerralvo, i., Mex.	42	24°00′N	109°59′W
Cervantes, Phil. (sĕr-vän′tās)	75a	16°59′N	120°42′E
Cessnock, Austl.	81	32°58′S	151°15′E
Cetinje, Serb. (tsĕt′ĭn-yĕ)	50	42°23′N	18°55′E
Ceylon *see* Sri Lanka, nation, Asia	73	8°45′N	82°30′E
Chad, nation, Afr.	85	17°48′N	19°00′E
Chad, Lake, l., Afr.	85	13°55′N	13°40′E
Chadron, Ne., U.S. (chăd′rŭn)	38	42°50′N	103°10′W
Chāgal Hills, hills, Afg.	68	29°15′N	63°28′E
Chahar, hist. reg., China (chä-här)	77	44°25′N	115°00′E
Chalkída, Grc.	55	38°28′N	23°38′E
Chālūs, Iran	68	36°38′N	51°26′E
Chaman, Pak. (chŭm-än′)	69	30°58′N	66°21′E
Chambal, r., India (chŭm-bäl′)	69	24°30′N	75°30′E
Champaign, Il., U.S. (shăm-pān′)	39	40°10′N	88°15′W
Champdāni, India	72a	22°48′N	88°21′E
Chandīgarh, India	69	30°51′N	77°13′E
Chandrapur, India	69	19°58′N	79°21′E
Chang *see* Yangtze, r., China	77	30°30′N	117°25′E
Changchun, China (chäŋ-chòn)	77	43°55′N	125°25′E
Changde, China (chäŋ-dŭ)	77	29°00′N	111°38′E
Changning, China (chäŋ-nĭŋ)	76	24°34′N	99°49′E
Changsha, China (chäŋ-shä)	77	28°20′N	113°00′E
Changzhou, China (chäŋ-jō)	77	31°47′N	119°56′E
Chaniá, Grc.	54	35°31′N	24°01′E
Channel Islands, is., Eur. (chăn′ĕl)	52	49°15′N	3°30′W
Chanthaburi, Thai.	74	12°37′N	102°04′E
Chanute, Ks., U.S. (shȧ-nōōt′)	39	37°41′N	95°27′W
Chany, l., Russia (chä′nĕ)	56	54°15′N	77°31′E
Chao′an, China (chou-än)	77	23°48′N	116°35′E
Chao Phraya, r., Thai.	74	16°13′N	99°33′E
Chaoyang, China	77	41°32′N	120°20′E
Chapala, Lago de, l., Mex. (lä′gô-dē-chä-pä′lä)	42	20°14′N	103°02′W
Chār Borjak, Afg.	71	30°17′N	62°03′E
Charjew, Turkmen.	59	38°52′N	63°37′E
Charleston, S.C., U.S.	39	32°47′N	79°56′W
Charleston, W.V., U.S.	39	38°20′N	81°35′W
Charleville, Austl. (chär′lē-vĭl)	81	26°16′S	146°28′E
Charlotte, N.C., U.S.	39	35°15′N	80°50′W
Charlotte Amalie, V.I.U.S. (shär-lŏt′ĕ ä-mä′lĭ-ä)	43	18°21′N	64°54′W
Charlottesville, Va., U.S. (shär′lŏtz-vĭl)	39	38°00′N	78°25′W
Charlotte Waters, Austl. (shär′lŏt)	80	26°00′S	134°50′E
Chārsadda, Pak. (chŭr-sä′dä)	69a	34°17′N	71°43′E
Charters Towers, Austl. (chär′tẽrz)	81	20°03′S	146°20′E
Chatham Islands, is., N.Z.	2	44°00′S	178°00′W
Chatham Strait, strt., Ak., U.S.	40	57°00′N	134°40′W
Chattanooga, Tn., U.S. (chăt-ȧ-nōō′gȧ)	39	35°01′N	85°15′W
Chau-phu, Viet.	74	10°49′N	104°57′E
Chechnya, prov., Russia	58	43°30′N	45°50′E
Cheduba Island, i., Mya.	74	18°45′N	93°01′E
Chefoo *see* Yantai, China	77	37°32′N	121°22′E
Chelyabinsk, Russia (chĕl-yä-bênsk′)	56	55°10′N	61°25′E
Chelyuskin, Mys, c., Russia (chĕl-yòs′-kĭn)	57	77°45′N	104°45′E
Chën, Gora, mtn., Russia	57	65°13′N	142°12′E
Chenāb, r., Asia (chê-näb)	69	30°30′N	71°30′E
Chengde, China (chŭŋ-dŭ)	77	40°50′N	117°50′E
Chengdu, China (chŭŋ-dōō)	76	30°30′N	104°10′E
Chennai (Madras), India	69	13°08′N	80°15′E
Cherbourg, Fr. (shär-bòr′)	50	49°39′N	1°43′W
Cherdyn′, Russia (chĕr-dyēn′)	56	60°25′N	56°32′E
Cheremkhovo, Russia (chĕr′yĕm-kô-vō)	57	52°58′N	103°18′E
Cherepanovo, Russia (chĕr′yĕ pä-nô′vō)	56	54°13′N	83°22′E
Cherepovets, Russia (chĕr-yĕ-pô′vyĕtz)	56	59°08′N	37°59′E
Chergui, i., Tun.	54	34°50′N	11°40′E
Chergui, Chott ech, l., Alg. (chĕr gē)	54	34°12′N	0°10′W
Cherkessk, Russia	58	44°14′N	42°04′E
Cherlak, Russia (chĭr-läk′)	56	54°04′N	74°28′E
Chernivtsi, Ukr.	56	48°18′N	25°56′E
Cherskogo, Khrebet, mts., Russia	57	67°15′N	140°00′E
Chesapeake Bay, b., U.S.	39	38°20′N	76°15′W
Chëshskaya Guba, b., Russia	56	67°25′N	46°00′E
Chesnokovka, Russia (chĕs-nŏ-kôf′kȧ)	56	53°28′N	83°41′E
Chetumal, Bahía de, b., N.A. (bä-ē-ä dĕ chĕt-ōō-mäl′)	42	18°07′N	88°05′W
Cheyenne, Wy., U.S. (shī-ĕn′)	38	41°10′N	104°49′W
Chhattisgarh, state, India	69	23°00′N	83°00′E
Chhindwāra, India	72	22°08′N	78°57′E
Chiang Mai, Thai.	74	18°38′N	98°44′E
Chiang Rai, Thai.	74	19°53′N	99°48′E
Chiapas, state, Mex. (chê-ä′päs)	42	17°10′N	93°00′W
Chiatura, Geor.	58	42°17′N	43°17′E
Chiba, Japan (chē′bä)	77	35°37′N	140°08′E
Chicago, Il., U.S. (shĭ-kô-gō) (chĭ-kä′gō)	39	41°49′N	87°37′W
Chichagof, i., Ak., U.S. (chê-chä′gôf)	40	57°50′N	137°00′W
Chickasha, Ok., U.S. (chĭk′ȧ-shā)	38	35°04′N	97°56′W
Chico, r., Phil.	75a	17°33′N	121°24′E
Chieti, Italy (kyĕ′tē)	54	42°22′N	14°22′E
Chifeng, China (chr-fŭŋ)	77	42°18′N	118°52′E
Chignik, Ak., U.S. (chĭg′nĭk)	40	56°14′N	158°12′W
Chignik Bay, b., Ak., U.S.	40	56°18′N	157°22′W
Chigu Co, l., China (chr-gōō tswo)	72	28°55′N	91°47′E
Chihuahua, Mex. (chê-wä′wä)	42	28°37′N	106°06′W
Chihuahua, state, Mex.	42	29°00′N	107°30′W
Chikishlyar, Turkmen. (chê-kêsh-lyär′)	59	37°40′N	53°50′E
Chile, nation, S.A. (chē′lā)	46	35°00′S	72°00′W
Chilibre, Pan. (chē-lē′brĕ)	42a	9°09′N	79°37′W
Chilka, l., India	72	19°26′N	85°42′E
Chiloé, Isla de, i., Chile	46	42°30′S	73°55′W
Chilpancingo de los Bravo, Mex.	42	17°32′N	99°30′W
Chilung, Tai. (chĭ′lung)	77	25°02′N	121°48′E
Chimbay, Uzb. (chĭm-bī′)	59	43°00′N	59°44′E
Chimborazo, mtn., Ec. (chêm-bô-rä′zō)	46	1°35′S	78°45′W
China, nation, Asia (chī′nȧ)	76	36°45′N	93°00′E
Chindwin, r., Mya. (chĭn-dwĭn)	69	23°30′N	94°34′E
Chíos, Grc. (kē′ôs)	55	38°23′N	26°09′E
Chíos, i., Grc.	55	38°20′N	25°45′E
Chirala, India	73	15°52′N	80°22′E
Chirchik, Uzb. (chĭr-chēk′)	59	41°28′N	69°18′E
Chirikof, i., Ak., U.S. (chĭ′rĭ-kôf)	40	55°50′N	155°35′W
Chirpan, Blg.	55	42°12′N	25°19′E
Chişinău, Mol.	56	47°02′N	28°52′E
Chistopol′, Russia (chĭs-tô′pôl-y′)	56	55°21′N	50°37′E
Chita, Russia (chê-tä′)	57	52°09′N	113°39′E
Chitina, Ak., U.S. (chĭ-tē′nȧ)	40	61°28′N	144°35′W
Chitorgarh, India	72	24°59′N	74°42′E
Chitrāl, Pak. (chē-träl′)	69	35°58′N	71°48′E
Chittagong, Bngl. (chĭt-ȧ-gông′)	69	22°26′N	90°51′E
Chŏngjin, Kor., N. (chŭng-jĭn′)	77	41°48′N	129°46′E
Chongqing, China (chôŋ-chyĭŋ)	76	29°38′N	107°30′E
Chongqing, prov., China	76	30°00′N	108°00′E
Chota Nagpur, plat., India	72	23°40′N	82°50′E
Choybalsan, Mong.	77	47°50′N	114°15′E
Christchurch, N.Z. (krīst′chûrch)	81a	43°30′S	172°38′E
Christiansted, V.I.U.S.	43b	17°45′N	64°44′W
Christmas Island, dep., Oc.	74	10°35′S	105°40′E
Chuguchak, hist. reg., China (chōō′gōō-chäk′)	76	46°09′N	83°58′E
Chukotskiy Poluostrov, pen., Russia	57	66°12′N	175°00′W
Chukotskoye Nagor′ye, mts., Russia	57	66°00′N	166°00′E
Chumikan, Russia (chōō-mē-kän′)	57	54°47′N	135°09′E
Chungking *see* Chongqing, China	76	29°38′N	107°30′E
Churu, India	72	28°22′N	75°00′E
Chusovoy, Russia (chōō-sô-vôy′)	56	58°18′N	57°50′E
Chust, Uzb. (chòst)	59	41°05′N	71°28′E
Chuuk (Truk), is., Micron.	78c	7°25′N	151°47′E
Chuxiong, China (chōō-shyôŋ)	76	25°19′N	101°34′E
Cide, Tur. (jē′dĕ)	55	41°50′N	33°00′E
Ciego de Avila, Cuba (syā′gō dā ä′vê-lä)	43	21°50′N	78°45′W
Cienfuegos, Cuba (syĕn-fwā′gōs)	43	22°10′N	80°30′W
Cincinnati, Oh., U.S. (sĭn-sĭ-nát′ĭ)	39	39°08′N	84°30′W
Cirebon, Indon.	74	6°50′S	108°33′E
Ciri Grande, r., Pan. (sē′rē-grà′n′dĕ)	42a	8°55′N	80°04′W
Ciudad Camargo, Mex.	42	27°42′N	105°10′W
Ciudad Chetumal, Mex.	42	18°30′N	88°17′W
Ciudad del Carmen, Mex. (syōō-dä′d-dĕl-kà′r-mĕn)	42	18°39′N	91°49′W
Ciudad García, Mex. (syōō-dhädh′gär-sē′ä)	42	22°39′N	103°02′W
Ciudad Guzmán, Mex. (syōō-dhädh′góz-män)	42	19°40′N	103°29′W
Ciudad Juárez, Mex. (syōō-dhädh hwä′räz)	42	31°44′N	106°28′W
Ciudad Mante, Mex. (syōō-dä′d-màn′tĕ)	42	22°34′N	98°58′W
Ciudad Obregón, Mex. (syōō-dhädh-ô-brĕ-gô′n)	42	27°40′N	109°58′W
Ciudad Rodrigo, Spain (thyōō-dhädh′rô-drē′gō)	54	40°38′N	6°34′W

118 Ciudad Victoria – Djibouti PRONOUNCING INDEX

PLACE (Pronunciation)	PAGE	LAT.	LONG.
Ciudad Victoria, Mex. (syōō-dhädh′vēk-tō′rē-ä)	42	23°43′N	99°09′W
Clarksburg, W.V., U.S. (klärkz′bûrg)	39	39°15′N	80°20′W
Cleburne, Tx., U.S. (klē′bŭrn)	38	32°21′N	97°23′W
Clermont, Austl. (klĕr′mŏnt)	81	23°02′S	147°46′E
Clermont-Ferrand, Fr. (klĕr-môn′fĕr-rän′)	50	45°47′N	3°03′E
Cleveland, Oh., U.S.	39	41°30′N	81°42′W
Clocolan, S. Afr.	87c	28°56′S	27°35′E
Cloncurry, Austl. (klŏn-kŭr′ē)	80	20°58′S	140°42′E
Clovis, N.M., U.S. (klō′vĭs)	38	34°24′N	103°11′W
Cluj-Napoca, Rom.	50	46°46′N	23°34′E
Coahuila, state, Mex. (kō-ä-wē′lä)	42	27°30′N	103°00′W
Coamo, P.R. (kō-ä′mō)	43b	18°05′N	66°21′W
Coast Mountains, mts., N.A. (kōst)	36	54°10′N	128°00′W
Coast Ranges, mts., U.S.	38	41°28′N	123°30′W
Coats Land, reg., Ant.	82	74°00′S	30°00′W
Coatzacoalcos, Mex.	42	18°09′N	94°26′W
Cobán, Guat. (kō-bän′)	42	15°28′N	90°19′W
Cobar, Austl.	81	31°28′S	145°50′E
Cobh, Ire. (kōv)	50	51°52′N	8°09′W
Coco, r., N.A.	43	14°55′N	83°45′W
Coco, Isla del, i., C.R. (ē′s-lä-dĕl-kō-kō)	42	5°33′N	87°02′W
Cocoli, Pan. (kŏ-kô′lē)	42a	8°58′N	79°36′W
Cocos (Keeling) Islands, is., Oc. (kō′kōs) (kē′lĭng)	3	11°50′S	90°50′E
Coco Solito, Pan. (kô-kô-sō-lē′tō)	42a	9°21′N	79°53′W
Cod, Cape, pen., Ma., U.S.	39	41°42′N	70°15′W
Coeur d'Alene, Id., U.S. (kûr dȧ-lān′)	38	47°43′N	116°35′W
Coffeyville, Ks., U.S. (kŏf′ĭ-vĭl)	39	37°01′N	95°38′W
Coimbatore, India (kō-ēm-bȧ-tōr′)	69	11°03′N	76°56′E
Coimbra, Port. (kō-ēm′brä)	50	40°14′N	8°23′W
Coligny, S. Afr.	87c	26°20′S	26°18′E
Colima, Mex. (kōlē′mä)	42	19°13′N	103°45′W
Colima, Nevado de, mtn., Mex. (nĕ-vä′dō-dĕ-kō-lē′mä)	42	19°30′N	103°38′W
College, Ak., U.S.	40	64°43′N	147°50′W
Collie, Austl. (kŏl′ē)	80	33°20′S	116°20′E
Cologne, Ger.	50	50°56′N	6°57′E
Colombia, nation, S.A.	46	3°30′N	72°30′W
Colombo, Sri L. (kō-lŏm′bō)	73	6°58′N	79°52′E
Colón, Pan. (kŏ-lô′n)	43	9°22′N	79°54′W
Colón, Archipiélago de, is., Ec.	46	0°10′S	87°45′W
Colorado, state, U.S.	38	39°30′N	106°55′W
Colorado, r., N.A.	36	36°00′N	113°30′W
Colorado Springs, Co., U.S. (kŏl-ō-rä′dō)	38	38°49′N	104°48′W
Columbia, Mo., U.S.	39	38°55′N	92°19′W
Columbia, S.C., U.S.	39	34°00′N	81°00′W
Columbia, r., N.A.	36	46°00′N	120°00′W
Columbus, Ga., U.S. (kō-lŭm′bŭs)	39	32°29′N	84°56′W
Columbus, Oh., U.S.	39	40°00′N	83°00′W
Colville, r., Ak., U.S.	40	69°00′N	156°25′W
Comayagua, Hond. (kō-mä-yä′gwä)	42	14°24′N	87°36′W
Comilla, Bngl. (kō-mĭl′ä)	69	23°33′N	91°17′E
Comitán, Mex. (kō-mē-tän′)	42	16°16′N	92°09′W
Como, Italy (kō′mō)	54	45°48′N	9°03′E
Como, Lago di, l., Italy (lä′gō-dē-kō′mō)	54	46°00′N	9°30′E
Comorin, Cape, c., India (kŏ′mō-rĭn)	73	8°05′N	78°05′E
Comoros, nation, Afr.	85	12°30′S	42°45′E
Conakry, Gui. (kō-nȧ-krē′)	85	9°31′N	13°43′W
Concepción, Chile	46	36°51′S	72°59′W
Concepcion, Phil.	75a	15°19′N	120°40′E
Concepción, r., Mex.	42	30°25′N	112°20′W
Concepción del Oro, Mex. (kōn-sĕp-syōn′ dĕl ō′rō)	42	24°39′N	101°24′W
Conchos, r., Mex.	42	29°30′N	105°00′W
Concord, N.H., U.S.	39	43°10′N	71°30′W
Congo, nation, Afr. (kŏn′gō)	85	3°00′S	13°48′E
Congo (Zaire), r., Afr. (kŏn′gō)	86	2°00′S	17°00′E
Congo, Democratic Republic of the (Zaire), nation, Afr.	85	1°00′S	22°15′E
Congo Basin, basin, D.R.C.	86	2°47′N	20°58′E
Connecticut, state, U.S. (kŏ-nĕt′ĭ-kŭt)	39	41°40′N	73°10′W
Con Son, is., Viet.	74	8°30′N	106°28′E
Constanţa, Rom. (kōn-stän′tsȧ)	50	44°12′N	28°36′E
Cooch Behār, India (kōoch bē-här′)	69	26°25′N	89°34′E
Cook Inlet, b., Ak., U.S.	40	60°50′N	151°38′W
Cook Islands, dep., Oc.	2	20°00′S	158°00′W
Cooktown, Austl. (kōok′toun)	81	15°40′S	145°20′E
Coolgardie, Austl. (kōōl-gär′dē)	80	31°00′S	121°25′E
Cooma, Austl. (kōō′mȧ)	81	36°22′S	149°10′E
Coonamble, Austl. (kōō-năm′b'l)	81	31°00′S	148°30′E
Coonoor, India	73	10°22′N	76°15′E
Cooper Center, Ak., U.S.	40	61°54′N	15°30′W
Copenhagen (København), Den.	50	55°43′N	12°27′E
Copper, r., Ak., U.S. (kŏp′ēr)	40	62°38′N	145°00′W
Corabia, Rom. (kō-rä′bĭ-ȧ)	55	43°45′N	24°29′E
Coral Sea, sea, Oc. (kŏr′ăl)	81	13°30′S	150°00′E
Córdoba, Arg. (kôr′dō-vä)	46	30°20′S	64°03′W
Córdoba, Mex. (kô′r-dō-bä)	42	18°53′N	96°54′W
Corfu *see* Kérkyra, i., Grc.	52	39°33′N	19°36′E
Corinth *see* Kórinthos, Grc.	50	37°56′N	22°54′E
Cork, Ire. (kôrk)	50	51°54′N	8°25′W
Cornelis, r., S. Afr. (kôr-nē′lĭs)	87c	27°48′S	29°15′E
Corno, Monte, mtn., Italy (kôr′nō)	54	42°28′N	13°37′E
Coromandel Coast, cst., India (kŏr-ō-man′dĕl)	69	13°30′N	80°30′E
Corpus Christi, Tx., U.S. (kôr′pŭs krĭstē)	38	27°48′N	97°24′W
Corregidor Island, i., Phil. (kō-rā-hē-dōr′)	75a	14°21′N	120°25′E
Corrientes, Cabo, c., Mex.	42	20°25′N	105°41′W
Corsica, i., Fr. (kō′r-sē-kä)	52	42°10′N	8°55′E
Corsicana, Tx., U.S. (kôr-sĭ-kăn′ȧ)	38	32°06′N	96°28′W
Corum, Tur. (chō-rōōm′)	68	40°34′N	34°45′E
Corvallis, Or., U.S. (kŏr-văl′ĭs)	38	44°34′N	123°17′W
Cosenza, Italy (kō-zĕnt′sä)	55	39°18′N	16°15′E
Costa Rica, nation, N.A. (kŏs′tȧ rē′kȧ)	43	10°30′N	84°30′W
Cotabato, Phil. (kō-tä-bä′tō)	75	7°06′N	124°13′E
Cote d'Ivoire (Ivory Coast), nation, Afr.	85	7°43′N	6°30′W
Council Bluffs, Ia., U.S. (koun′sĭl blŭf)	39	41°16′N	95°53′W
Covington, Ky., U.S.	39	39°05′N	84°31′W
Cox's Bāzār, Bngl.	72	21°32′N	92°00′E
Cozumel, Isla de, i., Mex. (ē′s-lä-dĕ-kŏ-zōō-mĕ′l)	42	20°26′N	87°10′W
Craiova, Rom. (krȧ-yō′vȧ)	55	44°18′N	23°50′E
Crecy, S. Afr. (krĕ-sē)	87c	24°38′S	28°52′E
Cremona, Italy (krā-mō′nä)	54	45°09′N	10°02′E
Crete, i., Grc.	52	35°15′N	24°30′E
Crna Gora (Montenegro), state, Serb.	55	42°55′N	18°52′E
Croatia, nation, Eur.	50	45°24′N	15°18′E
Cross Sound, strt., Ak., U.S. (krŏs)	40	58°12′N	137°20′W
Crown Mountain, mtn., V.I.U.S.	43c	18°22′N	64°58′W
Croydon, Austl. (kroi′dŭn)	81	18°15′S	142°15′E
Crozet, Îles, is., Afr. (krō-zĕ′)	3	46°20′S	51°30′E
Cruz, Cabo, c., Cuba (kȧ′-bō-krōōz)	43	19°50′N	77°45′W
Cuba, nation, N.A. (kū′bȧ)	43	22°00′N	79°00′W
Cuddalore, India (kŭd ȧ-lōr′)	69	11°49′N	79°46′E
Cuddapah, India (kŭd′ȧ-pä)	69	14°31′N	78°52′E
Cue, Austl. (kū)	80	27°30′S	118°10′E
Cuenca, Spain	54	40°05′N	2°07′W
Cuernavaca, Mex. (kwĕr-nä-vä′kä)	42	18°55′N	99°15′W
Cuevas del Almanzora, Spain (kwĕ′väs-dĕl-äl-män-zō-rä)	54	37°19′N	1°54′W
Culebra, i., P.R. (kōō-lā′brä)	43b	18°19′N	65°32′W
Culfa, Azer.	58	38°58′N	45°38′E
Culiacán, Mex. (kōō-lyä-kȧ′n)	42	24°45′N	107°30′W
Culion, Phil. (kōō-lē-ōn′)	74	11°43′N	119°58′E
Cullera, Spain (kōō-lyā′rä)	54	39°12′N	0°15′W
Cullinan, S. Afr. (kō′lĭ-nȧn)	87c	25°41′S	28°32′E
Cumberland, Md., U.S.	39	39°40′N	78°40′W
Cunnamulla, Austl. (kŭn-ȧ-mŭl-ȧ)	81	28°00′S	145°55′E
Cupula, Pico, mtn., Mex. (pē′kō-kōō′pōō-lä)	42	24°45′N	111°10′W
Curaçao, i., Neth. Ant. (kōō-rä-sä′ō)	43	12°12′N	68°58′W
Cuttack, India (kŭ-tăk′)	69	20°38′N	85°53′E
Cuyo Islands, is., Phil. (kōō′yō)	74	10°54′N	120°08′E
Cyclades *see* Kikládes, is., Grc.	52	37°30′N	24°45′E
Cyprus, nation, Asia (sī′prŭs)	68	35°00′N	31°00′E
Czech Republic, nation, Eur.	50	50°00′N	15°00′E

D

PLACE (Pronunciation)	PAGE	LAT.	LONG.
Daba Shan, mts., China (dä-bä shän)	76	32°25′N	108°20′E
Dabie Shan, mts., China (dä-bĭē shän)	77	31°40′N	114°50′E
Dacca *see* Dhaka, Bngl.	69	23°45′N	90°29′E
Dādra & Nagar Haveli, India	69	20°00′N	73°00′E
Daet, mtn., Phil. (dä′āt)	75a	14°07′N	122°59′E
Dagupan, Phil. (dä-gōō′pän)	75a	16°02′N	120°20′E
Dahomey *see* Benin, nation, Afr.	85	8°00′N	2°00′E
Dahra, Libya	70	29°34′N	17°50′E
Dairen *see* Dalian, China	77	38°54′N	121°35′E
Dajarra, Austl. (dȧ-jăr′ȧ)	80	21°45′S	139°30′E
Dakar, Sen. (dȧ-kär′)	86	14°40′N	17°26′W
Dalälven, r., Swe.	52	60°26′N	15°50′E
Dalby, Austl. (dôl′bē)	81	27°10′S	151°15′E
Daleside, S. Afr. (dāl′sīd)	87c	26°30′S	28°03′E
Daley Waters, Austl. (dā lē)	80	16°15′N	133°30′E
Dali, China	76	26°00′N	100°08′E
Dali, China	76	35°00′N	109°38′E
Dalian, China (lů-dä)	77	38°54′N	121°35′E
Dall, i., Ak., U.S. (dăl)	40	54°50′N	133°10′W
Dallas, Tx., U.S.	38	32°45′N	96°48′W
Dalnerechensk, Russia	57	46°07′N	133°21′E
Damān, India	69	20°32′N	72°53′E
Damar, Pulau, i., Indon.	75	7°15′S	129°15′E
Damascus, Syria	68	33°30′N	36°18′E
Damāvand, Qolleh-ye, mtn., Iran	68	36°05′N	52°05′E
Dāmghān, Iran (dām-gān′)	68	35°50′N	54°15′E
Dampier, Selat, strt., Indon. (däm′pēr)	75	0°40′S	131°15′E
Da Nang, Viet.	74	16°08′N	108°22′E
Dandong, China (dän-dōŋ)	77	40°10′N	124°30′E
Dänizkänarı, Azer.	58	40°13′N	49°33′E
Danube, r., Eur.	52	43°00′N	24°00′E
Danville, Va., U.S.	39	36°35′N	79°24′W
Danzig *see* Gdańsk, Pol.	50	54°20′N	18°40′E
Daphnae, hist., Egypt	67a	30°43′N	32°12′E
Darāw, Egypt (dȧ-rä′ōō)	87b	24°24′N	32°56′E
Darbhanga, India (dŭr-bŭŋ′gä)	69	26°03′N	85°09′E
Dardanelles *see* Çanakkale Boğazi, strt., Tur.	55	40°05′N	25°50′E
Dar es Salaam, Tan. (där ĕs sȧ-läm′)	85	6°48′S	39°17′E
Dargai, Pak. (dŭr-gä′ē)	72	34°35′N	72°00′E
Darjeeling, India (dŭr-jē′lĭng)	69	27°05′N	88°16′E
Darling, r., Austl.	81	31°50′S	143°20′E
Darling Range, mts., Austl.	80	30°30′S	115°45′E
Darnley Bay, b., Ak., U.S. (därn′lē)	40	70°00′N	124°00′W
Daru, Pap. N. Gui. (dä′rōō)	75	9°04′S	143°21′E
Darwin, Austl. (där′wĭn)	80	12°25′S	131°00′E
Dashhowuz, Turkmen.	59	41°50′N	59°45′E
Dasht, r., Pak. (dŭsht)	68	25°30′N	62°30′E
Dasol Bay, b., Phil. (dä-sōl′)	75a	15°53′N	119°40′E
Dattapukur, India	72a	22°45′N	88°32′E
Datu, Tandjung, c., Asia	74	2°08′N	110°15′E
Dāvangere, India	73	14°30′N	75°55′E
Davao, Phil. (dä′vä-ō)	75	7°05′N	125°30′E
Davao Gulf, b., Phil.	75	6°30′N	125°45′E
Davenport, Ia., U.S. (dăv′ĕn-pōrt)	39	41°34′N	90°38′W
David, Pan. (dȧ-vēdh′)	43	8°27′N	82°27′W
Davis Strait, strt., N.A.	37	66°00′N	60°00′W
Dawāsir, Wādī ad, val., Sau. Ar.	68	20°48′N	44°07′E
Dawei, Mya.	74	14°04′N	98°19′E
Dawna Range, mts., Mya. (dô′nȧ)	74	17°02′N	98°01′E
Dawson Range, mts., Can.	40	62°15′N	138°10′W
Daxian, China (dä-shyĕn)	76	31°12′N	107°30′E
Dayr az Zawr, Syria (dȧ-ērēz-zōr′)	68	35°15′N	40°01′E
Dayton, Oh., U.S.	39	39°54′N	84°15′W
Daytona Beach, Fl., U.S. (dā-tō′nȧ)	39	29°11′N	81°02′W
Da Yunhe (Grand Canal), can., China (dä yŏn-hŭ)	77	35°00′N	117°00′E
Dead Sea, l., Asia	68	31°30′N	35°30′E
Deadwood, S.D., U.S. (dĕd′wŏod)	38	44°23′N	103°43′W
Death Valley, val., Ca., U.S.	38	36°30′N	117°00′W
Debao, China (dŭ-bou)	76	23°18′N	106°40′E
Debrecen, Hung. (dĕ′brĕ-tsĕn)	50	47°32′N	21°40′E
Decatur, Il., U.S.	39	39°50′N	88°59′W
Deccan, plat., India (dĕk′ăn)	69	19°05′N	76°40′E
Deganga, India	72a	22°41′N	88°41′E
Degeh Bur, Eth.	87a	8°10′N	43°25′E
Dehiwala-Mount Lavinia, Sri L.	73	6°47′N	79°55′E
Dehra Dūn, India (dā′rŭ)	69	30°09′N	78°07′E
Dej, Rom. (dāzh)	55	47°09′N	23°53′E
Delaware, state, U.S.	39	38°40′N	75°30′W
Delhi, India	69	28°54′N	77°13′E
Delhi, state, India	69	28°30′N	76°50′E
Delmas, S. Afr. (dĕl′mȧs)	87c	26°08′S	28°43′E
De-Longa, i., Russia	57	76°21′N	148°56′E
De Long Mountains, mts., Ak., U.S. (dē′lŏng)	40	68°38′N	162°30′W
Del Rio, Tx., U.S. (dĕl rē′ō)	38	29°21′N	100°52′W
Deming, N.M., U.S. (dĕm′ĭng)	38	32°15′N	107°45′W
Dempo, Gunung, mtn., Indon. (dĕm′pō)	74	4°04′S	103°11′E
Denham, Mount, mtn., Jam.	43	18°20′N	77°30′W
Deniliquin, Austl. (dĕ-nĭl′ĭ-kwĭn)	81	35°20′S	144°52′E
Denison, Tx., U.S.	38	33°45′N	97°02′W
Denizli, Tur. (dĕn-ĭz-lē′)	55	37°40′N	29°10′E
Denmark, nation, Eur.	50	56°14′N	8°30′E
Denmark Strait, strt., Eur.	37	66°30′N	27°00′W
Dennilton, S. Afr. (dĕn-ĭl-tŭn)	87c	25°18′S	29°13′E
Denpasar, Indon.	74	8°35′S	115°10′E
D'Entrecasteaux Islands, is., Pap. N. Gui. (dän-tr'-kȧs-tō′)	75	9°45′S	152°00′E
Denver, Co., U.S. (dĕn′vēr)	38	39°44′N	104°59′W
Deoli, India	72	25°52′N	75°23′E
Dera, Lach, r., Afr. (läk dā′rä)	87a	0°45′N	41°26′E
Dera Ghāzi Khān, Pak. (dā′rŭ gä-zē′ ĸan′)	69	30°09′N	70°39′E
Dera Ismāīl Khān, Pak. (dā′rŭ ĭs-mä-ēl′ ĸăn′)	72	31°55′N	70°51′E
Derby, Austl. (där′bē) (dûr′bē)	80	17°20′S	123°40′E
Derby, S. Afr. (där′bĭ)	87c	25°55′S	27°02′E
Derdepoort, S. Afr.	87c	24°39′S	26°21′E
Des Moines, Ia., U.S. (dē moin′)	39	41°35′N	93°37′W
Detroit, Mi., U.S. (dē-troit′)	39	42°22′N	83°10′W
Deva, Rom. (dā′vä)	55	45°52′N	22°52′E
Devils Lake, N.D., U.S.	38	48°10′N	98°55′W
Devon, S. Afr. (dĕv′ŭn)	87c	26°23′S	28°47′E
Devonport, Austl. (dĕv′ŭn-pôrt)	81	41°20′S	146°30′E
Dezfūl, Iran	68	32°14′N	48°37′E
Dezhnëva, Mys, c., Russia (dyĕzh′nyĭf)	66	68°00′N	172°00′W
Dhahran *see* Aẓ Ẓahrān, Sau. Ar.	68	26°13′N	50°00′E
Dhaka, Bngl. (dä′kä) (dăk′ȧ)	69	23°45′N	90°29′E
Dharamtar Creek, r., India	73b	18°49′N	72°54′E
Dharmavaram, India	73	14°32′N	77°43′E
Dhawalāgiri, mtn., Nepal	69	28°42′N	83°31′E
Dhībān, Jord.	67a	31°30′N	35°46′E
Dhule, India	69	20°58′N	74°43′E
Diablo Heights, Pan. (dyȧ′blō)	42a	8°58′N	79°34′W
Dian Chi, l., China (dĭĕn chē)	76	24°58′N	103°18′E
Dickinson, N.D., U.S. (dĭk′ĭn-sŭn)	38	46°52′N	102°49′W
Dien Bien Phu, Viet.	76	21°38′N	102°49′E
Digul, r., Indon.	75	7°00′S	140°27′E
Dijohan Point, c., Phil. (dē-kō-än)	75a	16°24′N	122°25′E
Dijon, Fr. (dē-zhŏn′)	50	47°21′N	5°02′E
Dikson, Russia (dĭk′sŏn)	56	73°30′N	80°35′E
Dili, E. Timor (dĭl′ē)	75	8°35′S	125°35′E
Di Linosa Island, i., Italy (dē-lē-nō′sä)	54	36°01′N	12°43′E
Dimashq *see* Damascus, Syria	68	33°31′N	36°18′E
Dimitrovo *see* Pernik, Blg.	55	42°36′N	23°04′E
Dimona, Isr.	67a	31°03′N	35°01′E
Dinagat Island, i., Phil.	75	10°15′N	126°15′E
Dinājpur, Bngl.	72	25°38′N	87°39′E
Dinara, mts., Serb. (dē′nä-rä)	55	43°50′N	16°15′E
Dindigul, India	73	10°25′N	78°03′E
Dingalan Bay, b., Phil. (dĭŋ-gä′län)	75a	15°19′N	121°33′E
Dingo, Austl. (dĭŋ′gō)	81	23°45′S	149°26′E
Diphu Pass, p., Asia (dĭ-pōō)	76	28°15′N	96°45′E
Dirranbandi, Austl. (dĭ-rȧ-băn′dē)	81	28°24′S	148°29′E
Disko, i., Grnld. (dĭs′kō)	37	70°00′N	54°00′W
Dispur, India	72	26°00′N	91°50′E
District of Columbia, state, U.S.	39	38°50′N	77°00′W
Disūq, Egypt (dē-sōōk′)	87b	31°07′N	30°41′E
Diu, India (dē′ōō)	69	20°48′N	70°58′E
Divilacan Bay, b., Phil. (dē-vē-lä′kän)	75a	17°26′N	122°25′E
Diyarbakir, Tur. (dē-yär-bĕk′ĭr)	68	38°00′N	40°10′E
Djedi, Oued, r., Alg.	54	34°18′N	4°39′E
Djerba, Île de, i., Tun.	54	33°53′N	11°26′E
Djibouti, Dji. (jē-bōō-tē′)	87a	11°34′N	43°00′E
Djibouti, nation, Afr.	87a	11°35′N	48°08′E

ăt; finăl; rāte; senåte; ärm; åsk; sofȧ; fâre; ch-choose; dh-as th in other; bē; ēvent; bĕt; recĕnt; cratēr; g-gō; gh-guttural g; bĭt; ĭ-short neutral; rīde; ĸ-guttural k as ch in German ich;

PRONOUNCING INDEX

Dnepropetrovsk – Fort Sandeman 119

PLACE (Pronunciation)	PAGE	LAT.	LONG.
Dnepropetrovsk *see* Dnipropetrovs'k, Ukr.	56	48°15'N	34°08'E
Dniprodzerzhyns'ke vodoskhovyshche, res., Ukr.	56	49°00'N	34°10'E
Dnipropetrovs'k, Ukr.	56	48°15'N	34°08'E
Dobbyn, Austl. (dôb'ĭn)	80	19°45'S	140°02'E
Doberai, Jazirah, pen., Indon.	75	1°25'S	133°15'E
Dobo, Indon.	75	6°00'S	134°18'E
Dobrich, Blg.	55	43°33'N	27°52'E
Dodge City, Ks., U.S. (dŏj)	38	37°44'N	100°01'W
Dodoma, Tan. (dō'dō-mȧ)	85	6°11'S	35°45'E
Doha *see* Ad Dawhah, Qatar	68	25°02'N	51°28'E
Dohad, India	72	22°52'N	74°18'E
Dominica, nation, N.A. (dŏ-mĭ-nē'kȧ)	43	15°30'N	60°45'W
Dominican Republic, nation, N.A. (dō-mĭn'ĭ-kăn)	43	19°00'N	70°45'W
Don, r., Russia	56	49°50'N	41°30'E
Dondra Head, c., Sri L.	73	5°52'N	80°52'E
Donets'k, Ukr.	56	48°00'N	37°35'E
Dong, r., China (dôŋ)	77	24°13'N	115°08'E
Dongara, Austl. (dôn-gă'rȧ)	80	29°15'S	115°00'E
Donggala, Indon. (dôn-gä'lä)	74	0°45'S	119°32'E
Dong Hoi, Viet. (dông-hô-ē')	74	17°25'N	106°42'E
Dongon Point, c., Phil. (dŏng-ôn')	75a	12°43'N	120°35'E
Dongting Hu, l., China (dôŋ-tĭŋ hōō)	77	29°10'N	112°30'E
Donostia-San Sebastián, Spain	50	43°19'N	1°59'W
Doolow, Som.	87a	4°10'N	42°05'E
Doonerak, Mount, mtn., Ak., U.S. (dōō'nē-răk)	40	68°00'N	150°34'W
Dordogne, r., Fr. (dôr-dôn'yĕ)	52	44°53'N	0°16'E
Dörgön Nuur, l., Mong.	76	47°47'N	94°01'E
Dörtyol, Tur. (dûrt'yôl)	55	36°50'N	36°20'E
Dothan, Al., U.S. (dō'thăn)	39	31°13'N	85°23'W
Douglas, Ak., U.S. (dŭg'lăs)	40	58°18'N	134°35'W
Douglas, Az., U.S.	38	31°20'N	109°30'W
Dover, S. Afr.	87c	27°05'S	27°44'E
Dover, Eng., U.K.	50	51°08'N	1°19'E
Dover, De., U.S. (dō vẽr)	39	39°10'N	75°30'W
Dover, Strait of, strt., Eur.	52	50°50'N	1°15'W
Dovre Fjell, mts., Nor. (dŏv'rĕ fyĕl')	52	62°03'N	8°36'E
Drake Passage, strt., (drāk păs'ĭj)	47	57°00'S	65°00'W
Dráma, Grc. (drä'mä)	55	41°09'N	24°10'E
Drava, r., Eur. (drä'vä)	52	45°45'N	17°30'E
Dresden, Ger. (drās'dĕn)	50	51°05'N	13°45'E
Driefontein, S. Afr.	87c	25°53'S	29°10'E
Drina, r., Serb. (drē'nä)	55	44°09'N	19°30'E
Drobeta-Turnu Severin, Rom.	55	43°54'N	24°49'E
Duarte, Pico, mtn., Dom. Rep. (dŭ'ärtĕh pēcô)	43	19°00'N	71°00'W
Dubai *see* Dubayy, U.A.E.	68	25°18'N	55°26'E
Dubayy, U.A.E.	68	25°18'N	55°26'E
Dubbo, Austl. (dŭb'ō)	81	32°20'S	148°42'E
Dublin, Ire.	50	53°20'N	6°15'W
Dubrovnik, Cro. (dŏŏ'brôv-nĕk) (rä-gōō'sä)	50	42°40'N	18°10'E
Dubuque, Ia., U.S. (dŏŏ-bŭk')	39	42°30'N	90°43'W
Duchess, Austl. (dŭch'ĕs)	80	21°30'S	139°55'E
Ducie Island, i., Pit. (dü-sē')	2	25°30'S	126°20'W
Dudinka, Russia (dōō-dĭn'kȧ)	56	69°15'N	85°42'E
Duero, r., Eur.	52	41°30'N	4°30'W
Dukhān, Qatar	71	25°25'N	50°48'E
Dulce, Golfo, b., C.R. (gōl'fô dōōl'sä)	43	8°25'N	83°13'W
Duluth, Mn., U.S. (dŏŏ-lōōth')	39	46°50'N	92°07'W
Dumai, Indon.	67b	1°39'N	101°30'E
Dumali Point, c., Phil. (dōō-mä'lē)	75a	13°07'N	121°42'E
Dum-Dum, India	72a	22°37'N	88°25'E
Dumjor, India	72a	22°37'N	88°14'E
Dundee, Scot., U.K.	50	56°30'N	2°55'W
Dunhua, China (dŏŏn-hwä)	77	43°18'N	128°10'E
Duolun, China (dwô-lōōn)	77	42°12'N	116°15'E
Dupax, Phil. (dōō'päks)	75a	16°16'N	121°06'E
Durango, Mex. (dōō-rä'n-gô)	42	24°02'N	104°42'W
Durango, state, Mex.	42	25°00'N	106°00'W
Durban, S. Afr. (dûr'băn)	85	29°48'S	31°00'E
Đurđevac, Cro.	55	46°03'N	17°03'E
Durham, N.C., U.S.	39	36°00'N	78°55'W
Durrës, Alb. (dŏŏr'ĕs)	55	41°19'N	19°27'E
Dushanbe, Taj.	59	38°30'N	68°45'E
Duyun, China (dōō-yŏŏn)	76	26°18'N	107°40'E
Dwārka, India	72	22°18'N	68°59'E
Dykhtau, Gora, mtn., Russia	58	43°03'N	43°08'E
Dzamïn Üüd, Mong.	77	44°38'N	111°32'E
Dzavhan, r., Mong.	76	48°19'N	94°08'E
Dzhalal-Abad, Kyrg. (jȧ-läl'ȧ-bät')	59	40°56'N	73°00'E
Dzhambul *see* Zhambyl, Kaz.	59	42°51'N	71°29'E
Dzhizak, Uzb. (dzhĕ'zäk)	59	40°13'N	67°58'E
Dzhugdzhur Khrebet, mts., Russia (jŏŏg-jōōr')	57	56°15'N	137°00'E
Dzungaria, reg., China (dzôŋ-gä'rĭ-ȧ)	76	44°39'N	86°13'E
Dzungarian Gate, p., Asia	76	45°00'N	88°00'E

E

PLACE (Pronunciation)	PAGE	LAT.	LONG.
Eagle Pass, Tx., U.S.	38	28°49'N	100°30'W
East, Mount, mtn., Pan.	42a	9°09'N	79°46'W
East Cape *see* Dezhnëva, Mys, c., Russia	66	68°00'N	172°00'W
East China Sea, sea, Asia	77	30°28'N	125°52'E
Easter Island *see* Pascua, Isla de, i., Chile	89	26°50'S	109°00'W
Eastern Ghāts, mts., India	69	13°50'N	78°45'E
Eastern Turkestan, hist. reg., China (tôr-kĕ-stän')(tûr-kĕ-stăn')	76	39°40'N	78°20'E
East Pakistan *see* Bangladesh, nation, Asia	69	24°15'N	90°00'E
East Saint Louis, Il., U.S.	39	38°38'N	90°10'W
East Siberian Sea, sea, Russia (sī-bĭr'y'n)	57	73°00'N	153°28'E
East Timor, nation, Asia	75	9°00'S	125°30'E
Eau Claire, Wi., U.S. (ō klâr')	39	44°47'N	91°32'W
Ebro, r., Spain (ā'brō)	52	42°00'N	2°00'W
Echague, Phil. (ā-chä'gwä)	75a	16°43'N	121°40'E
Echmiadzin, Arm.	58	40°10'N	44°18'E
Echuca, Austl. (ē-chŏŏ'kȧ)	81	36°10'S	144°47'E
Écija, Spain (ā'thē-hä)	54	37°20'N	5°07'W
Ecuador, nation, S.A. (ĕk'wȧ-dôr)	46	0°00'N	78°30'W
Edenville, S. Afr. (ē'd'n-vĭl)	87c	27°33'S	27°42'E
Édessa, Grc.	55	40°48'N	22°04'E
Edinburgh, Scot., U.K. (ĕd''n-bŭr-ô)	50	55°57'N	3°10'W
Edmonton, Can.	36	53°33'N	113°28'W
Edremit, Tur. (ĕd-rĕ-mēt')	55	39°35'N	27°00'E
Edward, l., Afr.	85	0°25'S	29°40'E
Egadi, Isole, is., Italy (ĕ'sō-lĕ-ĕ'gä-dē)	54	38°01'N	12°00'E
Egegik, Ak., U.S. (ĕg'ĕ-jĭt)	40	58°10'N	157°22'W
Egiyn, r., Mong.	76	49°41'N	100°40'E
Egypt, nation, Afr. (ē'jĭpt)	85	26°58'N	27°01'E
Eivissa, i., Spain	52	38°55'N	1°24'E
Elands, r., S. Afr.	87c	25°11'S	28°52'E
Elat, Isr.	68	29°34'N	34°57'E
Elazığ, Tur. (ĕl-ä'zēz)	68	38°40'N	39°00'E
Elba, Isola d', i., Italy (ē-sō lä-d-ĕl'bȧ)	54	42°42'N	10°25'E
Elbansan, Alb. (ĕl-bä-sän')	55	41°08'N	20°05'E
Elbe (Labe), r., Eur. (ĕl'bĕ)(lä'bĕ)	52	52°30'N	11°30'E
El Beyadh, Alg.	54	33°42'N	1°06'E
Elbistan, Tur. (ĕl-bē-stän')	55	38°20'N	37°10'E
El'brus, Gora, mtn., Russia (ĕl'brŏŏs')	56	43°20'N	42°25'E
Elbrus, Mount *see* El'brus, Gora, mtn., Russia	56	43°20'N	42°25'E
Elburz Mountains, mts., Iran (ĕl'bŏŏrz')	68	36°30'N	51°00'E
El Dorado, Ar., U.S. (ĕl dô-rä'dō)	39	33°13'N	92°39'W
Eleuthera, i., Bah. (ē-lū'thēr-ȧ)	43	25°05'N	76°10'W
El Grara, Alg.	54	32°50'N	4°26'E
El Kef, Tun. (kĕf')	54	36°14'N	8°42'E
Elko, Nv., U.S. (ĕl'kō)	38	40°51'N	115°46'W
Ellesmere Island, i., Can.	37	81°00'N	80°00'W
Ellice Islands *see* Tuvalu, nation, Oc.	3	5°20'S	174°00'E
Ellsworth Mountains, mts., Ant.	82	77°00'S	90°00'W
El Mahdia, Tun. (mä-dēȧ)(mä'dē-ȧ)	54	35°30'N	11°09'E
El Paso, Tx., U.S. (pas'ō)	38	31°47'N	106°27'W
El Qala, Alg.	54	36°52'N	8°23'E
Elsa, Can.	40	63°55'N	135°25'W
El Salvador, nation, N.A.	42	14°00'N	89°30'W
Elūru, India	69	16°44'N	80°09'E
Elvas, Port. (ĕl'väzh)	54	38°53'N	7°11'W
Ely, Nv., U.S.	38	39°16'N	114°53'W
Emāmshahr, Iran	68	36°25'N	55°01'E
Emira Island, i., Pap. N. Gui. (ā-mē-rä')	75	1°40'S	150°28'E
Emporia, Ks., U.S. (ĕm-pō'rĭ-ȧ)	38	38°24'N	96°11'W
Empty Quarter *see* Ar Rub' al Khālī, des., Asia	68	20°00'N	51°00'E
Encantada, Cerro de la, mtn., Mex. (sĕ'r-rô-dĕ-lä-ĕn-kän-tä'dä)	42	31°58'N	115°15'W
Encanto, Cape, c., Phil. (ĕn-kän'tō)	75a	15°44'N	121°46'E
Endau, r., Malay.	67b	2°29'N	103°40'E
Enderbury, i., Kir. (ĕn'dĕr-bûrĭ)	88	2°00'S	171°00'W
Enderby Land, reg., Ant. (ĕn'dĕr bĭ)	82	72°00'S	52°00'E
Endicott Mountains, mts., Ak., U.S.	40	67°30'N	153°45'W
Engaño, Cabo, c., Dom. Rep. (kȧ'-bô- ĕn-gä-nô)	43	18°40'N	68°30'W
Enggano, Pulau, i., Indon. (ĕng-gä'nō)	74	5°22'S	102°18'E
England, state, U.K. (ĭŋ'glănd)	50	51°35'N	1°40'W
English Channel, strt., Eur.	52	49°45'N	3°06'W
Enid, Ok., U.S. (ē'nĭd)	38	36°25'N	97°52'W
Enkeldoring, S. Afr. (ĕŋ'k'l-dôr-ĭng)	87c	25°24'S	28°43'E
Ensenada, Mex. (ĕn-sĕ-nä'dä)	42	32°00'N	116°30'W
Enshi, China (ŭn-shr)	76	30°18'N	109°25'E
Eolie, Isole, is., Italy (ĕ'sō-lĕ-ĕ-ô'lyĕ)	54	38°43'N	14°43'E
Epi, Vanuatu (ā'pē)	81	16°59'S	168°29'E
Episkopi, Cyp.	67a	34°38'N	32°55'E
Equatorial Guinea, nation, Afr.	85	2°00'N	7°15'E
Erciyeş Dağı, mtn., Tur.	55	38°30'N	35°36'E
Ereğli, Tur. (ĕ-rā'ĭ-le)	55	37°40'N	34°00'E
Ereğli, Tur.	55	41°15'N	31°25'E
Erie, Lake, l., N.A.	36	42°15'N	81°25'W
Erie, Pa., U.S.	39	42°05'N	80°05'W
Erimo Saki, c., Japan (ā'rē-mō sä-kē)	77	41°53'N	143°20'E
Eritrea, nation, Afr. (ā-rē-trā'ȧ)	85	16°15'N	38°30'E
Ernākulam, India	69	9°58'N	76°23'E
Erode, India	73	11°20'N	77°45'E
Erzgebirge, mts., Eur. (ĕrts'gĕ-bē'gĕ)	52	50°29'N	12°40'E
Erzincan, Tur. (ĕr-zĭn-jän')	68	39°50'N	39°30'E
Erzurum, Tur. (ĕrz'rōōm')	68	39°55'N	41°10'E
Esashi, Japan (ĕs'ä-shē)	77	41°50'N	140°10'E
Escanaba, Mi., U.S. (ĕs-kȧ-nô'bȧ)	39	45°44'N	87°05'W
Escarpada Point, Phil.	74	18°40'N	122°45'E
Escuinapa, Mex. (ĕs-kwē-nä'pä)	42	22°49'N	105°44'W
Eşfahān, Iran	68	32°38'N	51°30'E
Eskifjördur, Ice. (ĕs'kē-fyûr'dōōr)	50	65°04'N	14°01'W
Eskişehir, Tur. (ĕs-kē-shĕ'h'r)	68	39°40'N	30°20'E
Esperance, Austl. (ĕs'pē-răns)	80	33°45'S	122°07'E
Essen, Ger. (ĕs'sĕn)	50	51°26'N	6°59'E
Estonia, nation, Eur.	56	59°10'N	25°00'E
Etah, Grnld. (ē'tȧ)	37	78°20'N	72°42'W
Ethiopia, nation, Afr. (ē-thē-ō'pē-ȧ)	85	7°53'N	37°55'E
Etna, Mount, vol., Italy	52	37°48'N	15°00'E
Etolin Strait, strt., Ak., U.S. (ĕt ō lĭn)	40	60°35'N	165°40'W
Eucla, Austl. (ū'klȧ)	80	31°45'S	128°50'E
Eugene, Or., U.S. (û-jēn')	38	44°02'N	123°06'W
Euphrates, r., Asia (û-frā'tēz)	68	36°00'N	40°00'E
Eureka, Ca., U.S. (û-rē'kȧ)	38	40°45'N	124°10'W
Europe, cont. (ū'rŭp)	52	50°00'N	15°00'E
Evanston, Il., U.S. (ĕv'ăn-stŭn)	39	42°03'N	87°41'W
Evansville, In., U.S. (ĕv'ănz-vĭl)	39	38°00'N	87°30'W
Evaton, S. Afr. (ĕv'ȧ-tŏn)	87d	26°32'S	27°53'E
Everest, Mount, mtn., Asia (ĕv'ēr-ĕst)	69	28°00'N	86°57'E
Everett, Wa., U.S. (ĕv'ēr-ĕt)	38	47°59'N	122°11'W
Everglades, The, sw., Fl., U.S.	39	25°35'N	80°55'W
Évora, Port. (ĕv'ô-rä)	54	38°35'N	7°54'W
Évvoia, i., Grc.	55	38°38'N	23°45'E
Eyl, Som.	87a	7°53'N	49°45'E
Eyre, Austl. (âr)	80	32°15'S	126°20'E
Eyre, l., Austl.	80	28°43'S	137°50'E

F

PLACE (Pronunciation)	PAGE	LAT.	LONG.
Faddeya, i., Russia (fȧd-yā')	57	76°12'N	145°00'E
Fafen, r., Eth.	87a	8°15'N	42°40'E
Fā'id, Egypt (fä-yēd')	87d	30°19'N	32°18'E
Fairbanks, Ak., U.S. (fâr'bănks)	40	64°50'N	147°48'W
Fairweather, Mount, mtn., N.A. (fâr-wĕdh'ēr)	40	59°12'N	137°22'W
Faisalabad, Pak.	69	31°29'N	73°06'E
Faizābād, India	69	26°50'N	82°17'E
Fajardo, P.R.	43b	18°20'N	65°40'W
Fakfak, Indon.	75	2°56'S	132°25'E
Falkland Islands, dep., S.A. (fôk'lănd)	46	50°45'S	61°00'W
Fall River, Ma., U.S.	39	41°42'N	71°07'W
False Divi Point, c., India	73	15°45'N	80°50'E
Famagusta, Cyp. (fä-mä-gōōs'tä)	55	35°08'N	33°59'E
Farāh, Afg. (fä-rä')	68	32°15'N	62°13'E
Farasān, Jaza'ir, is., Sau. Ar.	68	16°45'N	41°08'E
Fargo, N.D., U.S. (fär'gō)	38	46°53'N	96°48'W
Fārigh, Wādī al, r., Libya (wädē ĕl fä-rēg')	55	30°10'N	19°34'E
Fāriskūr, Egypt (fä-rēs-kōōr')	87b	31°19'N	31°46'E
Faro, Port.	54	37°01'N	7°57'W
Faroe Islands, is., Eur.	52	62°00'N	5°45'W
Farrukhābād, India (fŭ-rŏŏk-hä-bäd')	69	27°29'N	79°35'E
Fartak, Ra's, c., Yemen	68	15°43'N	52°17'E
Farvel, Kap, c., Grnld.	37	60°00'N	44°00'W
Fatsa, Tur. (fät'sä)	55	40°50'N	37°30'E
Fazilka, India	72	30°30'N	74°02'E
Felanitx, Spain (fā-lä-nēch')	54	39°29'N	3°09'E
Fengdu, China (fŭŋ-dōō)	76	29°58'N	107°50'E
Fengjie, China (fŭŋ-jyĕ)	76	31°02'N	109°30'E
Fengxiang, China (fŭŋ-shyäŋ)	76	34°25'N	107°20'E
Fengzhen, China (fŭŋ-jŭn)	77	40°28'N	113°20'E
Fennimore Pass, strt., Ak., U.S. (fĕn-ĭ-mōr)	40a	51°40'N	175°38'W
Fenyang, China	77	37°20'N	111°48'E
Ferdows, Iran	68	34°00'N	58°13'E
Fergana, Uzb.	59	40°23'N	71°46'E
Fergus Falls, Mn., U.S. (fûr'gŭs)	39	46°17'N	96°03'W
Ferrara, Italy (fĕr-rä'rä)	54	44°50'N	11°37'E
Ferrol, Spain	50	43°30'N	8°12'W
Fethiye, Tur. (fĕt-hē'yĕ)	55	36°40'N	29°05'E
Ficksburg, S. Afr. (fĭks'bûrg)	87c	28°53'S	27°53'E
Fiji, nation, Oc. (fē'jē)	3	18°40'S	175°00'E
Filchner Ice Shelf, ice, Ant. (fĭlk'nēr)	82	80°00'S	35°00'W
Finland, nation, Eur. (fĭn'lănd)	50	62°45'N	26°13'E
Finland, Gulf of, b., Eur. (fĭn'lănd)	52	59°35'N	23°35'E
Firenze *see* Florence, Italy	50	43°47'N	11°15'E
Firozpur, India	69	30°58'N	74°39'E
Fisterra, Cabo de, c., Spain	52	42°52'N	9°48'W
Fitzroy Crossing, Austl.	80	18°08'S	126°00'E
Fiume *see* Rijeka, Cro.	54	45°22'N	14°24'E
Flagstaff, Az., U.S. (flăg-stäf)	38	35°15'N	111°40'W
Flamingo Point, c., V.I.U.S.	43c	18°19'N	65°00'W
Flint, Mi., U.S.	39	43°00'N	83°45'W
Florence, Italy	50	43°47'N	11°15'E
Florence, Al., U.S. (flŏr'ĕns)	39	34°46'N	87°40'W
Flores, i., Indon.	74	8°14'S	121°08'E
Flores, Laut (Flores Sea), sea, Indon.	74	7°09'S	120°30'E
Florida, state, U.S. (flŏr'ĭ-dȧ)	39	30°30'N	84°40'W
Florida, Straits of, strt., N.A.	43	24°10'N	81°00'W
Florina, Grc. (flô-rē'nä)	55	40°48'N	21°24'E
Fly, r., Pap. N. Gui. (flī)	75	8°00'S	141°45'E
Fochville, S. Afr. (fŏk'vĭl)	87c	26°29'S	27°29'E
Foggia, Italy (fôd'jä)	55	41°30'N	15°34'E
Folādī, Koh-e, mtn., Afg.	69	34°38'N	67°32'E
Fond du Lac, Wi., U.S. (fŏn dū lăk')	39	43°47'N	88°29'W
Fonseca, Golfo de, b., N.A. (gôl-fô-dĕ-fôn-sā'kä)	42	13°09'N	87°55'W
Fontur, c., Ice.	52	66°21'N	14°02'W
Foraker, Mount, mtn., Ak., U.S. (fŏr'ȧ-kēr)	40	62°40'N	152°40'W
Forbes, Austl. (fôrbz)	81	33°24'S	148°05'E
Forel, Mont, mtn., Grnld.	37	65°50'N	37°41'W
Forli, Italy (fôr-lē')	54	44°13'N	12°03'E
Formentera, Isla de, i., Spain (ē's-lä-dĕ-fôr-mĕn-tā'rä)	54	38°43'N	1°25'E
Formosa Strait *see* Taiwan Strait, strt., Asia	77	24°30'N	120°00'E
Forsayth, Austl. (fôr-sīth')	81	18°33'S	143°42'E
Fortaleza, Braz. (fôr'tä-lā'zȧ)	46	3°35'S	38°31'W
Fort Collins, Co., U.S. (kŏl'ĭns)	38	40°36'N	105°04'W
Fort-de-France, Mart. (dē fräns)	43	14°37'N	61°06'W
Fort Dodge, Ia., U.S. (dŏj)	39	42°31'N	94°10'W
Forth, Firth of, b., Scot., U.K. (fûrth ôv fōrth)	52	56°04'N	3°03'W
Fort Sandeman, Pak. (săn'da-măn)	69	31°28'N	69°29'E

ng-sing; ŋ-baŋk; N-nasalized n; nŏd; cŏmmit; ōld; ôbey; ôrder; oi-boil; fōōd; ŏŏ-as oo in foot; ou-out; s-soft; sh-dish; th-thin; pūre; ûnite; ûrn; stŭd; circŭs; ü-as in French tu; '-indeterminate vowel.

120 Fort Scott – Gydanskiy Poluostrov PRONOUNCING INDEX

PLACE (Pronunciation)	PAGE	LAT.	LONG.
Fort Scott, Ks., U.S. (skŏt)	39	37°50'N	94°43'W
Fort-Shevchenko, Kaz. (shĕv-chĕn´kô)	59	44°30'N	50°18'E
Fort Smith, Ar., U.S. (smĭth)	39	35°23'N	94°24'W
Fort Wayne, In., U.S. (wān)	39	41°00'N	85°10'W
Fort Worth, Tx., U.S. (wûrth)	38	32°45'N	97°20'W
Foshan, China	77	23°02'N	113°07'E
Fouriesburg, S. Afr. (fō´rēz-bûrg)	87c	28°38'S	28°13'E
Four Mountains, Islands of the, is., Ak., U.S.	40a	52°58'N	170°40'W
Fox Islands, is., Ak., U.S. (fŏks)	40a	53°04'N	167°30'W
Framnes Mountains, mts., Ant.	82	67°50'S	62°35'E
France, nation, Eur. (fråns)	50	46°39'N	0°47'E
Frankfort, S. Afr.	87c	27°17'S	28°30'E
Frankfort, Ky., U.S.	39	38°10'N	84°55'W
Frankfurt am Main, Ger.	50	50°07'N	8°40'E
Franz Josef Land *see* Zemlya Frantsa-Iosifa, is., Russia	56	81°32'N	40°00'E
Frederick, Md., U.S. (frĕd´ẽr-ĭk)	39	39°25'N	77°25'W
Freeport, Il., U.S. (frē´pōrt)	39	42°19'N	89°30'W
Freetown, S.L. (frē´toun)	85	8°30'N	13°15'W
Fremantle, Austl. (frē´măn-t'l)	80	32°03'S	116°05'E
French Guiana, dep., S.A. (gē-ä´nä)	46	4°20'N	53°00'W
French Polynesia, dep., Oc.	2	15°00'S	140°00'W
Fresnillo, Mex. (frås-nēl´yô)	42	23°10'N	102°52'W
Fresno, Ca., U.S.	38	36°44'N	119°46'W
Frunze *see* Bishkek, Kyrg.	59	42°49'N	74°42'E
Fuerte, Río del, r., Mex. (rē´ō-dĕl-fo͞o-ĕ´r-tĕ)	42	26°15'N	108°50'W
Fuhai, China	76	47°01'N	87°07'E
Fujian, prov., China (fo͞o-jyĕn)	77	25°40'N	117°30'E
Fujin, China (fo͞o-jyĭn)	77	47°13'N	132°11'E
Fuji San, mtn., Japan (fo͞o´jē sän)	77	35°23'N	138°44'E
Fujiyama *see* Fuji San, mtn., Japan	77	35°23'N	138°44'E
Fukui, Japan (fo͞o´ko͞o-ē)	77	36°05'N	136°14'E
Fukuoka, Japan (fo͞o´kō-ō´kȧ)	77	33°35'N	130°23'E
Fuling, China (fo͞o-lĭŋ)	76	29°40'N	107°30'E
Fundy, Bay of, b., Can. (fŭn´dĭ)	36	45°00'N	66°00'W
Furgun, mtn., Iran	68	28°47'N	57°00'E
Fushun, China (fo͞o´sho͞on´)	77	41°50'N	124°00'E
Fuwah, Egypt (fo͞o´wȧ)	87b	31°13'N	30°35'E
Fuyang, China (fo͞o-yäŋ)	77	32°53'N	115°48'E
Fuyu, China (fo͞o-yo͞o)	77	45°20'N	125°00'E
Fuzhou, China (fo͞o-jō)	77	26°02'N	119°18'E

G

PLACE (Pronunciation)	PAGE	LAT.	LONG.
Gaalkacyo, Som.	87a	7°00'N	47°30'E
Gabés, Golfe de, b., Tun.	85	32°22'N	10°59'E
Gabon, nation, Afr. (gȧ-bôɴ´)	85	0°30'S	10°45'E
Gaborone, Bots.	85	24°28'S	25°59'E
Gachsārān, Iran	71	30°12'N	50°47'E
Gadsden, Al., U.S. (gădz´dĕn)	39	34°00'N	86°00'W
Gagra, Geor.	58	43°20'N	40°15'E
Gaillard Cut, reg., Pan. (gä-ēl-yä´rd)	42a	9°03'N	79°42'W
Gainesville, Fl., U.S. (gānz´vĭl)	39	29°40'N	82°20'W
Gairdner, Lake, l., Austl. (gârd´nẽr)	80	32°20'S	136°30'E
Galapagos Islands *see* Colón, Archipiélago de, is., Ec.	46	0°10'S	87°45'W
Galaţi, Rom.	50	45°25'N	28°05'E
Galdhøpiggen, mtn., Nor.	50	61°37'N	8°17'E
Galera, Cerro, mtn., Pan. (sĕ´r-rō-gä-lĕ´rä)	42a	8°55'N	79°38'W
Galesburg, Il., U.S. (gālz´bûrg)	39	40°56'N	90°21'W
Galilee, Sea of, l., Isr.	67a	32°53'N	35°45'E
Galle, Sri L. (gäl)	73	6°13'N	80°10'E
Gallipoli *see* Gelibolu, Tur. (gäl-lē´pō-lē)	55	40°25'N	26°40'E
Gallup, N.M., U.S. (găl´ŭp)	38	35°30'N	108°45'W
Galveston, Tx., U.S. (găl´vĕs-tŭn)	39	29°18'N	94°48'W
Galway, Ire.	50	53°16'N	9°05'W
Gamba, China (gäm-bä)	72	28°23'N	89°42'E
Gambia, The, nation, Afr.	85	13°38'N	19°38'W
Gandak, r., India	72	26°37'N	84°22'E
Gandhinagar, India	72	23°30'N	72°47'E
Gangdisê Shan (Trans Himalayas), mts., China	76	30°25'N	83°43'E
Ganges, r., Asia (găn´jēz)	69	24°00'N	89°30'E
Ganges, Mouths of the, mth., Asia (găn´jēz)	69	21°18'N	88°40'E
Gangtok, India	69	27°15'N	88°30'E
Gansu, prov., China (gän-so͞o)	76	38°50'N	101°10'E
Ganzhou, China (gän-jō)	77	25°50'N	114°30'E
Gaoyou Hu, l., China (kä´ō-yo͞o´ho͞o)	77	32°42'N	118°40'E
Gapan, Phil. (gä-pän)	75a	15°18'N	120°56'E
Gar, China	76	31°11'N	80°35'E
Garda, Lago di, l., Italy (lä-gō-dē-gär´dä)	54	45°43'N	10°26'E
Garden Reach, India	72a	22°33'N	88°17'E
Gardeyz, Afg.	72	33°43'N	69°09'E
Gareloi, i., Ak., U.S. (gär-lo͞o-ä´)	40a	51°40'N	178°48'W
Garm, Taj.	59	39°12'N	70°28'E
Garonne, r., Fr. (gȧ-rôn)	52	44°00'N	1°00'E
Garulia, India	72a	22°48'N	88°23'E
Gary, In., U.S. (gā´rĭ)	39	41°35'N	87°21'W
Gasan, Phil. (gä-sän´)	75a	13°19'N	121°52'E
Gasan-Kuli, Turkmen.	59	37°25'N	53°55'E
Gata, Cabo de, c., Spain (kȧ´bô-dĕ-gä´tä)	54	36°42'N	2°00'W
Gata, Sierra de, mts., Spain (syĕr´rȧ dā gȧ´tä)	54	40°12'N	6°39'W
Gátes, Akrotírion, c., Cyp.	67a	34°30'N	33°15'E
Gates of the Arctic National Park, rec., Ak., U.S.	40	67°45'N	153°30'W
Gatun, r., Pan.	42a	9°21'N	79°40'W
Gatun Locks, trans., Pan.	42a	9°16'N	79°57'W
Gauhāti, India	69	26°09'N	91°51'E
Gávdos, i., Grc. (gäv´dôs)	55	34°48'N	24°08'E
Gāvkhūnī, Bātlāq-e, l., Iran	68	31°40'N	52°48'E
Gävle, Swe. (yĕv´lĕ)	50	60°40'N	17°07'E
Gawler, Austl. (gô´lẽr)	80	34°35'S	138°47'E
Gaya, India (gŭ´yä)(gī´ȧ)	69	24°53'N	85°00'E
Gaza, Gaza	68	31°30'N	34°29'E
Gaziantep, Tur. (gä-zē-än´tĕp)	68	37°10'N	37°30'E
Gdańsk, Pol. (g´dänsk)	50	54°20'N	18°40'E
Gediz, r., Tur.	55	38°44'N	28°45'E
Geelong, Austl. (jē-lông´)	81	38°06'S	144°13'E
Gelibolu, Tur. (gĕ-lĭb´ō-lō)	55	40°25'N	26°40'E
Gemas, Malay. (jĕm´ås)	67b	2°35'N	102°37'E
Gemlik, Tur. (gĕm´lĭk)	55	40°30'N	29°10'E
Genale (Jubba), r., Afr.	87a	5°15'N	41°00'E
Geneva (Genève), Switz.	50	46°14'N	6°04'E
Genève *see* Geneva, Switz.	50	46°14'N	6°04'E
Genoa, Italy	50	44°23'N	9°52'E
Genova, Golfo di, b., Italy (gôl-fō-dē-jĕn´ō-vä)	52	44°10'N	8°45'E
Georgetown, Guy. (jôrj´toun)	46	7°45'N	58°04'W
George Town, Malay.	74	5°21'N	100°09'E
Georgia, nation, Asia	56	42°17'N	43°00'E
Georgia, state, U.S. (jôr´ji-ă)	39	32°40'N	83°50'W
Geraldton, Austl. (jĕr´ăld-tŭn)	80	28°40'S	114°35'E
Germany, nation, Eur. (jûr´mȧ-nĭ)	50	51°00'N	10°00'E
Gerona, Phil. (hā-rō´nä)	75a	15°36'N	120°36'E
Ghāghra, r., India	69	26°00'N	83°00'E
Ghana, nation, Afr. (gän´ä)	85	8°00'N	2°00'W
Gharo, Pak.	72	24°50'N	68°35'E
Ghazzah *see* Gaza, Gaza	68	31°30'N	34°29'E
Gheorgheni, Rom.	55	46°48'N	25°30'E
Ghorīān, Afg.	71	34°21'N	61°30'E
Gibraltar, dep., Eur. (gĭ-bräl-tä´r)	50	36°08'N	5°22'W
Gibraltar, Strait of, strt.,	52	35°55'N	5°45'W
Gifu, Japan (gē´fo͞o)	77	35°25'N	136°45'E
Gijón, Spain (hē-hōn´)	50	43°33'N	5°37'W
Gilbert Islands, is., Kir.	89	0°30'S	174°00'E
Gilgit, Pak. (gĭl´gĭt)	69	35°58'N	73°48'E
Giluwe, Mount, mtn., Pap. N. Gui.	75	6°04'S	144°00'E
Giresun, Tur. (ghēr´ĕ-sōn´)	68	40°55'N	38°20'E
Giridih, India (jē-rē-dē)	69	24°12'N	86°18'E
Girona, Spain	54	41°55'N	2°48'E
Gironde, r., Fr. (zhē-rôɴd´)	52	45°31'N	1°00'W
Giza *see* Al Jīzah, Egypt	87b	30°01'N	31°12'E
Gizhiga, Russia (gē´zhi-gȧ)	57	61°59'N	160°46'E
Gizo, Sol. Is.	78e	8°06'S	156°51'E
Gjirokastër, Alb.	55	40°04'N	20°10'E
Gladdeklipkop, S. Afr.	87c	24°17'S	29°36'E
Gladstone, Austl. (glăd´stōn)	81	23°45'S	152°00'E
Gladstone, Austl.	80	33°15'S	138°20'E
Glåma, r., Nor.	52	61°30'N	10°30'E
Glasgow, Scot., U.K. (glås´gō)	50	55°54'N	4°25'W
Glazov, Russia (glä´zôf)	56	58°05'N	52°52'E
Glendale, Ca., U.S.	38	34°09'N	118°15'W
Glendive, Mt., U.S. (glĕn´dīv)	38	47°08'N	104°41'W
Glen Innes, Austl. (ĭn´ĕs)	81	29°45'S	152°02'E
Globe, Az., U.S. (glōb)	38	33°20'N	110°50'W
Goa, state, India (gō´ȧ)	69	15°45'N	74°00'E
Gobi , des., Asia (gō´be)	76	43°29'N	103°15'E
Godāvari, r., India (gō-dä´vŭ-rē)	69	19°00'N	78°30'E
Godhavn, Grnld. (gōdh´håvn)	37	69°15'N	53°30'W
Godthåb, Grnld. (gŏt´hōb)	37	64°10'N	51°32'W
Gold Hill, mtn., Pan.	42a	9°03'N	79°08'W
Golets-Purpula, Gora, mtn., Russia	57	59°08'N	115°22'E
Golo Island, i., Phil. (gō´lō)	75a	13°38'N	120°17'E
Gómez Palacio, Mex. (pä-lä´syō)	42	25°35'N	103°30'W
Gonaïves, Haiti (gō-nȧ-ēv´)	43	19°25'N	72°45'W
Gonâve, Île de la, i., Haiti (gō-nàv´)	43	18°50'N	73°30'W
Gonda, India	72	27°13'N	82°00'E
Gondal, India	72	22°02'N	70°47'E
Gongga Shan, mtn., China (gōŋ-gä shän)	76	29°16'N	101°46'E
Good Hope, Cape of, c., S. Afr. (kāp ov go͝od hōp)	85	34°21'S	18°29'E
Gorakhpur, India (gō´rŭk-po͞or)	69	26°45'N	82°39'E
Gorgān, Iran	68	36°44'N	54°30'E
Gorgona, Isola di, Italy (gôr-gō´nä)	54	43°27'N	9°55'E
Gor'kiy *see* Nizhniy Novgorod, Russia	56	56°15'N	44°05'E
Gor'kovskoye, res., Russia	56	56°38'N	43°40'E
Gorno-Altaysk, Russia (gôr´nŭ´ŭl-tīsk´)	56	51°58'N	85°58'E
Gorodok, Russia	57	50°30'N	103°58'E
Gorontalo, Indon. (gō-rōn-tä´lo)	75	0°40'N	123°04'E
Göteborg, Swe. (yû´tĕ-bôrgh)	50	57°39'N	11°56'E
Gothenburg *see* Göteborg, Swe.	50	57°39'N	11°56'E
Gotland, i., Swe.	52	57°35'N	17°35'E
Gough, i., St. Hel. (gŏf)	2	40°00'S	10°00'W
Goukou, China (gō-kō)	77	48°45'N	121°42'E
Goulburn, Austl. (gōl´bŭrn)	81	34°47'S	149°40'E
Gradačac, Bos. (gra-dä´chats)	55	44°50'N	18°28'E
Grafton, Austl. (graf´tŭn)	81	29°38'S	153°05'E
Grampian Mountains, mts., Scot., U.K. (grăm´pĭ-ăn)	52	56°30'N	4°55'W
Granada, Nic. (grä-nä´dhä)	42	11°55'N	85°58'W
Granada, Spain (grä-nä´dä)	54	37°13'N	3°37'W
Gran Chaco, reg., S.A. (grån´chå´kŏ)	46	25°30'S	62°15'W
Grand Bahama, i., Bah.	43	26°35'N	78°30'W
Grand Canal *see* Da Yunhe, can., China	77	35°00'N	117°00'E
Grand Canyon, p., Az., U.S.	38	35°50'N	113°16'W
Grand Cayman, i., Cay. Is. (kā´măn)	43	19°15'N	81°15'W
Grande de Santiago, Río, r., Mex. (rēō-grä´n-dĕ-dĕ-sän-tyȧ´gô)	42	20°30'N	104°00'W
Grand Forks, N.D., U.S.	38	47°55'N	97°05'W
Grand Island, Ne., U.S. (ī´lȧnd)	38	40°56'N	98°20'W
Grand Junction, Co., U.S. (jŭngk´shŭn)	38	39°05'N	108°35'W
Grand Rapids, Mi., U.S. (răp´ĭdz)	39	43°00'N	85°45'W
Grand Turk, T./C. Is. (tûrk)	43	21°30'N	71°10'W
Grande, Rio, r., N.A. (grän´dā)	36	26°50'N	99°10'W
Grass Cay, i., V.I.U.S.	43c	18°22'N	64°50'W
Graz, Aus. (gräts)	50	47°05'N	15°26'E
Great Abaco, i., Bah. (ä´bä-kō)	43	26°30'N	77°05'W
Great Australian Bight, b., Austl. (ôs-trā´lĭ-ăn bĭt)	80	33°30'S	127°00'E
Great Barrier Reef, rf., Austl. (bȧ-rĭ-ẽr rēf)	81	16°43'S	146°34'E
Great Basin, basin, U.S. (grāt bā´s'n)	38	40°08'N	117°10'W
Great Bear Lake, l., Can. (bâr)	36	66°10'N	119°53'W
Great Bitter Lake, l., Egypt	87b	30°24'N	32°27'E
Great Dividing Range, mts., Austl. (dĭ-vī-dĭng rānj)	81	35°16'S	146°38'E
Greater Antilles, is., N.A.	43	20°30'N	79°15'W
Greater Khingan Range, mts., China (dä hĭŋ-gän lĭŋ)	77	46°30'N	120°00'E
Greater Sunda Islands, is., Asia	74	4°00'S	108°00'E
Great Falls, Mt., U.S. (fôlz)	38	47°30'N	111°15'W
Great Inagua, i., Bah. (ē-nä´gwä)	43	21°00'N	73°15'W
Great Indian Desert, des., Asia	69	27°35'N	71°37'E
Great Nicobar Island, i., India (nĭk-ō-bär´)	74	7°00'N	94°18'E
Great Plains, pl., N.A. (plāns)	37	45°00'N	104°00'W
Great Salt Lake, l., Ut., U.S.	38	41°19'N	112°48'W
Great Sandy Desert, des., Austl. (săn´dē)	80	21°50'S	123°10'E
Great Sitkin, i., Ak., U.S. (sĭt-kĭn)	40a	52°18'N	176°22'W
Great Slave Lake, l., Can. (slāv)	36	61°37'N	114°58'W
Great Victoria Desert, des., Austl. (vĭk-tō´rĭ-ȧ)	80	29°45'S	124°30'E
Great Wall, hist., China	76	38°00'N	109°00'E
Greece, nation, Eur. (grēs)	50	39°00'N	21°30'E
Greeley, Co., U.S. (grē´lĭ)	38	40°25'N	104°41'W
Green Bay, Wi., U.S.	39	44°30'N	88°04'W
Greenland, dep., N.A. (grēn´lănd)	37	74°00'N	40°00'W
Greenland Sea, sea,	93	77°00'N	1°00'W
Greensboro, N.C., U.S.	39	36°04'N	79°45'W
Greenville, Ms., U.S.	39	33°25'N	91°00'W
Greenville, S.C., U.S.	39	34°50'N	82°25'W
Grenada, nation, N.A.	43	12°02'N	61°15'W
Greylingstad, S. Afr. (grā-lĭng´shtȧt)	87c	26°40'S	29°13'E
Groblersdal, S. Afr.	87c	25°11'S	29°25'E
Groot Marico, S. Afr.	87c	25°36'S	26°23'E
Groot Marico, r., Afr.	87c	25°13'S	26°20'E
Groznyy, Russia (grŏz´nĭ)	56	43°20'N	45°40'E
Guadalajara, Mex. (gwä-dhä-lä-hä´rä)	42	20°41'N	103°21'W
Guadalajara, Spain (gwä-dä-lä-kä´rä)	54	40°37'N	3°10'W
Guadalcanal, i., Sol. Is.	81	9°48'S	158°43'E
Guadalquivir, Río, r., Spain (rē´ō-gwä-dhäl-kē-vēr´)	52	37°30'N	5°00'W
Guadalupe, i., Mex.	42	29°00'N	118°45'W
Guadalupe, Sierra de, mts., Spain (syĕr´rä dā gwä-dhä-lo͞o´pā)	54	39°30'N	5°25'W
Guadarrama, Sierra de, mts., Spain (gwä-dhär-rä´mä)	52	41°00'N	3°40'W
Guadeloupe, dep., N.A. (gwä-dē-lo͞op)	43	16°40'N	61°10'W
Guadiana, r., Eur. (gwä-dvä´nä)	52	39°00'N	6°00'W
Guam, i., Oc. (gwäm)	3	14°00'N	143°20'E
Guanabacoa, Cuba (gwä-nä-bä-kō´ä)	43	23°08'N	82°19'W
Guanacevi, Mex. (gwä-nä-sĕ-vē´)	42	25°30'N	105°45'W
Guanajuato, Mex. (gwä-nä-hwä´tō)	42	21°01'N	101°16'W
Guanajuato, state, Mex.	42	21°00'N	101°00'W
Guangdong, prov., China (gůäŋ-dōŋ)	77	23°45'N	113°15'E
Guangxi Zhuangzu, prov., China (gůäŋ-shyē)	76	24°00'N	108°30'E
Guangzhou, (Canton) China	76	23°07'N	113°15'W
Guantánamo, Cuba (gwän-tä´nä-mō)	43	20°10'N	75°10'W
Guatemala, Guat. (guä-tå-mä´lä)	42	14°37'N	90°32'W
Guatemala, nation, N.A.	42	15°45'N	91°45'W
Guayama, P.R. (gwä-yä´mä)	43b	18°00'N	66°08'W
Guayaquil, Ec. (gwī-ä-kēl´)	46	2°16'S	79°53'W
Guaymas, Mex. (gwȧ´y-mäs)	42	27°49'N	110°58'W
Gubakha, Russia (go͞o-bä´kå)	56	58°53'N	57°35'E
Gudermes, Russia	58	43°20'N	46°08'E
Guernsey, dep., Eur.	54	49°28'N	2°35'W
Guerrero, state, Mex.	42	17°45'N	100°15'W
Guiana Highlands, mts., S.A.	47	3°20'N	60°00'W
Guilin, China (gwā-lĭn)	77	25°18'N	110°22'E
Guinea, nation, Afr. (gĭn´ē)	85	10°48'N	12°28'W
Guinea, Gulf of, b., Afr.	85	2°00'N	1°00'E
Guinea-Bissau, nation, Afr. (gĭn´ē)	85	12°00'N	20°00'W
Guir, r., Mor.	54	31°55'N	2°48'W
Guiyang, China (gwā-yäŋ)	76	26°45'N	107°00'E
Guizhou, prov., China	76	27°00'N	106°10'E
Gujānwāla, Pak. (gōj-rän´va-lȧ)	69	32°08'N	74°14'E
Gujarat, India	69	22°54'N	72°00'E
Gulbarga, India (gōl-bûr´gȧ)	69	17°25'N	76°52'E
Gulja *see* Yining, China	76	43°58'N	80°40'E
Gumaca, Phil. (go͞o-mä-kä´)	75a	13°55'N	122°06'E
Guna, India	72	24°44'N	77°17'E
Guntūr, India (gōn´to͞or)	69	16°22'N	80°29'E
Guru Sikhar, mtn., India	72	29°42'N	72°50'E
Gur'yevsk, Russia (go͞or-yĭfsk´)	56	54°17'N	85°56'E
Guyana, nation, S.A. (gŭy´ănă)	46	7°45'N	59°00'W
Gwādar, Pak. (gwä´dŭr)	68	25°15'N	62°29'E
Gwalior, India	69	26°13'N	78°10'E
Gwardafuy, Gees, c., Som.	87a	11°55'N	51°30'E
Gyaring Co, l., China	72	30°37'N	88°33'E
Gydan, Khrebet (Kolymskiy), mts., Russia	57	61°45'N	155°00'E
Gydanskiy Poluostrov, pen., Russia	56	70°42'N	76°03'E

ăt; finăl; rāte; senåte; ärm; åsk; sofȧ; fâre; ch-choose; dh-as th in other; bē; ēvent; bĕt; recĕnt; cratẽr; g-gō; gh-guttural g; bĭt; ĭ-short neutral; rīde; ᴋ-guttural k as ch in German ich;

PRONOUNCING INDEX

Gympie – İskenderun Körfezi 121

PLACE (Pronunciation)	PAGE	LAT.	LONG.
Gympie, Austl. (gĭm′pê)	81	26°20′S	152°50′E
Gyöngyös, Hung. (dyûn′dyûsh)	55	47°47′N	19°55′E
Györ, Hung. (dyûr)	55	47°40′N	17°37′E
Gyzylarbat, Turkmen.	59	38°55′N	56°33′E

H

PLACE (Pronunciation)	PAGE	LAT.	LONG.
Ha'Arava (Wādī al Jayb), val., Asia	67a	30°33′N	35°10′E
Hābra, India	72a	22°49′N	88°38′E
Ḩadd, Ra's al, c., Oman	68	22°29′N	59°46′E
Ḩadera, Isr. (ка-dĕ′rä)	67a	32°26′N	34°55′E
Ḩadīdū, Yemen	68	12°40′N	53°50′E
Ḩadramawt, reg., Yemen	68	15°22′N	48°40′E
Hadūr Shu'ayb, mtn., Yemen	68	15°45′N	43°45′E
Haft Gel, Iran	71	31°27′N	49°27′E
Hafun, Ras, c., Som. (hä-fo͞on′)	87a	10°15′N	51°35′E
Hagerstown, Md., U.S.	39	39°40′N	77°45′W
Haifa, Isr. (hä′ē-fà)	68	32°48′N	35°00′E
Ḩā'il, Sau. Ar.	68	27°30′N	41°47′E
Hailar, China	77	49°10′N	118°40′E
Hailun, China (hä′ê-lo͞on′)	77	47°18′N	126°50′E
Hainan, prov., China	76	19°00′N	109°30′E
Hainan Dao, i., China (hī-nän dou)	77	19°00′N	111°10′E
Haines, Ak., U.S. (hānz)	40	59°10′N	135°38′W
Hai Phong, Viet. (hī′fông′)(hä′ēp-hŏng)	74	20°52′N	106°40′E
Haiti, nation, N.A. (hā′tĭ)	43	19°00′N	72°15′W
Hakodate, Japan (hä-kō-dä′t å)	77	41°46′N	140°42′E
Halcon, Mount, mtn., Phil. (häl-kōn′)	75a	13°19′N	120°55′E
Halkett, Cape, c., Ak., U.S.	40	70°50′N	151°15′W
Halls Creek, Austl. (hôlz)	80	18°15′S	127°45′E
Halmahera, i., Indon. (häl-mä-hā′rä)	75	0°45′N	128°45′E
Halmahera, Laut, Indon.	75	1°00′S	129°00′E
Hamadān, Iran (hŭ-mŭ-dän′)	68	34°45′N	48°07′E
Ḩamāh, Syria (hä′mä)	68	35°08′N	36°53′E
Hamburg, Ger. (häm′bo͝orgh)	50	53°34′N	10°02′E
Hamhŭng, Kor., N. (häm′hòng′)	77	39°57′N	127°35′E
Hami, China (hä-mē)	76	42°58′N	93°14′E
Hamilton, Austl. (hăm′ĭl-tŭn)	81	37°50′S	142°10′E
Hamilton, Oh., U.S.	39	39°22′N	84°33′W
Hammanskraal, S. Afr. (hä-mȧns-kräl′)	87c	25°24′S	28°17′E
Hammerfest, Nor. (hä′mĕr-fĕst)	50	70°38′N	23°59′E
Hammond, In., U.S. (hăm′ŭnd)	39	41°37′N	87°31′W
Han, r., China	77	31°40′N	112°04′E
Hancock, Mi., U.S. (hăn′kŏk)	39	47°08′N	88°37′W
Hangayn Nuruu, mts., Mong.	76	48°03′N	99°45′E
Hango, Fin. (häŋ′gû)	50	59°49′N	22°56′E
Hangzhou, China (häng′chō′)	77	30°17′N	120°12′E
Hannibal, Mo., U.S. (hăn′ĭ bȧl)	39	39°42′N	91°22′W
Hannover, Ger. (hän-ō′vĕr)	50	52°22′N	9°45′E
Hanoi, Viet. (hä-noi′)	74	21°04′N	105°50′E
Hans Lollick, i., V.I.U.S. (häns′lôl′ĭk)	43c	18°24′N	64°55′W
Hantengri Feng, mtn., Asia (hän-tŭŋ-rē fŭŋ)	76	42°10′N	80°20′E
Hanyang, China (han′yäng′)	77	30°30′N	114°10′E
Ḩaql, Sau. Ar.	67a	29°15′N	34°57′E
Harare, Zimb.	85	17°50′S	31°03′E
Harbin, China	77	45°40′N	126°30′E
Hardwār, India (hŭr′dvär)	69	29°56′N	78°06′E
Hargeysa, Som. (här-gā′ê-sȧ)	87a	9°20′N	43°57′E
Harīrūd, r., Asia	68	34°29′N	61°16′E
Harlingen, Tx., U.S.	38	26°12′N	97°42′W
Harrisburg, Pa., U.S.	39	40°15′N	76°50′W
Harrismith, S. Afr. (hă-rĭs′mĭth)	87c	28°17′S	29°08′E
Hartbeesfontein, S. Afr.	87c	26°46′S	26°25′E
Hartford, Ct., U.S.	39	41°45′N	72°40′W
Hārua, India	72a	22°36′N	88°40′E
Haryana, state, India	69	29°00′N	75°45′E
Hastings, Ne., U.S.	38	40°34′N	98°42′W
Hatay, Tur.	68	36°20′N	36°10′E
Hatteras, Cape, c., N.C., U.S. (hăt′ĕr-ȧs)	39	35°15′N	75°24′W
Hattiesburg, Ms., U.S. (hăt′ĭz-bûrg)	39	31°20′N	89°18′W
Hat Yai, Thai.	74	7°01′N	100°29′E
Hauptsrus, S. Afr.	87c	26°35′S	26°16′E
Haut Atlas, mts., Mor.	54	32°10′N	5°49′W
Havana, Cuba	43	23°08′N	82°23′W
Havre, Mt., U.S. (hăv′ĕr)	38	48°34′N	109°42′W
Hawai'ian Islands, is., Hi., U.S. (hä-wī′ȧn)	38c	22°00′N	158°00′W
Hawaii, state, U.S.	38c	20°00′N	157°40′W
Hayes, Mount, mtn., Ak., U.S. (hāz)	40	63°32′N	146°40′W
Heard Island, i., Austl. (hûrd)	3	53°10′S	74°35′E
Hebei, prov., China (hŭ-bā)	77	39°15′N	115°40′E
Hebrides, is., Scot., U.K.	52	57°00′N	6°30′W
Hebron *see* Al Khalīl, W.B.	67a	31°31′N	35°07′E
Hechuan, China (hŭ-chyüän)	76	30°00′N	106°20′E
Hefei, China (hŭ-fā)	77	31°51′N	117°15′E
Heidelberg, S. Afr.	87c	26°32′S	28°22′E
Heilbron, S. Afr. (hīl′brōn)	87c	27°17′S	27°58′E
Heilongjiang, prov., China (hā-lôŋ-jyäŋ)	77	46°36′N	128°07′E
Hejaz *see* Al Ḩijāz, reg., Sau. Ar.	68	23°45′N	39°08′E
Hekla, vol., Ice.	52	63°53′N	19°37′W
Helan Shan, mts., China (hŭ-län shän)	76	38°02′N	105°20′E
Helena, Ar., U.S. (hē-lē′nȧ)	39	34°33′N	90°35′W
Helena, Mt., U.S. (hĕ-lē′nȧ)	38	46°35′N	112°01′W
Hellín, Spain (ĕl-yēn′)	54	38°30′N	1°40′W
Helmand, r., Afg. (hĕl′mŭnd)	68	31°00′N	63°48′E
Helsingfors *see* Helsinki, Fin.	50	60°10′N	24°53′E
Helsinki, Fin. (hĕl′sēn-kê)	50	60°10′N	24°53′E
Henan, prov., China (hŭ-nän)	77	33°58′N	112°33′E
Hendrina, S. Afr. (hĕn-drē′nȧ)	87c	26°10′S	29°44′E
Hengyang, China	77	26°58′N	112°30′E
Hennenman, S. Afr.	87c	27°59′S	27°03′E
Hentiyn Nuruu, mts., Mong.	76	49°25′N	107°51′E
Henzada, Mya.	69	17°38′N	95°28′E
Herāt, Afg. (hĕ-rät′)	68	34°28′N	62°13′E
Hermanusdorings, S. Afr.	87c	24°08′S	27°46′E
Hermit Islands, is., Pap. N. Gui. (hûr′mĭt)	75	1°48′S	144°55′E
Hermosillo, Mex. (ĕr-mô-sē′l-yō)	42	29°00′N	110°57′W
Herzliyya, Isr.	67a	32°10′N	34°49′E
Heuningspruit, S. Afr.	87c	27°28′S	27°26′E
Heystekrand, S. Afr.	87c	25°16′S	27°14′E
Hibbing, Mn., U.S. (hĭb′ĭng)	39	47°26′N	92°58′W
Hidalgo, state, Mex.	42	20°45′N	99°30′W
Hidalgo del Parral, Mex. (ē-dä′l-gō-dĕl-pär-rà′l)	42	26°55′N	105°40′W
High Peak, mtn., Phil.	75a	15°38′N	120°05′E
Higuero, Punta, c., P.R.	43b	18°21′N	67°11′W
Himachal Pradesh, India	69	32°00′N	77°30′E
Himalayas, mts., Asia	69	29°30′N	85°02′E
Ḩimş, Syria	68	34°44′N	36°43′E
Hindu Kush, mts., Asia (hĭn′do͞o ko͞osh′)	69	35°15′N	68°44′E
Hindupur, India (hĭn′do͞o-po͞or)	73	13°52′N	77°34′E
Hirara, Japan	78d	24°48′N	125°17′E
Hirosaki, Japan (hē′rô-sä′kê)	77	40°31′N	140°38′E
Hiroshima, Japan (hē-rô-shē′mȧ)	77	34°22′N	132°25′E
Hisar, India	72	29°15′N	75°47′E
Hispaniola, i., N.A. (hĭ′spăn-ĭ-ō-là)	43	17°30′N	73°15′W
Hobart, Austl. (hō′bȧrt)	81	43°00′S	147°30′E
Hobyo, Som.	87a	5°24′N	48°28′E
Ho Chi Minh City, Viet.	74	10°46′N	106°34′E
Hodna, Chott el, l., Alg.	54	35°20′N	3°27′E
Hohhot, China (hŭ-ho͞o-tŭ)	77	41°05′N	111°50′E
Hokkaidō, i., Japan (hŏk′kī-dō)	65	43°30′N	142°45′E
Holguín, Cuba (ôl-gēn′)	43	20°55′N	76°15′W
Holy Cross, Ak., U.S. (hō′lĭ krôs)	40	62°10′N	159°40′W
Homer, Ak., U.S. (hō′mẽr)	40	59°42′N	151°30′W
Honduras, nation, N.A. (hŏn-do͞o′rȧs)	42	14°30′N	88°00′W
Honduras, Gulf of, b., N.A.	42	16°30′N	87°30′W
Hong Kong (Xianggang), China	77	21°45′N	115°00′E
Hongshui, r., China (hôŋ-shwā)	76	24°30′N	105°00′E
Hongze Hu, l., China	77	33°17′N	118°37′E
Honiara, Sol. Is.	81	9°26′S	159°57′E
Honolulu, Hi., U.S. (hŏn-ō-lo͞o′lo͞o)	38c	21°18′N	157°50′W
Honshū, i., Japan	77	36°00′N	138°00′E
Hood, Mount, mtn., Or., U.S.	38	45°20′N	121°43′W
Hood River, Or., U.S.	38	45°42′N	121°30′W
Hoogly, r., India (ho͞og′lĭ)	69	21°35′N	87°50′E
Hoonah, Ak., U.S. (ho͞o′nȧ)	40	58°05′N	135°25′W
Hooper Bay, Ak., U.S.	40	61°32′N	166°02′W
Hope, Ak., U.S. (hōp)	40	60°54′N	149°48′W
Hopetoun, Austl. (hōp′toun)	80	33°50′S	120°15′E
Hopkinsville, Ky., U.S. (hŏp′kĭns-vĭl)	39	36°50′N	87°28′W
Hoquiam, Wa., U.S. (hō′kwĭ-ăm)	38	47°00′N	123°53′W
Hormuz, Strait of, strt., Asia (hôr′mŭz′)	68	26°30′N	56°30′E
Horn, Cape *see* Hornos, Cabo de, c., Chile	46	56°00′S	67°00′W
Hornos, Cabo de, c., Chile	46	56°00′S	67°00′W
Horsham, Austl. (hôr′shăm) (hôrs′ăm)	81	36°42′S	142°17′E
Horton, r., Ak., U.S. (hôr′tŭn)	40	68°38′N	122°00′W
Hotan, China (hwô-tän)	76	37°11′N	79°50′E
Hotan, r., China	76	39°09′N	81°08′E
Hot Springs, Ak., U.S. (hŏt sprĭngs)	40	65°00′N	150°20′W
Hot Springs, Ar., U.S.	39	34°29′N	93°02′W
Houston, Tx., U.S.	39	29°46′N	95°21′W
Hovd, Mong.	76	48°08′N	91°40′E
Hovd Gol, r., Mong.	76	49°06′N	91°16′E
Hövsgöl Nuur, l., Mong.	76	51°11′N	99°11′E
Howland, i., Oc. (hou′lănd)	2	1°00′N	176°00′W
Howrah, India (hou′rä)	69	22°33′N	88°20′E
Hsawnhsup, Mya.	76	24°29′N	94°45′E
Huai, r., China (hwī)	77	32°07′N	114°38′E
Huang (Yellow), r., China (hûäng)	77	35°06′N	113°39′E
Huang, Old Beds of the, mth., China	76	40°28′N	106°34′E
Huangyuan, China (hûäŋ-yûän)	76	37°00′N	101°01′E
Huascarán, Nevados, mts., Peru (wäs-kä-rän′)	46	9°05′S	77°50′W
Hubei, prov., China (ho͞o-bā)	77	31°20′N	111°58′E
Hubli, India (ho͝ob′lê)	69	15°25′N	75°09′E
Hudson, r., U.S.	39	42°30′N	73°55′W
Hudson Bay, b., Can.	36	60°15′N	85°30′W
Hudson Strait, strt., Can.	36	63°25′N	74°05′W
Hue, Viet. (ü-ā′)	74	16°28′N	107°42′E
Huelva, Spain (wĕl′vä)	54	37°16′N	6°58′W
Huésca, Spain (wĕs-kä)	54	42°07′N	0°25′W
Hughenden, Austl. (hū′ĕn-dĕn)	81	20°58′S	144°13′E
Hughes, Austl. (hūz)	80	30°45′S	129°30′E
Huili, China	76	26°48′N	102°20′E
Huimin, China (ho͞oĭ mĭn)	77	37°29′N	117°32′E
Hukou, China (ho͞o-kō)	77	29°58′N	116°20′E
Hulan, China (ho͞o′län′)	77	45°58′N	126°32′E
Huludao, China (ho͞o-lo͞o-dou)	77	40°40′N	120°55′E
Hulun Nur, l., China (ho͞o-lòn nòr)	77	48°50′N	116°45′E
Humacao, P.R. (o͞o-mä-kä′ō)	43b	18°09′N	65°49′W
Hunan, prov., China (ho͞o′nän′)	77	28°08′N	111°25′E
Hunchun, China (hòn-chŭn)	77	42°53′N	130°34′E
Hungary, nation, Eur. (hŭŋ′gȧ-rĭ)	50	46°44′N	17°55′E
Hungerford, Austl. (hŭn′gẽr-fẽrd)	81	28°50′S	144°32′E
Huntington, W.V., U.S.	39	38°25′N	82°25′W
Huon Gulf, b., Pap. N. Gui.	75	7°15′S	147°45′E
Ḩuraydin, Wādī, r., Egypt	67a	30°55′N	34°12′E
Hurdiyo, Som.	87a	10°43′N	51°05′E
Huron, S.D., U.S.	38	44°22′N	98°15′W
Huron, Lake, l., N.A. (hū′rŏn)	36	45°15′N	82°40′W
Hurricane, Ak., U.S. (hŭr′ĭ-kān)	40	63°00′N	149°30′W
Hutchinson, Ks., U.S. (hŭch′ĭn-sŭn)	38	38°02′N	97°56′W
Hyargas Nuur, l., Mong.	76	48°00′N	92°32′E
Hydaburg, Ak., U.S. (hī-dă′bûrg)	40	55°12′N	132°49′W
Hyderābād, India (hī-dẽr-å-băd′)	69	17°29′N	78°28′E
Hyderābād, India	69	18°30′N	76°50′E
Hyderābād, Pak.	69	25°29′N	68°28′E

I

PLACE (Pronunciation)	PAGE	LAT.	LONG.
Iaşi, Rom. (yä′shê)	50	47°10′N	27°40′E
Iba, Phil. (ē′bä)	75a	15°20′N	119°59′E
Ibadan, Nig. (ê-bä′dän)	85	7°17′N	3°30′E
Ibb, Yemen	71	14°01′N	44°10′E
Ibiza *see* Eivissa, i., Spain	52	38°55′N	1°24′E
Ibrāhīm, Būr, b., Egypt	87d	29°57′N	32°33′E
Ibrāhīm, Jabal, mtn., Sau. Ar.	68	20°31′N	41°17′E
İçel, Tur.	68	37°00′N	34°40′E
Iceland, nation, Eur. (īs′lănd)	50	65°12′N	19°45′W
Icy Cape, c., Ak., U.S. (ī′sĭ)	40	70°20′N	161°40′W
Idaho, state, U.S. (ī′dȧ-hō)	38	44°00′N	115°10′W
Idaho Falls, Id., U.S.	38	43°30′N	112°01′W
Ider, r., Mong.	76	48°58′N	98°38′E
Idi, Indon. (ē′dê)	74	4°58′N	97°47′E
Idkū Lake, l., Egypt	87b	31°13′N	30°22′E
Idlib, Syria	70	35°55′N	36°38′E
Igarka, Russia (ê-gär′kȧ)	56	67°22′N	86°16′E
Iglesias, Italy (ē-lĕ′syôs)	54	39°20′N	8°34′E
Iguig, Phil. (ē-gēg′)	75a	17°46′N	121°44′E
Ilagan, Phil.	75a	17°09′N	121°52′E
Ílhavo, Port. (ēl′yȧ-vô)	54	40°36′N	8°41′W
Iliamna, Ak., U.S. (ē-lê-ăm′nȧ)	40	59°45′N	155°05′W
Iliamna, Ak., U.S.	40	60°18′N	153°25′W
Iliamna, l., Ak., U.S.	40	59°25′N	155°30′W
Ilimsk, Russia (ē-lyêmsk′)	57	56°47′N	103°43′E
Ilin Island, i., Phil. (ē-lyēn′)	75a	12°16′N	120°57′E
Illimani, Nevado, mtn., Bol. (nĕ-vá′dô-ēl-yê-mä′nê)	46	16°50′S	67°38′W
Illinois, state, U.S. (ĭl-ĭ-noi′) (ĭl-ĭ-noiz′)	39	40°25′N	90°40′W
Iloilo, Phil. (ē-lô-ē′lō)	74	10°49′N	122°33′E
Imbābah, Egypt (ēm-bä′bȧ)	87b	30°06′N	31°09′E
Immerpan, S. Afr. (ĭmẽr-pän)	87c	24°29′S	29°14′E
Imperia, Italy (êm-pā′rē-ä)	54	43°52′N	8°00′E
Imphāl, India (ĭmp′hŭl)	69	24°42′N	94°00′E
Ince Burun, c., Tur. (ĭn′jȧ)	55	42°00′N	35°00′E
Inch'ŏn, Kor., S. (ĭn′chŭn)	77	37°26′N	126°46′E
India, nation, Asia (ĭn′dĭ-ȧ)	69	23°00′N	77°30′E
Indiana, state, U.S.	39	39°50′N	86°45′W
Indianapolis, In., U.S. (ĭn-dĭ-ăn-ăp′ō-lĭs)	39	39°45′N	86°08′W
Indian Ocean, o.	5	10°00′S	70°00′E
Indio, r., Pan. (ē′n-dyô)	42a	9°13′N	79°28′W
Indochina, reg., Asia (ĭn-dô-chī′nȧ)	74	17°22′N	105°18′E
Indonesia, nation, Asia (ĭn′dô-nē-zhȧ)	74	4°38′S	118°45′E
Indore, India (ĭn-dōr′)	69	22°48′N	76°51′E
Indragiri, r., Indon. (ĭn-drȧ-jē′rê)	74	0°27′S	102°05′E
Indrāvati, r., India (ĭn-drŭ-vä′tê)	69	19°00′N	82°00′E
Indus, r., Asia	69	26°43′N	67°41′E
Inebolu, Tur. (ê-nȧ-bō′lo͞o)	55	41°50′N	33°40′E
Infanta, Phil. (ēn-fän′tä)	75a	14°44′N	121°39′E
Infanta, Phil.	75a	15°50′N	119°53′E
Ingham, Austl. (ĭng′ăm)	81	18°45′S	146°14′E
Ingushetia, prov., Russia	58	43°15′N	45°00′E
Inner Brass, i., V.I.U.S. (bräs)	43c	18°23′N	64°58′W
Inner Mongolia *see* Nei Monggol, prov., China	76	40°15′N	105°00′E
International Falls, Mn., U.S. (ĭn′tẽr-năsh′ŭn-ăl fôlz)	39	48°34′N	93°26′W
Inverell, Austl. (ĭn-vẽr-ĕl′)	81	29°50′S	151°32′E
Ioánnina, Grc. (yô-ä′nē-nȧ)	55	39°39′N	20°52′E
Ionian Islands, is., Grc. (ī-ō′nĭ-ăn)	55	39°10′N	20°05′E
Ionian Sea, sea, Eur.	52	38°59′N	18°48′E
Iori, r., Asia	58	41°03′N	46°17′E
Iowa, state, U.S. (ī′ô-wȧ)	39	42°05′N	94°20′W
Iowa City, Ia., U.S.	39	41°39′N	91°31′W
Ipoh, Malay.	74	4°45′N	101°05′E
Ipswich, Austl. (ĭps′wĭch)	81	27°40′S	152°50′E
Irákleio, Grc.	50	35°20′N	25°10′E
Iran, nation, Asia (ē-rän′)	68	31°15′N	53°30′E
Iran, Plateau of, plat., Iran	68	32°28′N	58°00′E
Iran Mountains, mts., Asia	74	2°30′N	114°30′E
Iraq, nation, Asia (ē-räk′)	68	32°00′N	42°30′E
Irbid, Jord. (ẽr-bēd′)	70	32°33′N	35°51′E
Irbit, Russia (ẽr-bēt′)	56	57°40′N	63°10′E
Ireland, nation, Eur. (īr-lănd)	50	53°33′N	8°00′W
Iriomote Jima, i., Japan (ērē′-ō-mō-tā)	77	24°20′N	123°30′E
Irish Sea, sea, Eur. (ī′rĭsh)	52	53°55′N	5°25′W
Irkutsk, Russia (ĭr-kōtsk′)	57	52°16′N	104°00′E
Iron Bottom Sound, strt., Sol. Is.	78e	9°15′S	160°00′E
Irrawaddy, r., Mya. (ĭr-ȧ-wäd′ê)	69	23°27′N	96°25′E
Irtysh, r., Asia (ĭr-tĭsh′)	56	59°00′N	69°00′E
Isaacs, Mount, mtn., Pan. (ē-sä-ä′ks)	42a	9°22′N	79°31′W
Isangi, D.R.C. (ê-säŋ′gê)	76	0°46′N	24°15′E
Isarog, Mount, mtn., Phil. (ē-sä-rô-g)	75a	13°40′N	123°23′E
Ischia, Isola d', i., Italy (dē′sh-kyä)	54	40°26′N	13°55′E
Ishigaki, Japan	78d	24°20′N	124°09′E
Ishim, Russia (ĭsh-ĕm′)	56	56°07′N	69°13′E
Ishim, r., Asia	56	53°17′N	67°45′E
Ishinomaki, Japan (ĭsh-nō-mä′kē)	77	38°22′N	141°22′E
Ishmant, Egypt	87b	29°17′N	31°15′E
İskenderun, Tur. (ĭs-kĕn′dẽr-o͞on)	68	36°45′N	36°15′E
İskenderun Körfezi, b., Tur.	55	36°22′N	35°25′E

ng-sing; ŋ-baŋk; ɴ-nasalized n; nŏd; cŏmmit; ōld; ôbey; ôrder; oi-boil; fo͞od; ò-as oo in foot; ou-out; s-soft; sh-dish; th-thin; pūre; ûnite; ûrn; stŭd; circŭs; ü-as in French tu; ′-indeterminate vowel.

PLACE (Pronunciation)	PAGE	LAT.	LONG.
İskilip, Tur. (ēs'kĭ-lēp')	55	40°40'N	34°30'E
Islāmābād, Pak.	69	33°55'N	73°05'E
Ismailia, Egypt (ēs-mä-ēl'ēà)	87b	30°35'N	32°17'E
Ismā'īlīyah Canal, can., Egypt	87b	30°25'N	31°45'E
Ismail Samani, pik, mtn., Taj.	59	38°57'N	72°01'E
Isparta, Tur. (ē-spär'tȧ)	68	37°50'N	30°40'E
Israel, nation, Asia	68	32°40'N	34°00'E
Issyk-Kul, Ozero, l., Kyrg.	59	42°13'N	76°12'E
İstanbul, Tur. (ē-stän-bōōl')	68	41°02'N	29°00'E
İstanbul Boğazı (Bosporus), strt., Tur.	68	41°10'N	29°10'E
Italy, nation, Eur. (ĭt'ȧ-lē)	50	43°58'N	11°14'E
Itarsi, India	72	22°43'N	77°45'E
Ithaca, N.Y., U.S.	39	42°25'N	76°30'W
Ivanovo, Russia (ē-vä'nô-vō)	56	57°02'N	41°54'E
Iviza *see* Eivissa, i., Spain	52	38°55'N	1°24'E
Ivory Coast *see* Cote d'Ivoire, nation, Afr.	85	7°43'N	6°30'W
Ivrea, Italy (ē-vrĕ'ä)	54	45°25'N	7°54'E
Ivvavik National Park, rec., Can.	40	69°10'N	139°30'W
Izberbash, Russia	58	42°33'N	47°52'E
Izhevsk, Russia (ē-zhyĕfsk')	56	56°50'N	53°15'E
İzmir, Tur. (ĭz-mēr')	68	38°25'N	27°05'E
İzmit, Tur. (ĭz-mēt')	55	40°45'N	29°45'E
Izu Shichitō, is., Japan	77	34°32'N	139°25'E

J

PLACE (Pronunciation)	PAGE	LAT.	LONG.
Jabalpur, India	69	23°18'N	79°59'E
Jackson, Mi., U.S.	39	42°15'N	84°25'W
Jackson, Ms., U.S.	39	32°17'N	90°10'W
Jackson, Tn., U.S.	39	35°37'N	88°49'W
Jacksonville, Fl., U.S.	39	30°20'N	81°40'W
Jacksonville, Il., U.S.	39	39°43'N	90°12'W
Jacobābad, Pak.	72	28°22'N	68°30'E
Jaen, Spain	54	37°45'N	3°48'W
Jaffna, Sri L. (jăf'nȧ)	73	9°44'N	80°09'E
Jahore Strait, strt., Asia	67b	1°22'N	103°37'E
Jahrom, Iran	68	28°30'N	53°28'E
Jaipur, India	69	27°00'N	75°50'E
Jaisalmer, India	72	27°00'N	70°54'E
Jajpur, India	69	20°49'N	86°37'E
Jakarta, Indon. (yä-kär'tä)	74	6°17'S	106°45'E
Jalālābād, Afg. (jŭ-lä-lȧ-bäd)	69a	34°25'N	70°27'E
Jalālah al Baḩrīyah, Jabal, mts., Egypt	87b	29°20'N	32°00'E
Jaleswar, Nepal	72	26°50'N	85°55'E
Jalgaon, India	72	21°08'N	75°33'E
Jalisco, state, Mex.	42	20°07'N	104°45'W
Jamaica, nation, N.A.	43	17°45'N	78°00'W
Jamālpur, Bngl.	72	24°56'N	89°58'E
Jambi, Indon. (mäm'bē)	74	1°45'S	103°28'E
James Bay, b., Can. (jāmz)	36	53°53'N	80°40'W
James Ross, i., Ant.	47	64°20'S	58°20'W
Jamestown, N.D., U.S.	38	46°54'N	98°42'W
Jamestown, N.Y., U.S. (jāmz'toun)	39	42°05'N	79°15'W
Jammu, India	69	32°50'N	74°52'E
Jammu and Kashmīr, state, India (kȧsh-mēr')	69	34°30'N	76°00'E
Jammu and Kashmīr, hist. reg., Asia (kȧsh-mēr')	69	39°10'N	75°05'E
Jāmnagar, India (jäm-nŭ'gŭr)	69	22°33'N	70°03'E
Jamshedpur, India (jŭm'shād-pōōr)	69	22°52'N	86°11'E
Janin, W.B.	67a	32°27'N	35°19'E
Japan, nation, Asia (jȧ-păn')	77	36°30'N	133°30'E
Japan, Sea of, sea, Asia (jȧ-păn')	77	40°08'N	132°55'E
Jarash, Jord.	67a	32°17'N	35°53'E
Jarud Qi, China (jya-lōō-tŭ shyē)	77	44°35'N	120°40'E
Jasin, Malay.	67b	2°19'N	102°26'E
Jāsk, Iran (jäsk)	68	25°46'N	57°48'E
Jason Bay, b., Malay.	67b	1°53'N	104°14'E
Java (Jawa), i., Indon.	74	8°35'S	111°11'E
Java Trench, deep,	74	9°45'S	107°30'E
Jawa, Laut (Java Sea), sea, Indon.	74	5°10'S	110°30'E
Jaya, Puncak, mtn., Indon.	75	4°00'S	137°00'E
Jayapura, Indon.	74	2°30'S	140°45'W
Jayb, Wādī al (Ha'Arava), val., Asia	67a	30°33'N	35°10'E
Jazzīn, Leb.	67a	33°34'N	35°37'E
Jefferson City, Mo., U.S.	39	38°34'N	92°10'W
Jehol, hist. reg., China (jē-hōl)	77	42°31'N	118°12'E
Jerez de la Frontera, Spain	54	36°42'N	6°09'W
Jericho, Austl. (jĕr'ĭ-kō)	81	23°38'S	146°24'E
Jericho, S. Afr. (jĕr-ĭkô)	87c	25°16'N	27°47'E
Jericho	67a	31°51'N	35°28'E
Jerome, Az., U.S. (jē-rōm')	38	34°45'N	112°10'W
Jersey, dep., Eur.	54	49°15'N	2°10'W
Jersey City, N.J., U.S.	39	40°43'N	74°05'W
Jerusalem, Isr. (jē-rōō'sȧ-lĕm)	68	31°46'N	35°14'E
Jhālawār, India	69	24°30'N	76°00'E
Jhang Maghiāna, Pak.	72	31°21'N	72°19'E
Jhānsi, India (jän'sē)	69	25°29'N	78°32'E
Jharkhand, state, India	69	23°30'N	85°00'E
Jhārsuguda, India	72	22°51'N	84°13'E
Jhelum, Pak.	69	32°59'N	73°43'E
Jhelum, r., Asia (jā'lŭm)	69	31°40'N	71°51'E
Jialing, r., China (jyä-lĭŋ)	76	32°30'N	105°30'E
Ji'an, China (jyē-än)	77	27°15'N	115°10'E
Jiangling, China (jyäŋ-lĭŋ)	77	30°30'N	112°10'E
Jiangsu, prov., China (jyäŋ-sōō)	77	33°45'N	120°30'E
Jiangxi, prov., China (jyäŋ-shyē)	77	28°15'N	116°00'E
Jiaoxian, China (jyou shyĕn)	77	36°18'N	120°01'E
Jiaxing, China (jyä-shyĭŋ)	77	30°45'N	120°50'E
Jiazhou Wan, b., China (jyä-jō wän)	77	36°10'N	119°55'E
Jiddah, Sau. Ar.	68	21°30'N	39°15'E
Jieyang, China (jyĕ-yäŋ)	77	23°38'N	116°20'E
Jiggalong, Austl. (jĭg'ȧ-lông)	80	23°20'S	120°45'E
Jijiga, Eth.	87a	9°15'N	42°48'E
Jilin, China (jyē-lĭn)	77	43°58'N	126°40'E
Jilin, prov., China	77	44°20'N	124°50'E
Jinan, China (jyē-nän)	77	36°40'N	117°01'E
Jinhua, China (jyĭn-hwä)	77	29°10'N	119°42'E
Jining, China (jyē-nĭŋ)	77	35°26'N	116°34'E
Jinta, China (jyĭn-tä)	76	40°11'N	98°45'E
Jinzhou, China (jyĭn-jō)	77	41°00'N	121°00'E
Jiujiang, China	77	29°43'N	116°00'E
Jiuquan, China (jyô-chyän)	76	39°46'N	98°26'E
Jodhpur, India (hōd'pōōr)	69	26°23'N	73°00'E
Johannesburg, S. Afr. (yô-hän'ĕs-bôrgh)	85	26°08'S	27°54'E
Johnson City, Tn., U.S.	39	36°17'N	82°23'W
Johnston, i., Oc. (jŏn'stŭn)	2	17°00'N	168°00'W
Johnstown, Pa., U.S.	39	40°20'N	78°50'W
Johor, r., Malay. (jŭ-hōr')	67b	1°39'N	103°52'E
Johor Baharu, Malay.	74	1°28'N	103°46'E
Joinville, i., Ant.	47	63°00'S	53°30'W
Jolo, Phil. (hō-lô)	74	5°59'N	121°05'E
Jolo Island, i., Phil.	74	5°55'N	121°15'E
Jomalig, i., Phil. (hô-mä'lēg)	75a	14°44'N	122°34'E
Jones, Phil. (jōnz)	75a	12°56'N	122°05'E
Jones, Phil.	75a	16°35'N	121°39'E
Jonesboro, Ar., U.S. (jōnz'bŭro)	39	35°49'N	90°42'W
Joplin, Mo., U.S. (jŏp'lĭn)	39	37°05'N	94°31'W
Jordan, nation, Asia (jôr'dăn)	68	30°15'N	38°00'E
Jordan, r., Asia	67a	32°05'N	35°35'E
Jorhāt, India (jôr-hät')	69	26°43'N	94°16'E
Juan Diaz, r., Pan. (κōōà'n-dē'äz)	42a	9°05'N	79°30'W
Juan Fernández, Islas de, is., Chile	47	33°30'S	79°00'W
Jubayl, Leb. (jōō-bĭl')	67a	34°07'N	35°38'E
Jubba (Genale), r., Afr.	87a	1°30'N	42°25'E
Júcar, r., Spain (hōō'kär)	54	39°10'N	1°22'W
Juchitán, Mex. (hōō-chē-tän')	42	16°15'N	95°00'W
Julian Alps, mts., Serb.	54	46°05'N	14°05'E
Julianehåb, Grnld.	37	60°07'N	46°20'W
Jullundur, India	69	31°29'N	75°39'E
Julpaiguri, India	72	26°35'N	88°48'E
Jumrah, Indon.	67b	1°48'N	101°04'E
Junāgādh, India (jŏ-nä'gŭd)	69	21°33'N	70°25'E
Junayfah, Egypt	87d	30°11'N	32°26'E
Junaynah, Ra's al, mtn., Egypt	67a	29°02'N	33°58'E
Juneau, Ak., U.S. (jōō'nō)	40	58°25'N	134°30'W
Juniyah, Leb. (jōō-nē'ē)	67a	33°59'N	35°38'E
Juticalpa, Hond. (hōō-tē-käl'pä)	42	14°35'N	86°17'W
Juventud, Isla de la, i., Cuba	43	21°40'N	82°45'W

K

PLACE (Pronunciation)	PAGE	LAT.	LONG.
K2 (Qogir Feng), mtn., Asia	69	36°06'N	76°38'E
Kabaena, Pulau, i., Indon. (kä-bȧ-ā'nȧ)	74	5°35'S	121°07'E
Kaboudia, Ra's, c., Tun.	54	35°17'N	11°28'E
Kābul, Afg. (kä'bōl)	69	34°39'N	69°14'E
Kabul, r., Asia (kä'bōl)	69	34°44'N	69°43'E
Kachuga, Russia (kȧ-chōō-gȧ)	57	54°09'N	105°43'E
Kaesŏng, Kor., N. (kä'ē-sŭng) (kĭ'jō)	77	38°00'N	126°35'E
Kagoshima, Japan (kä'gô-shē'mȧ)	77	31°35'N	130°31'E
Kahayan, r., Indon.	74	1°45'S	113°40'E
Kahramanmaraş, Tur.	68	37°40'N	36°50'W
Kai, Kepulauan, is., Indon.	75	5°35'S	132°45'E
Kaiang, Malay.	67b	3°00'N	101°47'E
Kaidu, r., China (kī-dōō)	76	42°35'N	84°04'E
Kaifeng, China (kī-fŭŋ)	77	34°48'N	114°22'E
Kai Kecil, i., Indon.	75	5°45'S	132°40'E
Kaimana, Indon.	75	3°32'S	133°47'E
Kaiyuh Mountains, mts., Ak., U.S. (kī-yōō')	40	64°25'N	157°38'W
Kajang, Gunong, mtn., Malay.	67b	2°47'N	104°05'E
Kakhovs'ke vodoskhovyshche, res., Ukr.	56	47°21'N	33°33'E
Kākināda, India	69	16°58'N	82°18'E
Kaktovik, Ak., U.S. (kăk-tō'vĭk)	40	70°08'N	143°51'W
Kaladan, r., Asia	76	21°07'N	93°04'E
Kalahari Desert, des., Afr. (kä-lä-hä'rē)	85	23°00'S	22°03'E
Kalamáta, Grc.	50	37°04'N	22°08'E
Kalamazoo, Mi., U.S. (kăl-ȧ-mȧ-zōō')	39	42°20'N	85°40'W
Kalaotoa, Pulau, i., Indon.	74	7°22'S	122°30'E
Kalar, mtn., Iran	68	31°43'N	51°41'E
Kalāt, Pak. (kŭ-lät')	69	29°05'N	66°36'E
Kalgan *see* Zhangjiakou, China	77	40°45'N	114°58'E
Kalgoorlie-Boulder, Austl. (kăl-gōōr'lē)	80	30°45'S	121°35'E
Kaliakra, Nos, c., Blg.	55	43°25'N	28°42'E
Kaliningrad, Russia	56	54°42'N	20°32'E
Kalispell, Mt., U.S. (kăl'ĭ-spĕl)	38	48°12'N	114°18'W
Kalsubai Mount, mtn., India	72	19°43'N	73°47'E
Kālu, r., India	73b	19°18'N	73°14'E
Kaluga, Russia (kȧ-lò'gä)	56	54°29'N	36°12'E
Kalwa, India	73b	19°12'N	72°59'E
Kalyān, India	72	19°16'N	73°07'E
Kama, r., Russia (kä'mä)	56	56°10'N	53°50'E
Kamarān, i., Yemen	68	15°19'N	41°47'E
Kāmārhāti, India	72a	22°41'N	88°23'E
Kamchatka, Poluostrov, pen., Russia	57	55°19'N	157°45'E
Kamen'-na-Obi, Russia (kä-mĭny'nŭ ô'bē)	56	53°43'N	81°28'E
Kāmet, mtn., Asia	72	30°50'N	79°42'E
Kampala, Ug. (käm-pä'lä)	86	0°19'N	32°25'E
Kampar, r., Indon. (käm'pär)	74	0°30'N	101°30'E
Kâmpóng Saôm, Camb.	74	10°40'N	103°50'E
Kâmpóng Thum, Camb. (kŏm'pŏng-tŏm)	74	12°41'N	104°29'E
Kâmpôt, Camb. (käm'pōt)	74	10°41'N	104°07'E
Kampuchea *see* Cambodia, nation, Asia	74	12°15'N	104°00'E
Kamskoye, res., Russia	56	59°08'N	56°30'E
Kamyshin, Russia (kä-mwēsh'ĭn)	56	50°08'N	45°20'E
Kamyshlov, Russia (kä-mēsh'lôf)	56	56°50'N	62°32'E
Kanaga, i., Ak., U.S. (kă-nă'gă)	40a	52°02'N	177°38'W
Kanā'is, Ra's al, c., Egypt	55	31°14'N	28°08'E
Kanatak, Ak., U.S. (kă-nă'tŏk)	40	57°35'N	155°48'W
Kanazawa, Japan (kä'nȧ-zä'wȧ)	77	36°34'N	136°38'E
Kānchenjunga, mtn., Asia (kĭn-chĭn-jòn'gȧ)	69	27°30'N	88°18'E
Kānchipuram, India	69	12°55'N	79°43'E
Kandahār, Afg.	69	31°43'N	65°58'E
Kandalaksha, Russia (kȧn-dȧ-lák'shȧ)	56	67°10'N	33°05'E
Kandiāro, Pak.	72	27°09'N	68°12'E
Kandla, India (kŭnd'lŭ)	72	23°00'N	70°20'E
Kandy, Sri L. (kän'dē)	73	7°18'N	80°42'E
Kangāvar, Iran (kŭn'gā-vär)	68	34°37'N	46°45'E
Kangean, Kepulauan, is., Indon. (käŋ'gē-än)	74	6°50'S	116°22'E
Kanggye, Kor., N. (käng'gyĕ)	77	40°55'N	126°40'E
Kanin, Poluostrov, pen., Russia	56	68°00'N	45°00'E
Kanivs'ke vodoskhovyshche, res., Ukr.	56	50°10'N	30°40'E
Kano, Nig. (kä'nō)	85	12°00'N	8°30'E
Kānpur, India (kän'pûr)	72	26°30'N	80°10'E
Kansas, state, U.S. (kăn'zȧs)	38	38°30'N	99°40'W
Kansas City, Ks., U.S.	39	39°06'N	94°39'W
Kansas City, Mo., U.S.	39	39°05'N	94°35'W
Kansk, Russia	57	56°14'N	95°43'E
Kantang, Thai. (kän'täng')	74	7°26'N	99°28'E
Kanton, i., Kir.	88	3°50'S	174°00'W
Kaohsiung, Tai. (kä-ô-syòng')	77	22°35'N	120°25'E
Kara, Russia (kärȧ)	56	68°42'N	65°30'E
Karabalā', Iraq (kŭr'bȧ-lä)	68	32°31'N	43°58'E
Kara-Bogaz-Gol, Zaliv, b., Turkmen. (kȧ-rä' bŭ-gäs')	59	41°30'N	53°40'E
Karachay-Cherkessia, prov., Russia	58	44°00'N	42°00'E
Karāchi, Pak.	69	24°59'N	68°56'E
Karaganda *see* Qaraghandy, Kaz.	59	49°42'N	73°18'E
Karakoram Pass, p., Asia	69	35°35'N	77°45'E
Karakoram Range, mts., India (kä'rä kō'rôm)	69	35°24'N	76°38'E
Karakorum, hist., Mong.	76	47°25'N	102°22'E
Kara-Kum, des., Turkmen.	59	40°00'N	57°00'E
Kara Kum Canal, can., Turkmen.	59	37°35'N	61°50'E
Karaman, Tur. (kä-rä-män')	55	37°10'N	33°00'E
Karamay, China (kär-äm-ā)	76	45°37'N	84°53'E
Kara Sea *see* Karskoye More, sea, Russia	56	74°00'N	68°00'E
Karashahr (Yanqui), China (kä-rä-shä-är) (yän-chyē)	76	42°14'N	86°28'E
Kargat, Russia (kär-gät')	56	55°17'N	80°07'E
Karghalik *see* Yecheng, China	76	37°54'N	77°25'E
Kargopol', Russia (kär-gō-pōl'')	56	61°30'N	38°50'E
Kariba, Lake, res., Afr.	85	17°15'S	27°55'E
Kārikāl, India (kä-rē-käl')	73	10°58'N	79°49'E
Karimata, Kepulauan, is., Indon. (kä-rē-mä'tȧ)	74	1°08'S	108°10'E
Karimata, Selat, strt., Indon.	74	1°00'S	107°10'E
Karimun Besar, i., Indon.	67b	1°10'N	103°28'E
Karimunjawa, Kepulauan, is., Indon. (kä'rē-mōōn-yä'vä)	74	5°36'S	110°15'E
Karin, Som. (kȧr'ĭn)	87a	10°43'N	45°50'E
Karkar Island, i., Pap. N. Gui. (kär'kär)	75	4°50'S	146°45'E
Karkheh, r., Iran	68	32°45'N	47°50'E
Karkūk, Iraq	68	35°28'N	44°22'E
Karlovac, Cro. (kär'lô-vȧts)	55	45°29'N	15°16'E
Karlstad, Swe. (kärl'städ)	50	59°25'N	13°28'E
Karluk, Ak., U.S. (kär'lŭk)	40	57°30'N	154°22'W
Karnataka, state, India	69	14°55'N	75°00'E
Kárpathos, i., Grc.	55	35°34'N	27°26'E
Kars, Tur. (kärs)	68	40°35'N	43°00'E
Karshi, Uzb. (kär'shē)	59	38°30'N	66°08'E
Karskiye Vorota, Proliv, strt., Russia	56	70°30'N	58°07'E
Karskoye More (Kara Sea), sea, Russia	56	74°00'N	68°00'E
Kartaly, Russia (kȧr'tá lē)	56	53°05'N	60°40'E
Karunagapalli, India	73	9°09'N	76°34'E
Kāshān, Iran (kā-shän')	68	33°52'N	51°15'E
Kashgar *see* Kashi, China	76	39°29'N	76°00'E
Kashi (Kashgar), China (kä-shr) (käsh-gär)	76	39°29'N	76°00'E
Kāshmar, Iran	71	35°12'N	58°27'E
Kashmīr *see* Jammu and Kashmīr, state, India	69	34°30'N	76°00'E
Kashmor, Pak.	72	28°33'N	69°34'E
Kaskanak, Ak., U.S. (kăs-kă'năk)	40	60°00'N	158°00'W
Kásos, i., Grc.	55	35°20'N	26°55'E
Kaspiysk, Russia	58	42°52'N	47°38'E
Kastamonu, Tur. (kä-stȧ-mō'nōō)	68	41°20'N	33°50'E
Kastoría, Grc. (kȧs-tō'rĭ-ȧ)	55	40°28'N	21°17'E
Kasūr, Pak.	72	31°10'N	74°29'E
Katanning, Austl. (kȧ-tăn'ĭng)	80	33°45'S	117°45'E
Katherine, Austl. (kăth'ĕr-ĭn)	80	14°15'S	132°20'E
Kāthiāwār, pen., India (kä'tyȧ-wär')	69	22°10'N	70°20'E
Kathmandu, Nepal (kăt-män-dōō')	69	27°49'N	85°21'E
Katihār, India	72	25°39'N	87°39'E
Katowice, Pol.	50	50°15'N	19°00'E
Katta-Kurgan, Uzb. (kȧ-tȧ-kôr-gän')	59	39°45'N	66°42'E
Kattegat, strt., Eur. (kăt'ĕ-gät)	52	56°57'N	11°25'E
Kaua'i, i., Hi., U.S.	38c	22°09'N	159°15'W

ăt; finăl; rāte; senâte; ärm; ȧsk; sofȧ; fâre; ch-choose; dh-as th in other; bē; ēvent; bĕt; recĕnt; cratĕr; g-gō; gh-guttural g; bĭt; ĭ-short neutral; rīde; κ-guttural k as ch in German ich;

PRONOUNCING INDEX

Kaunas – Lādnun 123

PLACE (Pronunciation)	PAGE	LAT.	LONG.
Kaunas, Lith. (kou′nȧs) (kŏv′nŏ)	56	54°42′N	23°54′E
Kavála, Grc. (kä-vä′lä)	55	40°55′N	24°24′E
Kavieng, Pap. N. Gui. (kä-vē-ĕng′)	75	2°44′S	151°02′E
Kavīr, Dasht-e, des., Iran (dŭsht-ē-ka-vēr′)	68	34°41′N	53°30′E
Kaxgar, r., China	76	39°30′N	75°00′E
Kayan, r., Indon.	74	1°45′N	115°38′E
Kayseri, Tur. (kī′sĕ-rē)	68	38°45′N	35°20′E
Kazach′ye, Russia	57	70°46′N	135°47′E
Kazakhstan, nation, Asia	56	48°45′N	59°00′E
Kazan′, Russia (kȧ-zän′)	56	55°50′N	49°18′E
Kāzerūn, Iran	68	29°37′N	51°44′E
Kebnekaise, mtn., Swe. (kĕp′nĕ-kȧ-ēs′ĕ)	52	67°53′N	18°10′E
Kecskemét, Hung. (kĕch′kĕ-māt)	55	46°52′N	19°42′E
Kedah, hist. reg., Malay. (kā′dä)	74	6°00′N	100°31′E
Kefallonía, i., Grc.	55	38°08′N	20°58′E
Ke Ga, Mui, c., Viet.	74	12°58′N	109°50′E
Kelafo, Eth.	87a	5°40′N	44°00′E
Kelang, Malay.	74	3°20′N	101°27′E
Kelang, r., Malay.	67b	3°00′N	101°40′E
Kelkit, r., Tur.	55	40°38′N	37°03′E
Keluang, Malay.	67b	2°01′N	103°19′E
Kem′, Russia (kĕm)	56	65°00′N	34°48′E
Kemerovo, Russia	56	55°31′N	86°05′E
Kempsey, Austl. (kĕmp′sē)	81	30°59′S	152°50′E
Kempton Park, S. Afr. (kĕmp′tŏn pärk)	87c	26°07′S	28°29′E
Ken, r., India	72	25°00′N	79°55′E
Kenai, Ak., U.S. (kĕ-nī′)	40	60°38′N	151°18′W
Kenai Fjords National Park, rec., Ak., U.S.	40	59°45′N	150°00′W
Kenai Mountains, mts., Ak., U.S.	40	60°00′N	150°00′W
Kenai Pen, Ak., U.S.	40	64°40′N	150°18′W
Kendal, S. Afr.	87c	26°03′S	28°58′E
Kenitra, Mor. (kĕ-nē′trȧ)	54	34°21′N	6°34′W
Kennedy, Mount, mtn., Can.	40	60°25′N	138°50′W
Keno Hill, Can.	40	63°58′N	135°18′W
Kenosha, Wi., U.S. (kĕ-nō′shȧ)	39	42°34′N	87°50′W
Kentucky, state, U.S. (kĕn-tŭk′ĭ)	39	37°30′N	87°35′W
Kenya, nation, Afr. (kēn′yȧ)	85	1°00′N	36°53′E
Kenya, Mount (Kirinyaga), mtn., Kenya	85	0°10′S	37°20′E
Keokuk, Ia., U.S. (kē′ô-kŭk)	39	40°24′N	91°34′W
Kerala, state, India	69	16°38′N	76°00′E
Kerang, Austl. (kē-răng′)	81	35°32′S	143°58′E
Kerch, Ukr.	56	45°20′N	36°26′E
Kerempe Burun, c., Tur.	55	42°00′N	33°20′E
Kerguélen, Îles, is., Afr. (kĕr′gå-lĕn)	3	49°50′S	69°30′E
Kerinci, Gunung, mtn., Indon.	74	1°45′S	101°18′E
Keriya *see* Yutian, China	76	36°55′N	81°39′E
Keriya, r., China (kē′rē-yä)	76	37°13′N	81°59′E
Kerkebet, Erit.	70	16°18′N	37°24′E
Kerki, Turkmen. (kĕr′kē)	59	37°52′N	65°15′E
Kérkyra, Grc.	55	39°36′N	19°56′E
Kérkyra, i., Grc.	54	39°33′N	19°36′E
Kermadec Islands, is., N.Z. (kĕr-măd′ĕk)	3	30°30′S	177°00′E
Kermān, Iran (kĕr-män′)	68	30°23′N	57°08′E
Kermānshāh *see* Bakhtarān, Iran	68	34°01′N	47°00′E
Kerulen, r., Asia (kĕr′ōō-lĕn)	77	47°52′N	113°22′E
Keshan, China (kŭ-shän)	77	48°00′N	126°30′E
Kesour, Monts des, mts., Alg.	54	32°51′N	0°30′W
Kestell, S. Afr. (kĕs′tĕl)	87c	28°19′N	28°43′E
Ketamputih, Indon.	67b	1°25′N	102°19′E
Ketapang, Indon. (kē-tȧ-päng′)	74	2°00′S	109°57′E
Key West, Fl., U.S. (kē wĕst′)	39	24°31′N	81°47′W
Khabarovo, Russia (kŭ-bȧr-ôvŏ)	56	69°31′N	60°41′E
Khabarovsk, Russia (ĸä-bä′rôfsk)	57	48°35′N	135°12′E
Khālāpur, India	73b	18°48′N	73°17′E
Khal′mer-Yu, Russia (kŭl-myĕr′-yōō′)	56	67°52′N	64°25′E
Khambhāt, Gulf of, b., India	69	21°20′N	72°27′E
Khammam, India	73	17°09′N	80°13′E
Khānābād, Afg.	72	36°43′N	69°11′E
Khandwa, India	72	21°53′N	76°22′E
Khanka, l., Asia (ĸän′kȧ)	57	45°09′N	133°28′E
Khānpur, Pak.	72	28°42′N	70°42′E
Khanty-Mansiysk, Russia (ĸŭn-te′mŭn-sēsk′)	56	61°02′N	69°01′E
Khān Yūnus, Gaza	67a	31°21′N	34°19′E
Kharagpur, India (kŭ-rŭg′pȯr)	69	22°26′N	87°21′E
Kharkiv, Ukr.	56	50°00′N	36°10′E
Kharkov *see* Kharkiv, Ukr.	56	50°00′N	36°10′E
Khartoum, Sudan	85	15°34′N	32°36′E
Khasavyurt, Russia	58	43°15′N	46°37′E
Khāsh, Iran	68	28°08′N	61°08′E
Khāsh, r., Afg.	68	32°30′N	64°27′E
Khasi Hills, hills, India	69	25°38′N	91°55′E
Khaskovo, Blg. (ĸȧs′kô-vô)	55	41°56′N	25°32′E
Khatanga, Russia (ĸȧ-tän′gȧ)	57	71°48′N	101°47′E
Khatangskiy Zaliv, b., Russia (kä-täŋ′g-skē)	57	73°45′N	108°30′E
Khaybār, Sau. Ar.	68	25°45′N	39°28′E
Kholmsk, Russia (ĸŭlmsk)	57	47°09′N	142°33′E
Khomeynīshahr, Iran	71	32°41′N	51°31′E
Khon Kaen, Thai.	74	16°37′N	102°41′E
Khorog, Taj.	59	37°30′N	71°36′E
Khorramābād, Iran	71	33°30′N	48°20′E
Khorramshahr, Iran (kô-ram′shär)	68	30°36′N	48°15′E
Khudzhand, Taj.	59	40°17′N	69°37′E
Khulna, Bngl.	69	22°50′N	89°38′E
Khūryān Mūryān, is., Oman	68	17°27′N	56°02′E
Khvoy, Iran	68	38°32′N	45°01′E
Khyber Pass, p., Asia (kī′bĕr)	69	34°28′N	71°18′E
Kiel, Ger. (kēl)	50	54°19′N	10°08′E
Kiev (Kyïv), Ukr.	56	50°27′N	30°30′E
Kikládes, is., Grc.	54	37°30′N	24°45′E
Kilbuck Mountains, mts., Ak., U.S. (kĭl-bŭk)	40	60°05′N	160°00′W
Kilimanjaro, mtn., Tan. (kyl-ē-măn-jä′rô)	85	3°09′S	37°19′E
Kilis, Tur. (kē′lēs)	55	36°50′N	37°20′E
Kinabalu, Gunong, mtn., Malay.	74	5°45′N	115°26′E
Kingston, Austl. (kĭngz′tŭn)	80	37°52′S	139°52′E
Kingston, Jam.	43	18°00′N	76°45′W
Kingston, N.Y., U.S.	39	42°00′N	74°00′W
Kingston upon Hull, Eng., U.K.	50	53°45′N	0°25′W
Kingstown, St. Vin. (kĭngz′toun)	43	13°10′N	61°14′W
Kinshasa, D.R.C.	85	4°18′S	15°18′E
Kirakira, Sol. Is.	78e	10°27′S	161°55′E
Kirensk, Russia (kē-rĕnsk′)	57	57°47′N	108°22′E
Kirgiz Range, mts., Asia	59	42°30′N	74°00′E
Kiribati, nation, Oc.	3	1°30′S	173°00′E
Kirin *see* Chilung, Tai.	77	25°02′N	121°48′E
Kiritimati, i., Kir.	2	2°20′N	157°40′W
Kirklareli, Tur. (kērk′lȧr-ĕ′lē)	55	41°44′N	27°15′E
Kirksville, Mo., U.S. (kûrks′vĭl)	39	40°12′N	92°35′W
Kirov, Russia	56	58°35′N	49°35′E
Kirovakan, Arm.	58	40°48′N	44°30′E
Kirovsk, Russia	56	67°40′N	33°58′E
Kırşehir, Tur. (kēr-shĕ′hēr)	68	39°10′N	34°00′E
Kīrthar Range, mts., Pak. (kĭr-tŭr)	69	27°00′N	67°10′E
Kiselëvsk, Russia (kē-sĭ-lyôfsk′)	56	54°00′N	86°39′E
Kishinev *see* Chişinău, Mol.	56	47°02′N	28°52′E
Kislovodsk, Russia	58	43°55′N	42°44′E
Kitakyūshū, Japan	77	33°53′N	130°50′E
Kızıl, r., Tur.	68	40°00′N	34°00′E
Kizlyarskiy Zaliv, b., Russia	58	44°33′N	46°55′E
Klamath Falls, Or., U.S.	38	42°13′N	121°49′W
Klawock, Ak., U.S. (klā′wăk)	40	55°32′N	133°10′W
Klerksdorp, S. Afr. (klĕrks′dôrp)	87c	26°52′S	26°40′E
Klerksraal, S. Afr. (klĕrks′kräl)	87c	26°15′N	27°10′E
Klip, r., S. Afr. (klĭp)	87c	27°18′N	29°25′E
Klipgat, S. Afr.	87c	25°26′S	27°57′E
Klyuchevskaya, vol., Russia (klyōō-chĕfskä′yä)	57	56°13′N	160°00′E
Knezha, Blg. (knyā′zhȧ)	55	43°27′N	24°03′E
Knob Peak, mtn., Phil. (nŏb)	75a	12°30′N	121°20′E
Knoxville, Tn., U.S.	39	35°58′N	83°55′W
Kōbe, Japan (kō′bĕ)	77	34°30′N	135°10′E
København *see* Copenhagen, Den.	50	55°43′N	12°27′E
Kobuk, r., Ak., U.S. (kō′bŭk)	40	66°58′N	158°48′W
Kobuk Valley National Park, rec., Ak., U.S.	40	67°20′N	159°00′W
Kochi, India	73	9°58′N	76°19′E
Kōchi, Japan (kō′chē)	77	33°35′N	133°32′E
Kodiak Island, i., Ak., U.S.	40	57°24′N	153°32′W
Kohīma, India (kô-ē′mȧ)	69	25°45′N	94°41′E
Kokand, Uzb. (kô-känt′)	59	40°27′N	71°07′E
Koko Nor (Qinghai Hu), l., China (kō′kô nor) (chyĭŋ-hī hōō)	76	37°26′N	98°30′E
Kokopo, Pap. N. Gui. (kô-kô′pō)	75	4°25′S	152°27′E
Kökshetaū, Kaz.	59	53°15′N	69°13′E
Kola Peninsula *see* Kol′skiy Poluostrov, pen., Russia	56	67°15′N	37°40′E
Kolār (Kolār Gold Fields), India (kôl-är′)	69	13°39′N	78°33′E
Kolguyev, i., Russia (kôl-gò′yĕf)	56	69°00′N	49°00′E
Kolhāpur, India	73	16°48′N	74°15′E
Kolkata (Calcutta), India	69	22°32′N	88°22′E
Kolpashevo, Russia (kŭl pȧ shô′vȧ)	56	58°16′N	82°43′E
Kol′skiy Poluostrov, pen., Russia	56	67°15′N	37°40′E
Kolyma, r., Russia	57	66°30′N	151°45′E
Kolymskiy Mountains *see* Gydan, Khrebet, mts., Russia	57	61°45′N	155°00′E
Komandorskiye Ostrova, is., Russia	67	55°40′N	167°13′E
Komsomol′sk-na-Amure, Russia	57	50°46′N	137°14′E
Koné, N. Cal.	78f	21°04′S	164°52′E
Königsberg *see* Kaliningrad, Russia	56	54°42′N	20°32′E
Konnagar, India	72a	22°41′N	88°22′E
Konqi, r., China (kôn-chyē)	76	41°09′N	87°46′E
Konya, Tur. (kōn′yȧ)	68	36°55′N	32°25′E
Koppeh Dāgh, mts., Asia	68	37°28′N	58°29′E
Koppies, S. Afr.	87c	27°15′S	27°35′E
Korea, North, nation, Asia	77	40°00′N	127°00′E
Korea, South, nation, Asia	77	36°30′N	128°00′E
Korean Archipelago, is., Kor., S.	77	34°05′N	125°35′E
Korea Strait, strt., Asia	77	33°30′N	128°30′E
Korinthiakós Kólpos, b., Grc.	55	38°15′N	22°33′E
Kórinthos, Grc. (kô-rēn′thôs) (kŏr′ĭnth)	50	37°56′N	22°54′E
Korla, China (kōr-lä)	76	41°37′N	86°03′E
Koro Sea, sea, Fiji	78g	18°00′S	179°50′E
Korsakov, Russia (kôr′sȧ-kôf′)	57	46°42′N	143°16′E
Koryakskiy Khrebet, mts., Russia	57	62°00′N	168°45′E
Kosciuszko, Mount, mtn., Austl.	81	36°26′S	148°20′E
Kosi, r., India (kô′sē)	72	26°00′N	86°20′E
Koster, S. Afr.	87c	25°52′S	26°52′E
Kostroma, Russia (kôs-trô-mȧ′)	56	57°46′N	40°55′E
Kota, India	69	25°17′N	75°49′E
Kota Baharu, Malay. (kō′tä bä′rōō)	74	6°15′N	102°23′E
Kotabaru, Indon.	74	3°22′S	116°15′E
Kota Kinabalu, Malay.	74	5°55′N	116°05′E
Kota Tinggi, Malay.	67b	1°43′N	103°54′E
Kotel′nyy, i., Russia (kô-tyĕl′nē)	57	74°51′N	134°09′E
Kotzebue Sound, strt., Ak., U.S.	40	67°00′N	164°28′W
Koumac, N. Cal.	78f	20°33′S	164°17′E
Kovno *see* Kaunas, Lith.	56	54°42′N	23°54′E
Koyuk, Ak., U.S. (kô-yōōk′)	40	65°00′N	161°18′W
Koyukuk, r., Ak., U.S. (kô-yōō′kȯk)	40	66°25′N	153°50′W
Kozáni, Grc.	55	40°16′N	21°51′E
Kozhikode, India	69	11°19′N	75°49′E
Kra, Isthmus of, isth., Asia	74	9°30′S	99°45′E
Krâchéh, Camb.	74	12°28′N	106°06′E
Kragujevac, Serb. (krä′gōō′yĕ-vȧts)	55	44°01′N	20°55′E
Kraków, Pol. (krä′kȯf)	50	50°05′N	20°00′E
Kraljevo, Serb. (kräl′ye-vô)	55	43°39′N	20°48′E
Kranj, Slvn. (krän′)	54	46°16′N	14°23′E
Krasnodar, Russia (kräs′nô-dȧr)	56	45°03′N	38°55′E
Krasnotur′insk, Russia (krŭs-nŭ-tōō-rensk′)	56	59°47′N	60°15′E
Krasnoufimsk, Russia (krŭs-nŭ-ōō-fēmsk′)	56	56°38′N	57°46′E
Krasnoyarsk, Russia (kräs-nô-yȧrsk′)	57	56°13′N	93°12′E
Krestovyy, Pereval, p., Geor.	58	42°32′N	44°28′E
Krishna, r., India	69	16°00′N	79°00′E
Krishnanagar, India	72	23°29′N	88°33′E
Kristiansand, Nor. (krĭs-tyȧn-sän′′)	50	58°09′N	7°59′E
Krivoy Rog *see* Kryvyi Rih, Ukr.	56	47°54′N	33°22′E
Krokodil, r., S. Afr. (krŏ′kŏ-dī)	87c	24°25′S	27°08′E
Krung Thep *see* Bangkok, Thai.	74	13°50′N	100°29′E
Kryvyi Rih, Ukr.	56	47°54′N	33°22′E
Ksar-el-Kebir, Mor.	54	35°01′N	5°48′W
Ksar-es-Souk, Mor.	54	31°58′N	4°25′W
Kuala Klawang, Malay.	67b	2°57′N	102°04′E
Kuala Lumpur, Malay. (kwä′lä lȯm-pōōr′)	74	3°08′N	101°42′E
Kuching, Malay. (kōō′chĭng)	74	1°30′N	110°26′E
Kudap, Indon.	67b	1°14′N	102°30′E
Kudat, Malay. (kōō-dät′)	74	6°56′N	116°48′E
Kudymkar, Russia (kōō-dĭm-kär′)	56	58°43′N	54°52′E
Kuibyshev	56	53°10′N	50°05′E
Kuji, Japan	77	40°11′N	141°46′E
Kula, Tur.	55	38°32′N	28°30′E
Kula Kangri, mtn., Bhu.	69	33°11′N	90°36′E
Kulunda, Russia (kȯ-lȯn′dȧ)	56	52°38′N	79°00′E
Kumamoto, Japan (kōō′mä-mō′tô)	77	32°49′N	130°40′E
Kumbakonam, India (kȯm′bŭ-kō′nŭm)	69	10°59′N	79°25′E
Kumta, India	73	14°19′N	75°28′E
Kumul *see* Hami, China	76	42°58′N	93°14′E
Kunashir (Kunashiri), i., Russia (kōō-nŭ-shēr′)	77	44°00′N	145°45′E
Kundur, i., Indon.	67b	0°49′N	103°20′E
Kungur, Russia (kȯn-gōōr′)	56	57°27′N	56°53′E
Kunlun Shan, mts., China (kōōn-lōōn shän)	76	35°26′N	83°09′E
Kunming, China (kōōn-mĭŋ)	76	25°10′N	102°50′E
Kunsan, Kor., S. (kȯn′sän′)	77	35°54′N	126°46′E
Kupang, Indon.	75	10°14′S	123°37′E
Kupino, Russia (kōō-pĭ′nô)	56	54°00′N	77°47′E
Kuqa, China (kōō-chyä)	76	41°34′N	82°44′E
Kurdistan, hist. reg., Asia (kûrd′ĭ-stăn)	68	37°40′N	43°30′E
Kure, Japan (kōō′rĕ)	77	34°17′N	132°35′E
Kurgan, Russia (kȯr-găn′)	56	55°28′N	65°14′E
Kurgan-Tyube, Taj. (kȯr-gän′ tyȯ′bĕ)	59	38°00′N	68°49′E
Kurla, neigh., India	73b	19°03′N	72°53′E
Kurnool, India (kȯr-nōōl′)	69	16°00′N	78°04′E
Kursk, Russia (kȯrsk)	56	51°44′N	36°08′E
Kurume, Japan (kōō′rȯ-mĕ)	77	33°10′N	130°30′E
Kushiro, Japan (kōō′shē-rō)	77	43°00′N	144°22′E
Kushva, Russia (kōōsh′vȧ)	56	58°18′N	59°51′E
Kuskokwim, r., Ak., U.S.	40	61°32′N	160°36′W
Kuskokwim Bay, b., Ak., U.S. (kŭs′kô-kwĭm)	40	59°25′N	163°14′W
Kuskokwim Mountains, mts., Ak., U.S.	40	62°08′N	158°00′W
Kuskovak, Ak., U.S. (kŭs-kō′văk)	40	60°10′N	162°50′W
Kütahya, Tur. (kû-tä′hyȧ)	68	39°20′N	29°50′E
Kutch, Gulf of, b., India	69	22°45′N	68°33′E
Kutch, Rann of, sw., Asia	69	23°59′N	69°13′E
Kutulik, Russia (kȯ tōō′lyĭk)	57	53°12′N	102°51′E
Kuwait *see* Al Kuwayt, Kuw.	68	29°04′N	47°59′E
Kuwait, nation, Asia	68	29°00′N	48°45′E
Kuybyshevskoye, res., Russia	56	53°40′N	49°00′E
Kuznetsk Basin, basin, Russia	56	56°30′N	86°15′E
Kyakhta, Russia (kyäĸ′ta)	57	51°00′N	107°30′E
Kyaukpyu, Mya. (chouk′pyoo′)	69	19°19′N	93°33′E
Kyïv *see* Kiev, Ukr.	56	50°27′N	30°30′E
Kyïvs′ke vodoskhovyshche, res., Ukr.	56	51°00′N	30°20′E
Kynuna, Austl. (kī-nōō′nȧ)	81	21°30′S	142°12′E
Kyŏngju, Kor., S. (kyŭng′yōō)	77	35°48′N	129°12′E
Kyōto, Japan (kyō′tō′)	77	35°00′N	135°46′E
Kyparissía, Grc.	55	37°17′N	21°43′E
Kyren, Russia (kĭ-rĕn′)	57	51°46′N	102°13′E
Kyrgyzstan, nation, Asia	56	41°45′N	74°38′E
Kýthira, i., Grc.	55	36°15′N	22°56′E
Kyūshū, i., Japan	77	33°00′N	131°00′E
Kyustendil, Blg. (kyȯs-tĕn-dĭl′)	55	42°16′N	22°39′E
Kyzyl, Russia (kĭ zĭl)	57	51°37′N	93°38′E
Kyzyl-Kum, des., Asia	56	42°47′N	64°45′E

L

PLACE (Pronunciation)	PAGE	LAT.	LONG.
Laas Caanood, Som.	87a	8°24′N	47°20′E
Labis, Malay. (läb′ĭs)	67b	2°23′N	103°01′E
Labo, Phil. (lä′bô)	75a	14°11′N	122°49′E
Labo, Mount, mtn., Phil.	75a	14°00′N	122°47′E
Labrador, reg., Can. (lăb′rȧ-dôr)	36	53°05′N	63°30′W
Labuan, Pulau, i., Malay. (lä-bȯ-än′)	74	5°28′N	115°11′E
Labuha, Indon.	75	0°43′S	127°35′E
Laccadive Islands *see* Lakshadweep, is., India	69	11°00′N	73°02′E
Laccadive Sea, sea, Asia	73	9°10′N	75°17′E
La Ceiba, Hond. (lȧ sēbä)	42	15°45′N	86°52′W
La Crosse, Wi., U.S.	39	43°48′N	91°14′W
Lādīz, Iran	71	28°56′N	61°19′E
Lādnun, India (läd′nȯn)	72	27°45′N	74°20′E

ng-sing; ŋ-baŋk; ɴ-nasalized n; nŏd; cŏmmit; ōld; ôbey; ôrder; oi-boil; fōōd; ȯ-as oo in foot; ou-out; s-soft; sh-dish; th-thin; pūre; ûnite; ûrn; stŭd; circŭs; ü-as in French tu; ′-indeterminate vowel.

124 Ladoga, Lake – Madīnat ash Sha'b PRONOUNCING INDEX

PLACE (Pronunciation)	PAGE	LAT.	LONG.
Ladoga, Lake *see* Ladozhskoye Ozero, l., Russia	56	60°59′N	31°30′E
Ladozhskoye Ozero, l., Russia (lȧ-dôsh′skô-yē ô′zĕ-rô)	56	60°59′N	31°30′E
Lae, Pap. N. Gui. (lä′å)	75	6°15′S	146°57′E
Lafayette, In., U.S.	39	40°25′N	86°55′W
Lafayette, La., U.S.	39	30°15′N	92°02′W
La Galite, i., Tun. (gä-lēt)	54	37°36′N	8°03′E
Lågan, r., Nor. (lô′ghĕn)	52	61°00′N	10°00′E
Lagarto, r., Pan. (lä-gä′r-tô)	42a	9°08′N	80°05′W
Lagonay, Phil.	75a	13°44′N	123°31′E
Lagos, Nig. (lä′gōs)	85	6°27′N	3°24′E
Lagos de Moreno, Mex. (lä′gōs dā mỏ-rā′nō)	42	21°21′N	101°55′W
La Grande, Or., U.S. (lȧ grănd′)	38	45°20′N	118°06′W
La Grange, Austl. (lä grānj)	80	18°40′S	122°00′E
La Grange, Ga., U.S. (lȧ-grānj′)	39	33°01′N	85°00′W
La Habana *see* Havana, Cuba	43	23°08′N	82°23′W
Lāhījān, Iran	71	37°12′N	50°01′E
Lahore, Pak. (lä-hōr′)	69	32°00′N	74°18′E
Laizhou Wan, b., China (lī-jō wän)	77	37°22′N	119°19′E
Lake Brown, Austl. (broun)	80	31°03′S	118°30′E
Lake Charles, La., U.S. (chärlz′)	39	30°15′N	93°14′W
Lake Clark National Park, rec., Ak., U.S.	40	60°30′N	153°15′W
Lakeland, Fl., U.S. (lāk′lănd)	39	28°02′N	81°58′W
Lakewood, Oh., U.S.	39	41°29′N	81°48′W
Lakshadweep, state, India	69	10°10′N	72°50′E
Lakshadweep, is., India	69	11°00′N	73°02′E
La Línea, Spain (lä lē′nå-ä)	54	36°11′N	5°22′W
Lalitpur, Nepal	69	27°23′N	85°24′E
La Marmora, Punta, mtn., Italy (lä-mä′r-mô-rä)	54	40°00′N	9°28′E
Lambasa, Fiji	78g	16°26′S	179°24′E
Lamía, Grc. (lȧ-mē′ȧ)	55	38°54′N	22°25′E
Lamon Bay, b., Phil. (lä-mōn′)	74	14°35′N	121°52′E
Lampazos, Mex. (läm-pä′zōs)	42	27°03′N	100°30′W
Lampedusa, i., Italy (läm-på-do͞o′sä)	54	35°29′N	12°58′E
Lanak La, p., China	76	34°40′N	79°50′E
Lancaster, Pa., U.S.	39	40°05′N	76°20′W
Lands End, c., Eng., U.K.	52	50°03′N	5°45′W
Langat, r., Malay.	67b	2°46′N	101°33′E
Langla Co, l., China (läŋ-lä tswo)	72	30°42′N	80°40′E
Langsa, Indon. (läng′sȧ)	74	4°33′N	97°52′E
Lang Son, Viet. (lăng′sŏn′)	74	21°52′N	106°42′E
Langzhong, China (läŋ-jỏŋ)	76	31°40′N	106°05′E
Länkäran, Azer. (lĕn-kỏ-rän′)	56	38°52′N	48°58′E
Lansing, Mi., U.S.	39	42°45′N	84°35′W
Lanzhou, China (län-jō)	76	35°55′N	103°55′E
Laoag, Phil. (lä-wäg′)	74	18°13′N	120°38′E
Laos, nation, Asia (lā-ōs) (lȧ-ōs′)	74	20°15′N	102°00′E
La Paz, Bol.	46	16°31′S	68°03′W
La Paz, Mex.	42	24°00′N	110°15′W
Lapland, hist. reg., Eur. (lăp′lȧnd)	50	68°20′N	22°00′E
Laptev Sea, sea, Russia (läp′tyĭf)	57	75°39′N	120°00′E
L'Aquila, Italy (lȧ′kē-lä)	54	42°22′N	13°24′E
Lār, Iran (lär)	68	27°31′N	54°12′E
Laramie, Wy., U.S. (lăr′ȧ-mĭ)	38	41°20′N	105°40′W
Laredo, Tx., U.S.	38	27°31′N	99°29′W
Lárisa, Grc. (lä′rḗ-sȧ)	55	39°38′N	22°25′E
Lārkāna, Pak.	72	27°40′N	68°12′E
Larnaka, Cyp.	55	34°55′N	33°37′E
Lárnakos, Kólpos, b., Cyp.	67a	36°50′N	33°45′E
La Rochelle, Fr. (lȧ rỏ-shĕl′)	50	46°10′N	1°09′W
La Sagra, mtn., Spain (lä sä′grä)	54	37°56′N	2°35′W
Las Cruces, N.M., U.S.	38	32°20′N	106°50′W
Lashio, Mya. (lăsh′ḗ-ō)	76	22°58′N	98°03′E
La Spezia, Italy (lä-spĕ′zyä)	50	44°07′N	9°48′E
Lass Qoray, Som.	87a	11°13′N	48°19′E
Las Tres Vírgenes, Volcán, vol., Mex. (vē′r-hĕ-nĕs)	42	26°00′N	111°45′W
Las Vegas, N.M., U.S.	38	35°36′N	105°13′W
Las Vegas, Nv., U.S. (läs vā′gäs)	38	36°12′N	115°10′W
Latakia *see* Al Lādhiqīyah, Syria	68	35°32′N	35°51′E
Lātūr, India (lä-to͞or′)	72	18°20′N	76°35′E
Latvia, nation, Eur.	56	57°28′N	24°29′E
Lau Group, is., Fiji	78g	18°20′S	178°30′W
Launceston, Austl. (lôn′sĕs-tŭn)	81	41°35′S	147°22′E
La Unión, Spain	54	37°38′N	0°50′W
Laura, Austl. (lôrȧ)	81	15°40′S	144°45′E
Laurel, Ms., U.S.	39	31°42′N	89°07′W
Laurentian Highlands, hills, Can. (lô′rĕn-tĭ-ȧn)	37	49°00′N	74°50′W
Lauria, Italy (lou′rḗ-ä)	55	40°03′N	15°02′E
Lausanne, Switz. (lō-zȧn′)	50	46°32′N	6°35′E
Laut, Pulau, i., Indon.	74	3°39′S	116°07′E
Laut Kecil, Kepulauan, is., Indon.	74	4°44′S	115°43′E
Lautoka, Fiji	78g	17°37′S	177°27′E
Laverton, Austl. (lā′vĕr-tŭn)	80	28°45′S	122°30′E
Lawrence, Ks., U.S.	39	38°57′N	95°13′W
Lawton, Ok., U.S. (lô′tŭn)	38	34°36′N	98°25′W
Lawz, Jabal al, mtn., Sau. Ar.	68	28°46′N	35°37′E
Layang Layang, Malay. (lä-yäng′ lä-yäng′)	67b	1°49′N	103°28′E
Lead, S.D., U.S. (lēd)	38	44°22′N	103°47′W
Leavenworth, Ks., U.S. (lĕv′ĕn-wûrth)	39	39°19′N	94°54′W
Lebam, r., Malay.	67b	1°35′N	104°09′E
Lebanon, nation, Asia	68	34°00′N	34°00′E
Lecce, Italy (lĕt′chā)	55	40°22′N	18°11′E
Leeds, Eng., U.K.	50	53°48′N	1°33′W
Leeward Islands, is., N.A. (lē′wĕrd)	43	17°00′N	62°15′W
Lefkáda, i., Grc.	55	38°42′N	20°22′E
Legazpi, Phil. (lå-gäs′pḗ)	75	13°09′N	123°44′E
Leghorn *see* Livorno, Italy	50	43°32′N	11°18′E
Leh, India (lā)	72	34°10′N	77°40′E

PLACE (Pronunciation)	PAGE	LAT.	LONG.
Le Havre, Fr. (lĕ ȧv′r′)	50	49°31′N	0°07′E
Leicester, Eng., U.K. (lĕs′tĕr)	50	52°37′N	1°08′W
Leipzig, Ger. (līp′tsĭk)	50	51°20′N	12°24′E
Leizhou Bandao, pen., China (lā-jō bän-dou)	76	20°42′N	109°10′E
Lemery, Phil. (lā-mā-rē′)	75a	13°51′S	120°55′E
Lena, r., Russia	57	68°00′N	123°00′E
Lenik, r., Malay.	67b	1°59′N	102°51′E
Leningrad *see* Saint Petersburg, Russia	56	59°57′N	30°20′E
Leninogorsk, Kaz.	59	50°29′N	83°25′E
Lenīnsk, Kaz.	59	45°39′N	63°19′E
Leninsk-Kuznetski, Russia (lyĕ-nĕnsk′ko͞oz-nyĕt′skĭ)	56	54°28′N	86°48′E
León, Mex. (lå-ōn′)	42	21°08′N	101°41′W
León, Nic. (lĕ-ô′n)	42	12°28′N	86°53′W
León, Spain (lĕ-ô′n)	54	42°38′N	5°33′W
Lerdo, Mex. (lĕr′dō)	42	25°31′N	103°30′W
Lerwick, Scot., U.K. (lĕr′ĭk) (lûr′wĭk)	50	60°08′N	1°27′W
Lesbos *see* Lésvos, i., Grc.	52	39°15′N	25°40′E
Leshan, China (lŭ-shän)	76	29°40′N	103°40′E
Leslie, S. Afr.	87c	26°23′S	28°57′E
Lesotho, nation, Afr. (lĕsō′thỏ)	85	29°45′S	28°07′E
Lesser Antilles, is.,	43	12°15′N	65°00′W
Lesser Caucasus, mts., Asia	58	41°00′N	44°35′E
Lesser Khingan Range, mts., China	77	49°50′N	129°26′E
Lesser Sunda Islands, is., Indon.	74	9°00′S	120°00′E
Lésvos, i., Grc.	52	39°15′N	25°40′E
Levuka, Fiji	78g	17°41′S	178°50′E
Lewiston, Id., U.S. (lū′ĭs-tŭn)	38	46°24′N	116°59′W
Lewiston, Me., U.S.	39	44°05′N	70°14′W
Lewistown, Mt., U.S.	38	47°05′N	109°25′W
Lexington, Ky., U.S. (lĕk′sĭng-tŭn)	39	38°05′N	84°30′W
Leyte, i., Phil. (lā′tā)	75	10°35′N	125°35′E
Lhasa, China (läs′ä)	76	29°41′N	91°12′E
Lianyungang, China (lĭĕn-yỏn-gäŋ)	77	34°35′N	119°09′E
Liao, r., China	77	43°37′N	120°05′E
Liaodong Bandao, pen., China (lĭou-dỏŋ bän-dou)	77	39°45′N	122°22′E
Liaoning, prov., China	77	41°31′N	122°11′E
Liaoyang, China (lyä′ō-yäng′)	77	41°18′N	123°10′E
Liberia, nation, Afr. (lī-bē′rĭ-ȧ)	85	6°30′N	9°55′W
Libreville, Gabon (lē-br′vēl′)	85	0°23′N	9°27′E
Libya, nation, Afr. (lĭb′ḗ-ä)	85	27°38′N	15°00′E
Libyan Desert, des., Afr. (lĭb′ḗ-ăn)	85	28°23′N	23°34′E
Libyan Plateau, plat., Afr.	70	30°58′N	26°20′E
Lichtenburg, S. Afr. (lĭk′tĕn-bĕrgh)	87c	26°09′S	26°10′E
Liebenbergsvlei, r., S. Afr.	87c	27°35′S	28°25′E
Liechtenstein, nation, Eur. (lēk′tĕn-shtīn)	50	47°10′N	10°00′E
Ligao, Phil. (lḗ-gä′ỏ)	75a	13°14′N	123°33′E
Ligurian Sea, sea, Eur. (lĭ-gū′rĭ-ăn)	54	43°42′N	8°32′E
Lijiang, China (lē-jyäŋ)	76	27°00′N	100°08′E
Lille, Fr. (lēl)	50	50°38′N	3°01′E
Lilongwe, Mwi. (lḗ-lỏ-än)	85	13°59′S	33°44′E
Lima, Oh., U.S. (lī′mȧ)	39	40°40′N	84°05′W
Lima, Peru (lē′mä)	46	12°06′S	76°55′W
Limassol, Cyp.	55	34°39′N	33°02′E
Limbdi, India	72	22°37′N	71°52′E
Limnos, i., Grc.	55	39°58′N	24°48′E
Limón, C.R. (lḗ-mōn′)	43	10°01′N	83°02′W
Limón, Bahía, b., Pan.	42a	9°21′N	79°58′W
Limpopo, r., Afr. (lĭm-pō′pō)	85	23°15′S	27°46′E
Linares, Mex.	42	24°53′N	99°34′W
Linares, Spain (lē-nä′rĕs)	54	38°07′N	3°38′W
Linchuan, China (lĭn-chůän)	77	27°58′N	116°18′E
Lincoln, Ne., U.S.	38	40°49′N	96°43′W
Lindesnes, c., Nor. (lĭn′ĕs-nĕs)	52	58°00′N	7°05′E
Lindley, S. Afr. (lĭnd′lḗ)	87c	27°52′S	27°55′E
Linfen, China	77	36°00′N	111°38′E
Linga, Kepulauan, is., Indon.	74	0°35′S	105°05′E
Lingayen, Phil. (lĭŋ′gä-yān′)	74	16°01′N	120°13′E
Lingayen Gulf, b., Phil.	75a	16°18′N	120°11′E
Linqing, China (lĭn-chyĭŋ)	77	36°49′N	115°42′E
Linyi, China (lĭn-yē)	77	35°04′N	118°21′E
Lion, Golfe du, b., Fin.	52	43°00′N	4°00′E
Lipa, Phil. (lḗ-pä′)	74	13°55′N	121°10′E
Lipetsk, Russia (lyē′pĕtsk)	56	52°26′N	39°34′E
Liping, China (lē-pĭŋ)	76	26°18′N	109°00′E
Lisboa *see* Lisbon, Port.	50	38°42′N	9°05′W
Lisbon (Lisboa), Port.	50	38°42′N	9°05′W
Lishui, China	77	28°28′N	120°00′E
Lismore, Austl. (lĭz′môr)	81	28°48′S	153°18′E
Litani, r., Leb.	67a	33°28′N	35°42′E
Lithgow, Austl. (lĭth′gō)	81	33°23′S	149°31′E
Lithuania, nation, Eur. (lĭth-ů-ā-′nĭ-ȧ)	56	55°42′N	23°30′E
Little America, hist., Ant.	82	78°30′S	161°30′W
Little Andaman, i., India (ăn-dȧ-măn′)	74	10°39′N	93°08′E
Little Bitter Lake, l., Egypt	87b	30°10′N	32°36′E
Little Hans Lollick, i., V.I.U.S. (häns lôl′lĭk)	43c	18°25′N	64°54′W
Little Rock, Ar., U.S. (rôk)	39	34°42′N	92°16′W
Liuzhou, China (lĭỏ-jō)	76	24°25′N	109°30′E
Livengood, Ak., U.S. (līv′ĕn-gỏd)	40	65°30′N	148°35′W
Liverpool, Eng., U.K.	50	53°25′N	2°52′W
Liverpool Bay, b., Can.	40	69°45′N	130°00′W
Livingston, Mt., U.S.	38	45°40′N	110°35′W
Livno, Bos. (lēv′nỏ)	55	43°50′N	17°03′E
Livorno, Italy (lḗ-vôr′nō) (lĕg′hỏrn)	50	43°32′N	11°18′E
Ljubljana, Slvn. (lyo͞o′blyä′na)	50	46°04′N	14°29′E
Llanes, Spain (lyä′nås)	54	43°25′N	4°41′W
Llanos, reg., S.A. (lyȧ′nōs)	46	4°00′N	71°15′W
Lleida, Spain	54	41°38′N	0°37′E
Lobo, Phil.	75a	13°39′N	121°14′E
Loc Ninh, Viet. (lōk′nĭng′)	74	12°00′N	106°30′E
Lod, Isr. (lôd)	67a	31°57′N	34°55′E
Lodhran, Pak.	72	29°40′N	71°39′E

PLACE (Pronunciation)	PAGE	LAT.	LONG.
Łódź, Pol.	50	51°46′N	19°30′E
Lofoten, is., Nor. (lỏ′fō-tĕn)	52	68°26′N	13°42′E
Logan, Mount, mtn., Can.	36	60°54′N	140°33′W
Logan, Ut., U.S.	38	41°46′N	111°51′W
Logansport, In., U.S. (lō′gănz-pōrt)	39	40°45′N	86°25′W
Logroño, Spain (lỏ-grō′nyō)	54	42°28′N	2°25′W
Loire, r., Fr.	52	47°30′N	2°00′E
Lokala Drift, Bots. (lỏ′kä-lȧ drĭft)	87c	24°00′S	26°38′E
Lom, Blg. (lōm)	55	43°48′N	23°15′E
Lomblen, Pulau, i., Indon. (lŏm-blĕn′)	75	8°08′S	123°45′E
Lombok, i., Indon. (lŏm-bŏk′)	74	9°15′S	116°15′E
Lomé, Togo	85	6°08′N	1°13′E
London, Eng., U.K.	50	51°30′N	0°07′W
Long, i., Bah.	43	23°25′N	75°10′W
Long Beach, Ca., U.S. (lông bēch)	38	33°46′N	118°12′W
Long Island, i., N.Y., U.S. (lông)	39	40°50′N	72°50′W
Long Island, i., Pap. N. Gui.	75	5°10′S	147°30′E
Longreach, Austl. (lông′rēch)	81	23°32′S	144°17′E
Longxi, China (lỏŋ-shyē)	76	35°00′N	104°40′E
Long Xuyen, Viet. (loung′ so͞o′yĕn)	74	10°31′N	105°28′E
Longzhou, China (lỏŋ-jō)	76	22°20′N	107°02′E
Looc, Phil. (lỏ-ōk′)	75a	12°16′N	121°59′E
Lopatka, Mys, c., Russia (lỏ-pät′kȧ)	67	51°00′N	156°52′E
Lopez Bay, b., Phil. (lō′pāz)	75a	14°04′N	122°00′E
Loralai, Pak. (lō-rŭ-lī′)	69	30°31′N	68°35′E
Lorca, Spain (lôr′kä)	54	37°39′N	1°40′W
Los Angeles, Ca., U.S.	38	34°03′N	118°14′W
Los Reyes, Mex.	42	19°35′N	102°29′W
Louangphrabang, Laos (lo͞o-ang′prä-bäng′)	74	19°47′N	102°15′E
Louisiana, state, U.S.	39	30°50′N	92°50′W
Louisville, Ky., U.S.	39	38°15′N	85°45′W
Lowell, Ma., U.S.	39	42°38′N	71°18′W
Lower California *see* Baja California, pen., Mex.	37	28°00′N	113°30′W
Loznica, Serb. (lōz′nḗ-tsȧ)	55	44°31′N	19°16′E
Luan, r., China	77	41°25′N	117°15′E
Luanda, Ang. (lo͞o-än′dä)	85	8°48′S	13°14′E
Luarca, Spain (lwä′kä)	54	43°33′N	6°30′W
Lubang, Phil. (lo͞o-bäng′)	75a	13°49′N	120°07′E
Lubang Islands, is., Phil.	74	13°47′N	119°56′E
Lubbock, Tx., U.S.	38	33°35′N	101°50′E
Lübeck, Ger. (lü′bĕk)	50	53°53′N	10°42′E
Lublin, Pol. (lyỏ′blḗn′)	50	51°14′N	22°33′E
Lubuagan, Phil. (lỏ-bwä-gä′n)	75a	17°24′N	121°11′E
Lubumbashi, D.R.C.	85	11°40′S	27°28′E
Lucca, Italy (lo͝ok′kä)	54	43°51′N	10°29′E
Lucena, Phil. (lo͞o-sā′nä)	75a	13°55′N	121°36′E
Lucena, Spain (lo͞o-thā′nä)	54	37°25′N	4°28′W
Lucipara, Kepulauan, is., Indon.	75	5°45′S	128°15′E
Lucknow, India (lŭk′nou)	69	26°54′N	80°58′E
Ludhiāna, India	69	31°00′N	75°52′E
Lugo, Spain (lo͞o′gō)	54	43°01′N	7°32′W
Lugoj, Rom.	55	45°51′N	21°56′E
Luhans'k, Ukr.	56	48°34′N	39°18′E
Lulong, China (lo͞o-lỏŋ)	77	39°54′N	118°53′E
Lumajangdong Co, l., China	72	34°00′N	81°47′E
Lün, Mong.	76	47°58′N	104°52′E
Luna, Phil. (lo͞o′nȧ)	75a	16°51′N	120°22′E
Luoyang, China (lwỏ-yäŋ)	77	34°45′N	112°32′E
Lusaka, Zam. (lỏ-sä′kȧ)	85	15°25′S	28°17′E
Lüshun, China (lü-shŭn)	77	38°49′N	121°15′E
Lūt, Dasht-e, des., Iran (dă′sht-ē-lo͞ot)	68	31°47′N	58°38′E
Luuq, Som.	87a	3°38′N	42°35′E
Luxembourg, Lux.	50	49°38′N	6°30′E
Luxembourg, nation, Eur.	50	49°30′N	6°22′E
Luzhou, China (lo͞o-jō)	76	28°58′N	105°25′E
Luzon, i., Phil. (lo͞o-zŏn′)	74	17°10′N	119°45′E
Luzon Strait, strt., Asia	65	20°40′N	121°00′E
L'viv, Ukr.	56	49°50′N	24°00′E
L'vov *see* L'viv, Ukr.	56	49°50′N	24°00′E
Lynchburg, Va., U.S. (lĭnch′bûrg)	39	37°23′N	79°08′W
Lynn, Ma., U.S. (lĭn)	39	42°28′N	70°57′W
Lyon, Fr. (lē-ôɴ′)	50	45°44′N	4°52′E

M

PLACE (Pronunciation)	PAGE	LAT.	LONG.
Ma'ān, Jord. (mä-än′)	68	30°12′N	35°45′E
Mabeskraal, S. Afr.	87c	25°12′S	26°47′E
Mabula, S. Afr. (mă′bo͞o-la)	87c	24°49′S	27°59′E
Macalelon, Phil. (mä-kä-lā-lōn′)	75a	13°46′N	122°09′E
Macau, China	77	22°00′N	113°00′E
Macedonia, hist. reg., Eur. (măs-ḗ-dō′nĭ-ȧ)	55	41°05′N	22°15′E
Macedonia, nation, Eur.	50	41°50′N	22°00′E
Machilīpatnam, India	69	16°22′N	81°10′E
Mackay, Austl. (mă-kī′)	81	21°15′S	149°08′E
Mackenzie, r., Can.	36	63°38′N	124°23′W
Mackenzie Bay, b., Can.	40	69°20′N	137°10′W
Macon, Ga., U.S. (mā′kŏn)	39	32°49′N	83°39′W
Macquarie Islands, is., Austl. (mȧ-kwŏr′ḗ)	3	54°36′S	158°45′E
Madagascar, nation, Afr. (măd-ȧ-găs′kȧr)	85	18°05′S	43°12′E
Madanapalle, India	73	13°06′N	78°09′E
Madang, Pap. N. Gui. (mä-däng′)	75	5°15′S	145°45′E
Madeira, r., S.A.	46	6°48′S	62°43′W
Madeira, Arquipélago da, is., Port.	86	33°26′N	16°44′W
Madgaon, India	73	15°09′N	73°58′E
Madhya Pradesh, state, India (mŭd′vŭ prŭ-dāsh′)	69	22°04′N	77°48′E
Madīnat ash Sha'b, Yemen	68	12°45′N	44°00′E

ăt; finăl; rāte; senåte; ärm; åsk; sofȧ; fâre; ch-choose; dh-as th in other; bē; ḗvent; bĕt; recĕnt; cratẽr; g-gō; gh-guttural g; bĭt; ĭ-short neutral; rīde; ᴋ-guttural k as ch in German ich;

PRONOUNCING INDEX Madison – Mildura 125

PLACE (Pronunciation)	PAGE	LAT.	LONG.
Madison, Wi., U.S.	39	43°05′N	89°23′W
Madrakah, Ra's al, c., Oman	68	18°53′N	57°48′E
Madras see Chennai, India	69	13°08′N	80°15′E
Madre, Sierra, mts., Phil.	75a	16°40′N	122°10′E
Madre del Sur, Sierra, mts., Mex. (sē-ĕ′r-rä-mä′drå dĕlsōōr′)	42	17°35′N	100°35′W
Madre Occidental, Sierra, mts., Mex.	42	29°30′N	107°30′W
Madre Oriental, Sierra, mts., Mex.	42	25°30′N	100°45′W
Madrid, Spain (mä-drē′d)	50	40°26′N	3°42′W
Madura, i., Indon. (må-dōō′rä)	74	6°45′S	113°30′E
Madurai, India (mä-dōō′rä)	69	9°57′N	78°04′E
Maebashi, Japan (mä-ĕ-bä′shē)	77	36°26′N	139°04′E
Maestra, Sierra, mts., Cuba (sē-ĕ′r-rä-mä-ās′trä)	43	20°05′N	77°05′W
Magadan, Russia (må-gå-dän′)	57	59°39′N	150°43′E
Magaliesburg, S. Afr.	87c	26°01′S	27°32′E
Magallanes, Estrecho de, strt., S.A.	46	52°30′S	68°45′W
Magat, r., Phil. (mä-gät′)	75a	16°45′N	121°16′E
Magdalena, Mex.	38	30°34′N	110°50′W
Magdalena, Bahía, b., Mex. (bä-ē′ä-mäg-dä-lā′nä)	42	24°30′N	114°00′W
Magdeburg, Ger. (mäg′dē-boŏrgh)	50	52°07′N	11°39′E
Magellan, Strait of see Magallanes, Estrecho de, strt., S.A.	46	52°30′S	68°45′W
Maggiore, Lago, l., Italy	54	46°03′N	8°25′E
Maghāghah, Egypt	87b	28°38′N	30°50′W
Maghniyya, Alg.	54	34°52′N	1°40′W
Magnitogorsk, Russia (måg-nyē′tô-gôrsk)	56	53°26′N	59°05′E
Magwe, Mya. (mŭg-wā′)	69	20°19′N	94°57′E
Mahābād, Iran	71	36°55′N	45°50′E
Mahakam, r., Indon.	74	0°30′S	116°15′E
Maḥaṭṭat al Qaṭrānah, Jord.	67a	31°15′N	36°04′E
Maḥaṭṭat 'Aqabat al Ḥijāzīyah, Jord.	67a	29°45′N	35°55′E
Maḥaṭṭat ar Ramlah, Jord.	67a	29°31′N	35°57′E
Maḥaṭṭat Jurf ad Darāwīsh, Jord.	67a	30°41′N	35°51′E
Mahd adh-Dhahab, Sau. Ar.	71	23°30′N	40°52′E
Mahe, India (mä-ā′)	69	11°42′N	75°39′E
Mahi, r., India	72	23°16′N	73°20′E
Māhīm Bay, b., India	73b	19°03′N	72°45′E
Maijdi, Bngl.	72	22°59′N	91°08′E
Maikop see Maykop, Russia	56	44°35′N	40°07′E
Maine, state, U.S. (mān)	39	45°25′N	69°50′W
Mainz, Ger. (mīnts)	50	49°59′N	8°16′E
Maitland, Austl. (māt′lănd)	81	32°45′S	151°40′E
Majene, Indon.	74	3°34′S	119°00′E
Majorca see Mallorca, i., Spain	52	39°18′N	2°22′E
Makarakomburu, Mount, mtn., Sol. Is.	78e	9°43′S	160°02′E
Makasar see Ujungpandang, Indon.	74	5°08′S	119°28′E
Makasar, Selat (Makassar Strait), strt., Indon.	74	2°00′S	118°07′E
Makkah see Mecca, Sau. Ar.	68	21°27′N	39°45′E
Makushin, Ak., U.S. (mā-kö′shĭn)	40	53°57′N	166°28′W
Makushino, Russia (må-kö-shēn′ō)	56	55°03′N	67°43′E
Malabar Coast, cst., India (măl′á-bär)	73	11°19′N	75°33′E
Malabar Point, c., India	73b	18°57′N	72°47′E
Malabon, Phil.	75a	14°39′N	120°57′E
Malacca, Strait of, strt., Asia (mȧ-lăk′ȧ)	74	4°15′N	99°44′E
Málaga, Spain	50	36°45′N	4°25′W
Malang, Indon.	74	8°06′S	112°50′E
Malatya, Tur. (må-lä′tyå)	68	38°30′N	38°15′E
Malawi, nation, Afr.	85	11°15′S	33°45′E
Malaya Vishera, Russia (vē-shā′rä)	56	58°51′N	32°13′E
Malay Peninsula, pen., Asia (mȧ-lā′) (mā′lā)	74	6°00′N	101°00′E
Malaysia, nation, Asia (må-lā′zhå)	74	4°10′N	101°22′E
Malbon, Austl. (măl′bŭn)	80	21°15′S	140°30′E
Malden, i., Kir.	2	4°20′S	154°30′W
Maldives, nation, Asia	64	4°30′N	71°30′E
Maléas, Ákra, c., Grc.	55	36°31′N	23°13′E
Mālegaon, India	72	20°35′N	74°30′E
Mali, nation, Afr.	85	15°45′N	0°15′W
Mallorca, i., Spain	52	39°30′N	3°00′E
Malmö, Swe.	50	55°36′N	13°00′E
Malmyzh, Russia (mål-mēzh′)	57	49°58′N	137°07′E
Malolos, Phil. (mä-lō′lōs)	75a	14°51′N	120°49′E
Malta, nation, Eur.	50	35°52′N	13°30′E
Maluku (Moluccas), is., Indon.	75	2°22′S	128°25′E
Maluku, Laut (Molucca Sea), sea, Indon.	75	0°15′N	125°41′E
Mālvan, India	73	16°08′N	73°32′E
Mamburao, Phil. (mäm-bōō′rä-ō)	75a	13°14′N	120°35′E
Mamnoli, India	73b	19°17′N	73°15′E
Manado, Indon.	75	1°29′N	124°50′E
Managua, Nic.	42	12°10′N	86°16′W
Manama see Al Manāmah, Bahr.	68	26°01′N	50°33′E
Manas, China	76	44°30′N	86°00′E
Manaus, Braz. (mä-nä′ōōzh)	46	3°01′S	60°00′W
Manchester, Eng., U.K.	50	53°28′N	2°14′W
Manchester, N.H., U.S.	39	43°00′N	71°30′W
Manchuria, hist. reg., China (măn-chōō′rē-ȧ)	77	48°00′N	124°58′E
Mandalay, Mya. (măn′dá-lā)	69	22°00′N	96°08′E
Mandan, N.D., U.S. (măn′dăn)	38	46°49′N	100°54′W
Mandau Siak, r., Indon.	67b	1°03′N	101°25′E
Mandeb, Bab-el-, strt., (băb′ĕl män-dĕb′)	68	13°17′N	42°49′E
Mandla, India	72	22°43′N	80°23′E
Mandve, India	73b	18°47′N	72°52′E
Māndvi, India (mŭnd′vē)	73b	19°29′N	72°53′E
Māndvi, India (mŭnd′vē)	69	22°54′N	69°23′E
Mandya, India	73	12°40′N	77°00′E
Mangalore, India (mŭŋ-gŭ-lōr′)	69	12°53′N	74°52′E
Mangatarem, Phil. (män′gȧ-tä′rĕm)	75a	15°48′N	120°18′E
Mangkalihat, Tanjung, c., Indon.	74	1°25′N	119°55′E
Mangole, Pulau, i., Indon.	75	1°35′S	126°22′E
Manhattan, Ks., U.S. (măn-hăt′ăn)	38	39°11′N	96°34′W
Manihiki Islands, is., Cook Is. (mä′nē-hē′kē)	89	9°40′S	158°00′W
Manila, Phil.	74	14°37′N	121°00′E
Manila Bay, b., Phil. (mȧ-nĭl′ȧ)	75a	14°38′N	120°46′E
Manisa, Tur. (mä′nē-sȧ)	55	38°40′N	27°30′E
Mankato, Mn., U.S.	39	44°10′N	93°59′W
Mannar, Sri L. (mȧ-när′)	73	9°48′N	80°03′E
Mannar, Gulf of, b., Asia	69	8°47′N	78°33′E
Manokwari, Indon. (mä-nŏk-wä′rē)	75	0°56′S	134°10′E
Manori, neigh., India	73b	19°13′N	72°43′E
Manresa, Spain (män-rā′sä)	54	41°44′N	1°52′E
Mantova, Italy (män′tō-vä) (măn′tů-ȧ)	54	45°09′N	10°47′E
Mantua see Mantova, Italy	54	45°09′N	10°47′E
Manua Islands, is., Am. Sam.	78a	14°13′S	169°35′W
Manui, Pulau, i., Indon. (mä-nōō′ē)	75	3°35′S	123°38′E
Manus Island, i., Pap. N. Gui. (mä′nōōs)	75	2°22′S	146°22′E
Manzala Lake, l., Egypt	87b	31°14′N	32°04′E
Manzanillo, Cuba (män′zä-nēl′yō)	43	20°20′N	77°05′W
Manzanillo, Mex.	42	19°02′N	104°21′W
Manzhouli, China (män-jō-lē)	77	49°25′N	117°15′E
Maó, Spain	54	39°52′N	4°15′E
Maoke, Pegunungan, mts., Indon.	75	4°00′S	138°00′E
Maoming, China	77	21°55′N	110°40′E
Mapia, Kepulauan, i., Indon.	75	0°57′N	134°22′E
Maputo, Moz.	85	26°50′S	32°30′E
Maracaibo, Ven. (mä-rä-kī′bō)	46	10°38′N	71°45′W
Maracaibo, Lago de, l., Ven. (lä′gô-dĕ-mä-rä-kī′bō)	46	9°55′N	72°13′W
Marāgheh, Iran	71	37°20′N	46°10′E
Marand, Iran	71	38°26′N	45°46′E
Marble Bar, Austl. (märb″l bär)	80	21°15′S	119°15′E
Marchena, Spain (mär-chā′nä)	54	37°20′N	5°25′W
Marcus, i., Japan (mär′kŭs)	89	24°00′N	155°00′E
Mardin, Tur. (mär-dēn′)	68	37°25′N	40°40′E
Margarita, Pan. (mär-gōō-rē′tä)	42a	9°20′N	79°55′W
Margherita Peak, mtn., Afr.	85	0°22′N	29°51′E
Mariana Islands, is., Oc.	5	16°00′N	145°30′E
Marianao, Cuba (mä-rē-ä-nä′ō)	43	23°05′N	82°26′W
Mariana Trench, deep,	89	12°00′N	144°00′E
Marias, Islas, is., Mex. (mä-rē′äs)	42	21°30′N	106°40′W
Maribor, Slvn. (mä′re-bôr)	50	46°33′N	15°37′E
Maricaban, i., Phil. (mä-rē-kä-bän′)	75a	13°40′N	120°44′E
Marikana, S. Afr. (mă′-rĭ-kä-nă)	87c	25°40′S	27°28′E
Marinduque Island, i., Phil. (mä-rēn-dōō′kā)	75a	13°14′N	121°45′E
Marinette, Wi., U.S. (măr-ĭ-nĕt′)	39	45°04′N	87°40′W
Marion, In., U.S.	39	40°35′N	85°45′W
Mariupol', Ukr.	56	47°07′N	37°32′E
Mariveles, Phil.	75a	14°27′N	120°29′E
Marj Uyan, Leb.	67a	33°21′N	35°36′E
Marka, Som.	87a	1°45′N	44°47′E
Markham, Mount, mtn., Ant.	82	82°59′S	159°30′E
Markovo, Russia (mär′kô-vô)	57	64°46′N	170°48′E
Markrāna, India	72	27°08′N	74°43′E
Marmara Denizi, sea, Tur.	68	40°40′N	28°00′E
Marquard, S. Afr.	87c	28°41′S	27°26′E
Marquesas Islands, is., Fr. Poly. (mär-kĕ′säs)	2	8°50′S	141°00′W
Marquette, Mi., U.S.	39	46°32′N	87°25′W
Marrakech, Mor. (mär-rä′kĕsh)	85	31°38′N	8°00′W
Marree, Austl. (măr′rē)	80	29°38′S	137°55′E
Marsala, Italy (mär-sä′lä)	54	37°48′N	12°28′E
Marseille, Fr. (mår-sâ′y′)	50	43°18′N	5°25′E
Marshall, Tx., U.S.	39	32°33′N	94°22′W
Marshall Islands, nation, Oc.	3	10°00′N	165°00′E
Martaban, Gulf of, b., Mya. (mär-tŭ-bän′)	74	16°34′N	96°58′E
Martapura, Indon.	74	3°19′S	114°45′E
Martinique, dep., N.A. (mår-tē-nēk′)	43	14°50′N	60°40′W
Martin Point, c., Ak., U.S.	40	70°10′N	142°00′W
Marve, neigh., India	73b	19°12′N	72°43′E
Mary, Turkmen. (mä′rē)	59	37°45′N	61°47′E
Maryborough, Austl. (mā′rĭ-bŭr-ô)	81	25°35′S	152°40′E
Maryborough, Austl.	81	37°00′S	143°50′E
Maryland, state, U.S. (mĕr′ĭ-lănd)	39	39°10′N	76°25′W
Masalembo-Besar, i., Indon.	74	5°40′S	114°28′E
Masan, Kor., S. (mä-sän′)	77	35°10′N	128°31′E
Masbate, Phil. (mäs-bä′tā)	75a	12°21′N	123°38′E
Masbate, i., Phil.	75	12°19′N	123°03′E
Mascarene Islands, is., Afr.	5	20°20′S	56°40′E
Mashhad, Iran	68	36°17′N	59°30′E
Māshkel, Hāmūn-i-, l., Asia (hä-mōōn′ē mäsh-kĕl′)	68	28°28′N	64°13′E
Masjed Soleymān, Iran	68	31°45′N	49°17′E
Mason City, Ia., U.S.	39	43°08′N	93°14′W
Massachusetts, state, U.S. (măs-ȧ-chōō′sĕts)	39	42°20′N	72°30′W
Massif Central, Fr. (må-sēf′ sän-trål′)	50	45°12′N	3°02′E
Matagalpa, Nic. (mä-tä-gäl′pä)	42	12°52′N	85°57′W
Matamoros, Mex.	42	25°52′N	97°30′W
Matanzas, Cuba (mä-tän′zäs)	43	23°05′N	81°35′W
Matara, Sri L. (mä-tä′rä)	73	5°59′N	80°35′E
Mataram, Indon.	74	8°45′S	116°15′E
Matehuala, Mex. (mä-tå-wä′lä)	42	23°38′N	100°39′W
Mateur, Tun. (må-tûr′)	54	37°09′N	9°43′E
Mātherān, India	73b	18°58′N	73°16′E
Mathura, India (mu-tŏ′rŭ)	69	27°39′N	77°39′E
Matochkin Shar, Russia (mä′tŏch-kĭn)	56	73°57′N	56°16′E
Maṭraḥ, Oman (mȧ-trä′)	68	23°36′N	58°27′E
Matsue, Japan (mät′sö-ĕ)	77	35°29′N	133°04′E
Matsuyama, Japan (mät′sö-yä′mä)	77	33°48′N	132°45′E
Mattoon, Il., U.S. (mă-tōōn′)	39	39°30′N	88°20′W
Mauban, Phil. (mä′ōō-bän′)	75a	14°11′N	121°44′E
Maui, i., Hi., U.S. (mä′ōō-ē)	38c	20°52′N	156°02′W
Mauna Loa, mtn., Hi., U.S. (mä′ó-nälō′ä)	38c	19°28′N	155°38′W
Mauritania, nation, Afr. (mô-rē-tā′nĭ-ȧ)	85	19°38′N	13°30′W
Mauritius, nation, Afr. (mô-rĭsh′ĭ-ŭs)	3	20°18′S	57°36′E
Mawlamyine, Mya.	74	16°30′N	97°39′E
Mayagüez, P.R. (mä-yä-gwāz′)	43	18°12′N	67°10′W
Mayd, i., Som.	87a	11°24′N	46°38′E
Maykop, Russia	56	44°35′N	40°07′E
Maymyo, Mya. (mī′myō)	76	22°14′N	96°32′E
Mayon Volcano, vol., Phil. (mä-yōn′)	75a	13°21′N	123°43′E
Mayotte, dep., Afr. (mȧ-yôt′)	85	13°07′S	45°32′E
Mayran, Laguna de, l., Mex. (lä-ō′nä-dĕ-mī-rän′)	42	25°40′N	102°35′W
Mayskiy, Russia	58	43°38′N	44°04′E
Mazār-i-Sharif, Afg. (mȧ-zär′-ē-shȧ-rēf′)	69	36°48′N	67°12′E
Mazatenango, Guat. (mä-zä-tå-näŋ′gō)	42	14°30′N	91°30′W
Mazatlán, Mex.	42	23°14′N	106°27′W
Mazhafah, Jabal, mtn., Sau. Ar.	67a	28°56′N	35°05′E
McAlester, Ok., U.S. (măk ăl′ĕs-tẽr)	39	34°55′N	95°45′W
McDonald Island, i., Austl.	82	53°00′S	72°45′E
McKinley, Mount, mtn., Ak., U.S. (mȧ-kĭn′lĭ)	40a	63°00′N	151°02′W
Mead, Lake, l., U.S.	38	36°20′N	114°14′W
Mecca (Makkah), Sau. Ar. (mĕk′ȧ)	68	21°27′N	39°45′E
Mechriyya, Alg.	54	33°30′N	0°13′W
Medan, Indon. (må-dän′)	74	3°35′N	98°35′E
Medellín, Col. (må-dhĕl-yēn′)	46	6°15′N	75°34′W
Medenine, Tun. (mä-dĕ-nēn′)	54	33°22′N	10°33′E
Medford, Or., U.S.	38	42°19′N	122°52′W
Medina see Al Madīnah, Sau. Ar.	68	24°26′N	39°42′E
Medina del Campo, Spain (må-dē′nä dĕl käm′pō)	54	41°18′N	4°54′W
Mediterranean Sea, sea (mĕd-ĭ-tẽr-ā′nē-ăn)	54	36°22′N	13°25′E
Medjerda, Oued, r., Afr.	54	36°43′N	9°54′E
Mednogorsk, Russia	56	51°27′N	57°22′E
Meekatharra, Austl. (mē-kȧ-thăr′ȧ)	80	26°30′S	118°38′E
Meerut, India (mē′rŏt)	69	28°59′N	77°43′E
Mehsāna, India	72	23°42′N	72°23′E
Meiling Pass, p., China (mā′lĭng′)	77	25°22′N	115°00′E
Mekong, r., Asia	74	18°00′N	104°30′E
Melaka, Malay.	74	2°11′N	102°15′E
Melaka, state, Malay.	67b	2°19′N	102°09′E
Melanesia, is., Oc.	88	13°00′S	164°00′E
Melbourne, Austl. (mĕl′bŭrn)	81	37°52′S	145°08′E
Melkrivier, S. Afr.	87c	24°01′S	28°23′E
Melville, i., Austl.	80	11°30′S	131°12′E
Memel, S. Afr. (mĕ′mĕl)	87c	27°42′S	29°35′E
Memphis, Tn., U.S. (mĕm′fĭs)	39	35°07′N	90°03′W
Memphis, hist., Egypt	87b	29°50′N	31°12′E
Mendocino, Cape, c., Ca., U.S. (mĕn′dō-sē′nō)	38	40°25′N	12°42′W
Mengzi, China	76	23°22′N	103°20′E
Menorca (Minorca), i., Spain (mĕ-nō′r-kä)	52	40°05′N	3°58′E
Mentawai, Kepulauan, is., Indon. (mĕn-tä-vī′)	74	1°08′S	98°10′E
Menzel Bourguiba, Tun.	54	37°12′N	9°51′E
Menzies, Austl. (mĕn′zēz)	80	29°45′S	122°15′E
Merano, Italy (må-rä′nō)	54	46°39′N	11°10′E
Merauke, Indon. (må-rou′kā)	75	8°32′S	140°17′E
Mergui, Mya. (mẽr-gē′)	74	12°29′N	98°39′E
Mergui Archipelago, is., Mya.	74	12°04′N	97°02′E
Mérida, Mex.	42	20°58′N	89°37′W
Meridian, Ms., U.S. (mē-rĭd-ĭ-ăn)	39	32°21′N	88°41′W
Meron, Hare, mtn., Isr.	67a	32°58′N	35°25′E
Mersing, Malay.	67b	2°25′N	103°51′E
Merta Road, India (mār′tŭ rōd)	72	26°50′N	73°54′E
Merzifon, Tur. (mĕr′ze-fōn)	68	40°50′N	35°30′E
Mesopotamia, hist. reg., Asia	71	34°00′N	44°00′E
Messina, Italy (mĕ-sē′nȧ)	50	38°11′N	15°34′E
Messina, Stretto di, strt., Italy (stĕ′t-tô dē)	55	38°10′N	15°34′E
Metlakatla, Ak., U.S. (mĕt-lȧ-kăt′lȧ)	40	55°08′N	131°35′W
Mexian, China	77	24°20′N	116°10′E
Mexicali, Mex. (māk-sē-kä′lē)	42	32°28′N	115°29′W
Mexico, nation, N.A.	42	23°45′N	104°00′W
Mexico, Gulf of, b., N.A.	42	25°15′N	93°45′W
Mexico City, Mex. (mĕk′sĭ-kō)	42	19°28′N	99°09′W
Meyerton, S. Afr. (mī′ẽr-tŭn)	87c	26°35′S	28°01′E
Meymaneh, Afg.	68	35°53′N	64°38′E
Mezen', Russia	56	65°50′N	44°05′E
Mia, Oued, r., Alg.	54	29°26′N	3°15′E
Miami, Az., U.S.	38	33°20′N	110°55′W
Miami, Fl., U.S.	39	25°45′N	80°11′W
Mīāneh, Iran	68	37°15′N	47°13′E
Miangas, Pulau, i., Indon.	75	5°30′N	127°00′E
Michelson, Mount, mtn., Ak., U.S. (mĭch′ĕl-sŭn)	40	69°11′N	144°12′W
Michigan, state, U.S. (mĭsh-ĭ-găn)	39	45°55′N	87°00′W
Michigan, Lake, l., U.S.	39	43°20′N	87°10′W
Micronesia, is., Oc.	88	11°00′N	159°00′E
Micronesia, Federated States of, nation, Oc.	3	5°00′N	152°00′E
Middle Andaman, i., India (ăn-dȧ-măn′)	74	12°44′N	93°21′E
Middleburg, S. Afr.	87c	25°47′S	29°30′E
Middlewit, S. Afr. (mĭd′l′wĭt)	87c	24°50′S	27°00′E
Midway Islands, is., Oc.	2	28°00′N	179°00′W
Milan (Milano), Italy (mē-lä′nō)	50	45°29′N	9°12′E
Milâs, Tur. (mē′läs)	55	37°10′N	27°25′E
Mildura, Austl. (mĭl-dū′rȧ)	81	34°10′S	142°18′E

126 Miles City – Nevada PRONOUNCING INDEX

PLACE (Pronunciation)	PAGE	LAT.	LONG.
Miles City, Mt., U.S. (mīlz)	38	46°24'N	105°50'W
Miling, Austl. (mĭl'ĭng)	80	30°30'S	116°25'E
Millstream, Austl. (mĭl'strēm)	80	21°45'S	117°10'E
Milos, i., Grc. (mē'lŏs)	55	36°45'N	24°35'E
Milwaukee, Wi., U.S.	39	43°03'N	87°55'W
Min, r., China (mēn)	77	26°03'N	118°30'E
Minas, Indon.	67b	0°52'N	101°29'E
Minatitlán, Mex. (mē-nä-tē-tlän')	42	17°59'N	94°33'W
Minch, The, strt., Scot., U.K.	52	58°04'N	6°04'W
Mindanao, i., Phil.	75	8°00'N	125°00'E
Mindanao Sea, sea, Phil.	75	8°55'N	124°00'E
Mindoro, i., Phil.	74	12°50'N	121°05'E
Mindoro Strait, strt., Phil.	75a	12°28'N	120°33'E
Mingäçevir, Azer.	58	40°45'N	47°03'E
Mingäçevir su anbarı, res., Azer.	58	40°50'N	46°50'E
Mingenew, Austl. (mĭn'gĕ-nů)	80	29°15'S	115°45'E
Minneapolis, Mn., U.S.	39	44°58'N	93°15'W
Minnesota, state, U.S. (mĭn-ē-sō'tá)	39	46°10'N	90°20'W
Minot, N.D., U.S.	38	48°13'N	101°17'W
Minsk, Bela. (mênsk)	56	53°54'N	27°35'E
Minūf, Egypt (mē-nōōf')	87b	30°26'N	30°55'E
Minusinsk, Russia (mē-nò-sênsk')	57	53°47'N	91°45'E
Miraflores Locks, trans., Pan.	42a	9°00'N	79°35'W
Mirbāṭ, Oman	68	16°58'N	54°42'E
Miri, Malay. (mē'rē)	74	4°13'N	113°56'E
Mīrpur Khās, Pak. (mēr'pōōr κäs)	72	25°36'N	69°10'E
Mirzāpur, India (mēr'zä-pōōr)	69	25°12'N	82°38'E
Miskolc, Hung. (mĭsh'kōlts)	50	48°07'N	20°50'E
Misool, Pulau, i., Indon. (mê-sōl')	75	2°00'S	130°05'E
Miṣr al Jadīdah, Egypt	87b	30°06'N	31°35'E
Mississippi, state, U.S. (mĭs-ĭ-sĭp'ē)	39	32°30'N	89°45'W
Mississippi, r., U.S.	39	32°00'N	91°30'W
Missoula, Mt., U.S. (mĭ-zōō'lá)	38	46°55'N	114°00'W
Missouri, state, U.S. (mĭ-sōō'rē)	39	38°00'N	93°40'W
Missouri, r., U.S.	39	40°40'N	96°00'W
Misti, Volcán, vol., Peru	46	16°04'S	71°20'W
Misty Fjords National Monument, rec., Ak., U.S.	40	51°00'N	131°00'W
Mitchell, S.D., U.S.	38	43°42'N	98°01'W
Mitchell, Mount, mtn., N.C., U.S.	39	35°47'N	82°15'W
Mīt Ghamr, Egypt	87b	30°43'N	31°20'E
Mitla Pass, p., Egypt	67a	30°03'N	32°40'E
Mizdah, Libya (mēz'dä)	70	31°29'N	13°09'E
Mizoram, state, India	69	23°25'N	92°45'E
Moa, Pulau, i., Indon.	75	8°30'S	128°30'E
Moberly, Mo., U.S. (mō'bēr-lĭ)	39	39°24'N	92°25'W
Mobile, Al., U.S. (mò-bēl')	39	30°42'N	88°03'W
Modena, Italy (mô'dē-nä)	54	44°38'N	10°54'E
Mogadishu (Muqdisho), Som.	87a	2°08'N	45°22'E
Mogaung, Mya. (mō-gä'ông)	69	25°30'N	96°52'E
Mogok, Mya. (mō-gōk')	69	23°14'N	96°38'E
Mogol, r., S. Afr. (mō-gŏl)	87c	24°12'S	27°55'E
Mohe, China (mwo-hŭ)	77	53°33'N	122°30'E
Mohenjo-Dero, hist., Pak.	69	27°20'N	68°10'E
Mojave Desert, des., Ca., U.S.	38	35°00'N	117°00'W
Mokp'o, Kor., S. (mōk'pō')	77	34°50'N	126°30'E
Moldavia see Moldova, nation, Eur.	56	48°00'N	28°00'E
Moldova, nation, Eur.	56	48°00'N	28°00'E
Molfetta, Italy (môl-fĕt'tä)	55	41°11'N	16°38'E
Moller, Port, Ak., U.S. (pōrt mōl'ēr)	40	56°18'N	161°30'W
Moluccas see Maluku, is., Indon.	75	2°22'S	128°25'E
Mombasa, Kenya (mōm-bä'sä)	85	4°03'S	39°40'E
Mompog Pass, strt., Phil. (mōm-pōg')	75a	13°35'N	122°09'E
Monaco, nation, Eur. (mŏn'á-kō)	50	43°43'N	7°47'E
Mona Passage, strt., N.A. (mō'nä)	43	18°00'N	68°10'W
Monastir, Tun. (mōn-ás-tēr')	54	35°49'N	10°56'E
Monclova, Mex. (mōn-klō'vä)	42	26°53'N	101°25'W
Monghyr, India (mōn-gēr')	69	25°23'N	86°34'E
Mongolia, nation, Asia (mŏŋ-gō'lĭ-á)	76	46°00'N	100°00'E
Monroe, La., U.S.	39	32°30'N	92°06'W
Montague, i., Ak., U.S.	40	60°10'N	147°00'W
Montana, state, U.S. (mŏn-tăn'á)	38	47°10'N	111°50'W
Montego Bay, Jam. (mŏn-tē'gō)	43	18°30'N	77°55'W
Montemorelos, Mex. (mōn'tā-mō-rā'lōs)	42	25°14'N	99°50'W
Monterey, Ca., U.S. (mŏn-tĕ-rā')	38	36°36'N	121°53'W
Monterrey, Mex. (mōn-tĕr-rā')	42	25°43'N	100°19'W
Monte Sant'Angelo, Italy (mô'n-tĕ săn ä'n-gzhĕ-lô)	55	41°43'N	15°59'E
Montevideo, Ur. (mōn'tā-vē-dhā'ō)	46	34°50'S	56°10'W
Montgomery, Al., U.S. (mŏnt-gŭm'ēr-ĭ)	39	32°23'N	86°17'W
Montijo, Bahía, b., Pan. (bä-ē'ä mōn-tē'hō)	43	7°36'N	81°11'W
Montpelier, Vt., U.S.	39	44°20'N	72°35'W
Montréal, Can. (mŏn-trē-ôl')	36	45°30'N	73°35'W
Montserrat, dep., N.A. (mŏnt-sĕ-răt')	43	16°48'N	63°15'W
Monywa, Mya. (mŏn'yōō-wà)	69	22°02'N	95°16'E
Mooi, r., S. Afr. (mō'ĭ)	87c	26°34'S	27°03'E
Moonta, Austl. (mōōn'tá)	80	34°05'S	137°42'E
Moora, Austl. (mòr'à)	80	30°35'S	116°12'E
Mora, India	73b	18°54'N	72°56'E
Morādābād, India (mō-rä-dä-bäd')	69	28°57'N	78°48'E
Moratuwa, Sri L.	73	6°35'N	79°59'E
Moray Firth, b., Scot., U.K. (mŭr'å)	52	57°41'N	3°55'W
Moree, Austl. (mō'rē)	81	29°20'S	149°50'E
Morelia, Mex. (mô-rā'lyä)	42	19°43'N	101°12'W
Morena, Sierra, mts., Spain (syĕr'rä mō-rā'nä)	52	38°15'N	5°45'W
Morga Range, mts., Afg.	69a	34°02'N	70°38'E
Morgenzon, S. Afr. (môr'gănt-sŏn)	87c	26°44'S	29°39'E
Morioka, Japan (mō'rē-ō'kà)	77	39°40'N	141°21'E
Morobe, Pap. N. Gui.	75	8°03'S	147°45'E
Morocco, nation, Afr. (mō-rŏk'ō)	85	32°00'N	7°00'W
Moroni, Com.	85	11°41'S	43°16'E
Morotai, i., Indon. (mō-rō-tä'ē)	75	2°12'N	128°30'E
Moscow (Moskva), Russia	56	55°45'N	37°37'E
Moscow, Id., U.S. (mŏs'kō)	38	46°44'N	116°57'W
Moses, r., S. Afr.	87c	25°17'S	29°04'E
Moskva see Moscow, Russia	56	55°45'N	37°37'E
Mosquitos, Gulfo de los, b., Pan. (gōō'l-fō-dĕ-lôs-mōs-kē'tōs)	43	9°17'N	80°59'W
Mostar, Bos. (môs'tär)	55	43°20'N	17°51'E
Motril, Spain (mō-trēl')	54	36°44'N	3°32'W
Mount Gambier, Austl. (găm'bēr)	80	37°30'S	140°53'E
Mount Isa, Austl. (ī'zá)	80	21°00'S	139°45'E
Mount Magnet, Austl. (măg-nĕt)	80	28°00'S	118°00'E
Mount Morgan, Austl. (môr-găn)	81	23°42'S	150°45'E
Moyen Atlas, mts., Mor.	54	32°49'N	5°28'W
Moyynqum, des., Kaz.	59	44°30'N	70°00'E
Mozambique, nation, Afr. (mō-zăm-bēk')	85	20°15'S	33°53'E
Mozambique Channel, strt., Afr. (mō-zăm-bek')	85	24°00'S	38°00'E
Muar, r., Malay.	67b	2°18'N	102°43'E
Muğla, Tur. (mōōg'lä)	68	37°10'N	28°20'E
Mukden see Shenyang, China	77	41°45'N	123°22'E
Mukhtuya, Russia (mòk-tōō'yà)	57	61°00'N	113°00'E
Mulhacén, mtn., Spain	54	37°04'N	3°18'W
Müller, Pegunungan, mts., Indon. (mül'ĕr)	74	0°22'N	113°05'E
Multān, Pak. (mò-tän')	69	30°17'N	71°13'E
Mumbai (Bombay), India	69	18°58'N	72°50'E
München see Munich, Ger.	50	48°08'N	11°35'E
Muncie, In., U.S. (mŭn'sĭ)	39	40°10'N	85°30'W
Mungana, Austl. (mŭn-găn'á)	81	17°15'S	144°18'E
Mungindi, Austl. (mŭn-gĭn'dē)	81	29°00'S	148°45'E
Munich, Ger.	50	48°08'N	11°35'E
Munku Sardyk, mtn., Asia (mòn'kò sär-dĭk')	57	51°45'N	100°30'E
Muñoz, Phil. (mōōn-nyōth')	75a	15°44'N	120°53'E
Muntok, Indon. (mòn-tŏk')	74	2°05'S	105°11'E
Muong Sing, Laos (mōō'ông-sĭng')	74	21°06'N	101°17'E
Murat, r., Tur. (mōō-råt')	68	39°00'N	42°00'E
Murcia, Spain (mōōr'thyä)	50	38°00'N	1°10'W
Mureş, r., Rom. (mōō'rĕsh)	55	46°02'N	21°50'E
Murgab, Taj.	59	38°10'N	73°59'E
Murgab, r., Asia (mōōr-gäb')	68	37°07'N	62°32'E
Murmansk, Russia (mōōr-mänsk')	56	69°00'N	33°20'E
Murom, Russia (mōō'rôm)	56	55°30'N	42°00'W
Muroran, Japan (mōō'rô-ràn)	77	42°21'N	141°05'E
Murray, r., Austl.	80	34°20'S	140°00'E
Murray Bridge, Austl.	80	35°10'S	139°35'E
Murshidābād, India (mòr'shē-dä-bäd')	72	24°08'N	88°11'E
Murwāra, India	69	23°54'N	80°23'E
Musan, Kor., N. (mò'sän)	77	41°11'N	129°10'E
Muscat, Oman (mŭs-kăt')	68	23°23'N	58°30'E
Muscat and Oman see Oman, nation, Asia	68	20°00'N	57°45'E
Musi, r., Indon. (mōō'sē)	74	2°40'S	103°42'E
Muskegon, Mi., U.S. (mŭs-kē'gŭn)	39	43°15'N	86°20'W
Muskogee, Ok., U.S. (mŭs-kō'gē)	39	35°44'N	95°21'W
Mussau Island, i., Pap. N. Gui. (mōō-sä'ōō)	75	1°30'S	149°32'E
Mustafakemalpaşa, Tur.	55	40°05'N	28°30'E
Musu Dan, c., Kor., N. (mò'sò dàn)	77	40°51'N	130°00'E
Muzaffargarh, Pak.	72	30°09'N	71°15'E
Muzaffarpur, India	72	26°13'N	85°20'E
Muztagata, mtn., China	76	38°20'N	75°28'E
Myanmar (Burma), nation, Asia	64	21°00'N	95°15'E
Myingyan, Mya. (myĭng-yŭn')	69	21°37'N	95°26'E
Myitkyina, Mya. (myĭ'chē-nà)	69	25°33'N	97°25'E
Mykolaïv, Ukr.	56	46°58'N	32°02'E
Mymensingh, Bngl.	69	24°48'N	90°28'E
Mysore, India (mī-sōr')	69	12°31'N	76°42'E
Mytilíni, Grc.	55	39°09'N	26°35'E

N

PLACE (Pronunciation)	PAGE	LAT.	LONG.
Naberezhnyye Chelny, Russia	56	55°42'N	52°19'E
Naboomspruit, S. Afr.	87c	24°32'S	28°43'E
Nābulus, W.B.	67a	32°13'N	35°16'E
Nadiād, India	72	22°45'N	72°51'E
Nadir, V.I.U.S.	43c	18°19'N	64°53'W
Nafīshah, Egypt	87d	30°34'N	32°15'E
Nafūd ad Daḥy, des., Sau. Ar.	68	22°15'N	44°15'E
Nag, Co, l., China	72	31°38'N	91°18'E
Naga, Phil. (nä'gä)	75	13°37'N	123°12'E
Nagaland, India	69	25°47'N	94°15'E
Nagano, Japan (nä'gà-nō)	77	36°42'N	138°12'E
Nagaoka, Japan (nä'gà-ō'kà)	77	37°22'N	138°49'E
Nāgappattinam, India	69	10°48'N	79°51'E
Nagasaki, Japan (nä'gà-sä'kē)	77	32°48'N	129°53'E
Nāgaur, India	72	27°19'N	73°41'E
Nagcarlan, Phil. (näg-kär-län')	75a	14°07'N	121°24'E
Nāgercoil, India	73	8°15'N	77°29'E
Nagorno Karabakh, hist. reg., Azer. (nu-gôr'nŭ-kŭ-rŭ-bäk')	58	40°10'N	46°50'E
Nagoya, Japan	77	35°09'N	136°53'E
Nāgpur, India (näg'pōōr)	69	21°12'N	79°09'E
Nagykanizsa, Hung. (nôd'y'kô'nē-shô)	55	46°27'N	17°00'E
Naha, Japan (nä'hä)	77	26°02'N	127°43'E
Nahariyya, Isr.	67a	33°01'N	35°06'E
Naic, Phil. (nä-ēk)	75a	14°20'N	120°46'E
Nā'īn, Iran	71	32°52'N	53°05'E
Nairobi, Kenya (nī-rō'bē)	85	1°17'S	36°49'E
Najd, hist. reg., Sau. Ar.	68	25°18'N	42°38'E
Najin, Kor., N. (nä'jĭn)	77	42°04'N	130°35'E
Najran, des., Sau. Ar. (nŭj-rän')	68	17°29'N	45°30'E
Nakhodka, Russia (nŭ-κôt'kŭ)	57	43°03'N	133°08'E
Nakhon Ratchasima, Thai.	74	14°56'N	102°14'E
Nakhon Sawan, Thai.	74	15°42'N	100°06'E
Nakhon Si Thammarat, Thai.	74	8°27'N	99°58'E
Namak, Daryacheh-ye, l., Iran	68	34°58'N	51°33'E
Namangan, Uzb. (nà-màn-gän')	59	41°08'N	71°59'E
Namatanai, Pap. N. Gui. (nä'mä-tà-nä'ē)	75	3°43'S	152°26'E
Nam Co, l., China (näm tswo)	76	30°30'N	91°10'E
Nam Dinh, Viet. (näm dēnκ')	74	20°30'N	106°10'E
Namibia, nation, Afr.	85	19°30'S	16°13'E
Namous, Oued en, r., Alg. (nà-mōōs')	54	31°48'N	0°19'W
Nampa, Id., U.S. (năm'pá)	38	43°35'N	116°35'W
Namp'o, Kor., N.	77	38°47'N	125°28'E
Nan, r., Thai.	74	18°11'N	100°29'E
Nanchang, China (nän'chäng')	77	28°38'N	115°48'E
Nancheng, China (nän-chŭŋ)	77	26°50'N	116°40'E
Nanchong, China (nän-chôŋ)	76	30°45'N	106°05'E
Nanda Devi, mtn., India (nän'dä dā'vē)	69	30°30'N	80°25'E
Nānded, India	72	19°13'N	77°21'E
Nandurbār, India	72	21°29'N	74°13'E
Nandyāl, India	73	15°54'N	78°09'E
Nanga Parbat, mtn., Pak.	72	35°20'N	74°35'E
Nangi, India	72a	22°30'N	88°14'E
Nanjing, China (nän-jyĭŋ)	77	32°04'N	118°46'E
Nanking see Nanjing, China	77	32°04'N	118°46'E
Nan Ling, mts., China	77	25°15'N	111°40'E
Nannine, Austl. (nă-nēn')	80	25°50'S	118°30'E
Nanning, China (nän'nĭng')	76	22°56'N	108°10'E
Nanping, China (nän-pĭŋ)	77	26°40'N	118°05'E
Nansei-shotō, is., Japan	77	27°30'N	127°00'E
Nantes, Fr. (näNt')	50	47°13'N	1°37'W
Nanyang, China	77	33°00'N	112°42'E
Napa, Ca., U.S. (năp'á)	38	38°20'N	122°17'W
Naples (Napoli), Italy	50	40°37'N	14°12'E
Napoli see Naples, Italy	50	40°37'N	14°12'E
Napoli, Golfo di, b., Italy	54	40°29'N	14°08'E
Nara, Japan (nä'rä)	77	34°41'N	135°50'E
Naracoorte, Austl. (nă-rà-kōōn'tē)	80	36°50'S	140°50'E
Naraspur, India	73	16°32'N	81°43'E
Narmada, r., India	69	22°30'N	75°30'E
Narodnaya, Gora, mtn., Russia (nà-rôd'nà-yà)	56	65°10'N	60°10'E
Narrandera, Austl. (nà-răn-dē'rà)	81	34°40'S	146°40'E
Narrogin, Austl. (năr'ō-gĭn)	80	33°00'S	117°15'E
Narvacan, Phil. (när-vä-kän')	75a	17°27'N	120°29'E
Narvik, Nor. (när'vêk)	50	68°21'N	17°18'E
Nar'yan-Mar, Russia (när'yán mär')	56	67°42'N	53°30'E
Narym, Russia (nä-rēm')	56	58°47'N	82°05'E
Nashua, N.H., U.S.	39	42°47'N	71°23'W
Nashville, Tn., U.S.	39	36°10'N	86°48'W
Našice, Cro. (nä'shē-tsĕ)	55	45°29'N	18°06'E
Nāsik, India (nä'sĭk)	69	20°02'N	73°49'E
Nasirabād, India	72	26°13'N	74°48'E
Nassau, Bah. (năs'ô)	43	25°05'N	77°20'W
Nasser, Lake, res., Egypt	70	23°50'N	32°50'E
Nasugbu, Phil. (nà-sòg-bōō')	75a	14°05'N	120°37'E
Natal, Braz. (nä-täl')	46	6°00'S	35°13'W
Natchez, Ms., U.S. (năch'ĕz)	39	31°35'N	91°20'W
Naṭrūn, Wādī an, val., Egypt	87b	30°33'N	30°12'E
Natuna Besar, i., Indon.	74	4°00'N	106°50'E
Nau, Cap de la, c., Spain	52	38°43'N	0°14'E
Naujan, Phil. (nä-ò-hän')	75a	13°19'N	121°17'E
Nauru, nation, Oc.	3	0°30'S	167°00'E
Nautla, Mex. (nä-ōōt'lä)	42	20°14'N	96°44'W
Navojoa, Mex. (nä-vô-kô'ä)	42	27°00'N	109°40'W
Nawābshāh, Pak. (nà-wäb'shä)	72	26°20'N	68°30'E
Naxçıvan Muxtar, state, Azer.	58	39°20'N	45°30'E
Náxos, i., Grc. (nák'sôs)	55	37°15'N	25°20'E
Nayarit, state, Mex. (nä-yä-rēt')	42	22°00'N	105°15'W
Nazas, r., Mex.	42	25°30'N	104°40'W
Naẓerat, Isr.	67a	32°43'N	35°19'E
N'Djamena, Chad	85	12°07'N	15°03'E
Néa Páfos, Cyp.	67a	34°46'N	32°27'E
Near Islands, is., Ak., U.S. (nēr)	40a	52°20'N	172°40'E
Nebitdag, Turkmen.	59	39°30'N	54°20'E
Nebraska, state, U.S. (nē-brăs'ká)	38	41°45'N	101°30'W
Negeri Sembilan, state, Malay. (nā'grē-sĕm-bē-län')	67b	2°46'N	101°54'E
Negev, des., Isr. (nĕ'gĕv)	67a	30°34'N	34°43'E
Negombo, Sri L.	73	7°39'N	79°49'E
Negro, r., S.A. (nā'grò)	46	0°18'S	63°21'W
Negros, i., Phil. (nā'grōs)	74	9°50'N	121°45'E
Nehbandān, Iran	71	31°32'N	60°02'E
Nei Monggol (Inner Mongolia), state, China	76	40°15'N	105°00'E
Nel'kan, Russia (nĕl-kän')	57	57°45'N	136°36'E
Nellore, India (nĕl-lōr')	69	14°28'N	79°59'E
Nelson, i., Ak., U.S.	40	60°38'N	164°42'W
Nelson, r., Can.	36	56°50'N	93°40'W
Nemuro, Japan (nā'mò-rō)	77	43°13'N	145°10'E
Nen, r., China (nŭn)	77	47°07'N	123°28'E
Nenana, Ak., U.S. (nå-nä'ná)	40	64°28'N	149°18'W
Nenjiang, China (nŭn-jyäŋ)	77	49°02'N	125°15'E
Nepal, nation, Asia (nĕ-pôl')	69	28°45'N	83°00'E
Nerchinsk, Russia (nyĕr'chēnsk)	57	51°47'N	116°17'E
Nerchinskiy Khrebet, mts., Russia	57	50°30'N	118°30'E
Nerchinskiy Zavod, Russia (nyĕr'chēn-skĭzá-vôt')	57	51°35'N	119°46'E
Netanya, Isr.	67a	32°19'N	34°52'E
Netherlands, nation, Eur. (nĕdh'ēr-lăndz)	50	53°01'N	3°57'E
Nevada, state, U.S. (nĕ vä'dà)	38	39°30'N	117°00'W

PRONOUNCING INDEX

Nevada, Sierra – Ozurgeti 127

PLACE (Pronunciation)	PAGE	LAT.	LONG.
Nevada, Sierra, mts., Spain (syĕr′rä nā-vä′dhä)	52	37°01′N	3°28′W
Nevada, Sierra, mts., U.S. (sē-ĕ′r-rä nė-vä′då)	38	39°20′N	120°05′W
Nevinnomiyssk, Russia	58	44°38′N	41°56′E
Nevis, i., St. K./N. (nē′vĭs)	43	17°05′N	62°38′W
Nevşehir, Tur. (nĕv-shĕ′hẽr)	55	38°40′N	34°35′E
Nev′yansk, Russia (nĕv-yänsk′)	56	57°29′N	60°14′E
Newark, N.J., U.S. (no͞o′ûrk)	39	40°44′N	74°10′W
New Bedford, Ma., U.S. (bĕd′fẽrd)	39	41°35′N	70°55′W
New Bern, N.C., U.S. (bûrn)	39	35°05′N	77°05′W
New Britain, i., Pap. N. Gui.	75	6°45′S	149°38′E
New Caledonia, dep., Oc.	81	21°28′S	164°40′E
Newcastle, Eng., U.K.	50	55°00′N	1°45′W
Newcastle Waters, Austl. (wô′tẽrz)	80	17°10′S	133°25′E
New Delhi, India (dĕl′hī)	69	28°43′N	77°18′E
Newenham, Cape, c., Ak., U.S. (nū-ĕn-hăm)	40	58°40′N	162°32′W
Newfoundland, i., Can.	36a	48°30′N	56°00′W
Newfoundland and Labrador, prov., Can.	36	48°15′N	56°53′W
New Georgia Group, is., Sol. Is.	78e	8°30′S	157°20′E
New Georgia Sound, strt., Sol. Is.	78e	8°00′S	158°10′E
New Guinea, i., (gĭne)	75	5°45′S	140°00′E
New Hampshire, state, U.S. (hămp′shĭr)	39	43°55′N	71°40′W
New Hanover, i., Pap. N. Gui.	75	2°37′S	150°15′E
New Haven, Ct., U.S. (hā′vĕn)	39	41°20′N	72°55′W
New Hebrides, is., Vanuatu	81	16°00′S	167°00′E
New Ireland, i., Pap. N. Gui. (īr′lănd)	75	3°15′S	152°30′E
New Jersey, state, U.S. (jûr′zĭ)	39	40°30′N	74°50′W
New Mexico, state, U.S. (mĕk′sĭ-kō)	38	34°30′N	107°10′W
New Norfolk, Austl. (nôr′fŏk)	81	42°50′S	147°17′E
New Orleans, La., U.S. (ôr′lê-ănz)	39	30°00′N	90°05′W
Newport, Ky., U.S.	39	39°05′N	84°30′W
Newport News, Va., U.S.	39	36°59′N	76°24′W
New Siberian Islands *see* Novosibirskiye Ostrova, is., Russia	57	74°00′N	140°30′E
New South Wales, state, Austl. (wālz)	81	32°45′S	146°14′E
New York, N.Y., U.S. (yôrk)	39	40°40′N	73°58′W
New York, state, U.S.	39	42°45′N	78°05′W
New Zealand, nation, Oc. (zē′lȧnd)	81a	42°00′S	175°00′E
Neyshābūr, Iran	68	36°06′N	58°45′E
Ngangla Ringco, l., China (ŋäŋ-lä rĭŋ-tswo)	72	31°42′N	82°53′E
Nha Trang, Viet. (nyä-träng′)	74	12°08′N	108°56′E
Niagara Falls, N.Y., U.S.	39	43°06′N	79°02′W
Niamey, Niger (nē-ä-må′)	85	13°31′N	2°07′E
Nias, Pulau, i., Indon. (nē′äs′)	74	0°58′N	97°43′E
Nicaragua, nation, N.A. (nĭk-ȧ-rä′gwȧ)	42	12°45′N	86°15′W
Nicaragua, Lago de, l., Nic. (lä′gô dĕ)	42	11°45′N	85°28′W
Nicastro, Italy (nė-käs′trō)	55	38°39′N	16°15′E
Nice, Fr. (nēs)	50	43°42′N	7°21′E
Nicobar Islands, is., India (nĭk-ō-bär′)	74	8°28′N	94°04′E
Nicosia, Cyp. (nē-kō-sē′ȧ)	68	35°10′N	33°22′E
Nietverdiend, S. Afr.	87c	25°02′S	26°10′E
Niğde, Tur. (nĭg′dĕ)	55	37°55′N	34°40′E
Nigel, S. Afr. (nī′jĕl)	87c	26°26′S	28°27′E
Niger, nation, Afr. (nī′jẽr)	85	18°02′N	8°30′E
Niger, r., Afr.	85	8°00′N	6°00′E
Nigeria, nation, Afr. (nī-jē′rĭ-ȧ)	85	8°57′N	6°30′E
Niigata, Japan (nē′ė-gä′tȧ)	77	37°47′N	139°04′E
Nikolayevsk-na-Amure, Russia	57	53°18′N	140°49′E
Nikol′sk, Russia (nė-kôlsk′)	56	59°30′N	45°40′E
Nikopol, Blg. (nė′kô-pŏl′)	55	43°41′N	24°52′E
Nile, r., Afr. (nīl)	85	27°30′N	31°00′E
Nileshwar, India	73	12°08′N	74°14′E
Nilgiri Hills, hills, India	73	12°05′N	76°22′E
Nīmach, India	72	24°32′N	74°51′E
Nîmes, Fr. (nēm)	50	43°49′N	4°22′E
Nineveh, Iraq (nĭn′ė-vȧ)	68	36°30′N	43°10′E
Ning′an, China (nĭŋ-än)	77	44°20′N	129°20′E
Ningbo, China (nĭŋ-bwo)	77	29°56′N	121°30′E
Ningde, China (nĭŋ-dŭ)	77	26°38′N	119°33′E
Ningwu, China (nĭng′wo͞o′)	77	39°00′N	112°12′E
Ningxia Huizu, prov., China (nĭŋ-shyä)	76	37°10′N	106°00′E
Ninh Binh, Viet. (nēn bēnk′)	74	20°22′N	106°00′E
Ninigo Group, is., Pap. N. Gui.	75	1°15′S	143°30′E
Nirmali, India	72	26°30′N	86°43′E
Niš, Serb.	50	43°19′N	21°54′E
Niue, dep., Oc. (nĭ′ȯ)	89	19°50′S	167°00′W
Nizāmābād, India	69	18°48′N	78°07′E
Nizhne-Angarsk, Russia (nyēzh′nyĭ-ŭngärsk′)	57	55°49′N	108°46′E
Nizhne-Kolymsk, Russia (kō-lėmsk′)	57	68°32′N	160°56′E
Nizhneudinsk, Russia (nėzh′nyĭ-o͞odēnsk′)	57	54°58′N	99°15′E
Nizhniy Novgorod (Gor′kiy), Russia	56	56°15′N	44°05′E
Nizhniy Tagil, Russia (tŭgēl′)	56	57°54′N	59°59′E
Nizhnyaya Tunguska, r., Russia	57	64°13′N	91°30′E
Noākhāli, Bngl.	69	22°52′N	91°08′E
Noatak, Ak., U.S. (nō-ȧ′tȧk)	40	67°22′N	163°28′W
Noatak, r., Ak., U.S.	40	67°58′N	162°15′W
Nogales, Mex.	42	31°15′N	111°00′W
Nogales, Az., U.S. (nō-gä′lĕs)	38	31°20′N	110°55′W
Nogal Valley, val., Som. (nō′găl)	87a	8°30′N	47°50′E
Nome, Ak., U.S. (nōm)	40a	64°30′N	165°20′W
Nordvik, Russia (nôrd′vėk)	57	73°57′N	111°15′E
Norfolk, Ne., U.S.	38	42°10′N	97°25′W
Norfolk, Va., U.S.	39	36°55′N	76°15′W
Norfolk, i., Oc.	89	27°10′S	166°50′E
Noril′sk, Russia (nô rēlsk′)	56	69°00′N	87°11′E
Normanton, Austl. (nôr′mȧn-tŭn)	81	17°45′S	141°10′E
Nornalup, Austl. (nôr-năl′ŭp)	80	35°00′S	117°00′E
Norrköping, Swe. (nôr′chûp′ĭng)	50	58°37′N	16°10′E
Norseman, Austl. (nôrs′măn)	80	32°15′S	122°00′E
Northam, Austl. (nôr-dhăm)	80	31°50′S	116°45′E
Northam, S. Afr. (nôr′thăm)	87c	24°52′S	27°16′E
North America, cont.	37	45°00′N	100°00′W
North American Basin, deep, (ȧ-mĕr′ĭ-kȧn)	4	23°45′N	62°45′W
Northampton, Austl. (nôr-thămp′tŭn)	80	28°22′S	114°45′E
North Andaman Island, i., India (ăn-dȧ-măn′)	74	13°15′N	93°30′E
North Borneo *see* Sabah, hist. reg., Malay.	74	5°10′N	116°25′E
North Carolina, state, U.S. (kăr-ō-lī′nȧ)	39	35°40′N	81°30′W
North Channel, strt., U.K.	52	55°15′N	7°56′W
North Dakota, state, U.S. (dȧ-kō′tȧ)	38	47°20′N	101°55′W
North Dum-Dum, India	72a	22°38′N	88°23′E
Northeast Cape, c., Ak., U.S. (nôrth-ēst)	40	63°15′N	169°04′W
Northern Dvina *see* Severnaya Dvina, r., Russia	56	63°00′N	42°40′E
Northern Ireland, state, U.K. (īr′lănd)	50	54°48′N	7°00′W
Northern Land *see* Severnaya Zemlya, is., Russia	57	79°33′N	101°15′E
Northern Mariana Islands, dep., Oc. (mä-rė-ä′nȧ)	3	17°20′N	145°00′E
Northern Territory, ter., Austl.	80	18°15′S	133°00′E
Northern Yukon National Park, rec., Can.	40	69°00′N	140°00′W
North Island, i., N.Z.	81a	37°20′S	173°30′E
North Magnetic Pole, pt. of. i.,	93	77°19′N	101°49′W
North Platte, Ne., U.S. (plăt)	38	41°08′N	100°45′W
North Pole, pt. of. i.,	93	90°00′N	0°00′
North Sea, Eur.	50	56°09′N	3°16′E
Norton Bay, b., Ak., U.S.	40	64°22′N	162°18′W
Norton Sound, strt., Ak., U.S.	40	63°48′N	164°50′W
Norway, nation, Eur. (nôr′wā)	50	63°48′N	11°17′E
Nouakchott, Maur.	85	18°06′N	15°57′W
Nouméa, N. Cal. (no͞o-mā′ä)	81	22°16′S	166°27′E
Novara, Italy (nō-vä′rä)	54	45°24′N	8°38′E
Nova Scotia, prov., Can. (skō′shȧ)	36	44°28′N	65°00′W
Novaya Sibir, i., Russia (sė-bēr′)	57	75°00′N	149°00′E
Novaya Zemlya, i., Russia (zĕm-lyȧ′)	56	72°00′N	54°46′E
Novi Pazar, Serb. (pȧ-zär′)	55	43°08′N	20°30′E
Novi Sad, Serb. (säd′)	50	45°15′N	19°53′E
Novokuznetsk, Russia (nō′vō-kȯ′z-nyĕ′tsk) (stȧ′lênsk)	56	53°43′N	86°59′E
Novomoskovsk, Russia (nô′vō-mōs-kôfsk′)	56	54°06′N	38°08′E
Novorossiysk, Russia (nô′vō-rô-sēsk′)	56	44°43′N	37°48′E
Novosibirsk, Russia (nô′vō-sė-bērsk′)	56	55°09′N	82°58′E
Novosibirskiye Ostrova (New Siberian Islands), is., Russia	57	74°00′N	140°30′E
Novyy Port, Russia (nô′vē)	56	67°19′N	72°28′E
Nubian Desert, des., Sudan (no͞o′bĭ-ăn)	85	21°13′N	33°09′E
Nueva Rosita, Mex. (nȯĕ′vä rô-sė′tä)	38	27°55′N	101°10′W
Nuevitas, Cuba (nwā-vē′täs)	43	21°35′N	77°15′W
Nuevo Laredo, Mex. (lä-rā′dhō)	42	27°29′N	99°30′W
Nuevo León, state, Mex. (lå-ōn′)	42	26°00′N	100°00′W
Nuevo San Juan, Pan. (nwĕ′vō sän ᴋo͞o-ä′n)	42a	9°14′N	79°43′W
Nulato, Ak., U.S. (no͞o-lä′tō)	40	64°40′N	158°18′W
Nullagine, Austl. (nŭ-lă′jēn)	80	22°00′S	120°07′E
Nullarbor Plain, pl., Austl. (nŭ-lär′bôr)	80	31°45′S	126°30′E
Numfoor, Pulau, i., Indon.	75	1°20′S	134°48′E
Nunyama, Russia (nûn-yä′mȧ)	40	65°49′N	170°32′W
Nurata, Uzb. (no͞or′ät′ȧ)	59	40°33′N	65°28′E
Nuremberg *see* Nürnberg, Ger.	50	49°28′N	11°07′E
Nürnberg, Ger. (nürn′bĕrgh)	50	49°28′N	11°07′E
Nushagak, r., Ak., U.S. (nū-shă-găk′)	40	59°28′N	157°40′W
Nushki, Pak. (nŭsh′kė)	69	29°30′N	66°02′E
Nuwaybi ′al Muzayyinah, Egypt	67a	28°59′N	34°40′E
Nyainqêntanglha Shan, mts., China (nyä-ĭn-chyŭn-täŋ-lä shän)	76	29°55′N	88°08′E
Nyasa, Lake, l., Afr. (nyä′sä)	85	10°45′S	34°30′E
Nyíregyháza, Hung. (nyē′rĕd-y′hä′zä)	55	47°58′N	21°45′E
Nymagee, Austl. (nī-mȧ-gē′)	81	32°17′S	146°18′E
Nyngan, Austl. (nĭŋ′gȧn)	81	31°31′S	147°25′E

O

PLACE (Pronunciation)	PAGE	LAT.	LONG.
O′ahu, i., Hi., U.S. (ō-ä′ho͞o) (ō-ā′hü)	38c	21°38′N	157°48′W
Oakland, Ca., U.S. (ōk′lănd)	38	37°48′N	122°16′W
Oaxaca, Mex.	42	17°03′N	96°42′W
Oaxaca, state, Mex. (wä-hä′kä)	42	16°45′N	97°00′W
Ob′, r., Russia	56	62°15′N	67°00′E
Obi, Kepulauan, is., Indon. (ō′bė)	75	1°25′S	128°15′E
Obi, Pulau, i., Indon.	75	1°30′S	127°45′E
Obock, Dji. (ō-bŏk′)	87a	11°55′N	43°15′E
Obskaya Guba, b., Russia	56	67°13′N	73°45′E
Ochamchira, Geor.	58	42°44′N	41°28′E
Ödemiş, Tur. (û′dĕ-mēsh)	55	38°12′N	28°00′E
Odendaalsrus, S. Afr. (ō′dĕn-däls-rûs′)	87c	27°52′S	26°41′E
Oder, r., Eur. (ō′dẽr)	52	52°40′N	14°19′E
Odesa, Ukr.	56	46°28′N	30°44′E
Odiongan, Phil. (ō-dė-ōŋ′gän)	75a	12°24′N	121°59′E
Odra *see* Oder, r., Eur. (ô′drȧ)	52	52°40′N	14°19′E
Ogaden Plateau, plat., Eth.	87a	6°45′N	44°53′E
Ogden, Ut., U.S.	38	41°14′N	111°58′W
Ogdensburg, N.Y., U.S.	39	44°40′N	75°30′W
Ogies, S. Afr.	87c	26°03′S	29°04′E
Ohio, state, U.S. (ō′hī′ō)	39	40°30′N	83°15′W
Ohio, r., U.S.	39	37°25′N	88°05′W
Ojinaga, Mex. (ō-ᴋė-nä′gä)	42	29°34′N	104°26′W
Okayama, Japan (ō′kȧ-yä′mȧ)	77	34°39′N	133°54′E
Okha, Russia (ŭ-ᴋā′)	57	53°44′N	143°12′E
Okhotsk, Russia (ō-ᴋôtsk′)	57	59°28′N	143°32′E
Okhotsk, Sea of, sea, Asia (ō-kôtsk′)	57	56°45′N	146°00′E
Okinawa, i., Japan	77	26°30′N	128°00′E
Oklahoma, state, U.S. (ô-klȧ-hō′mȧ)	38	36°00′N	98°20′W
Oklahoma City, Ok., U.S.	38	35°27′N	97°32′W
Oktemberyan, Arm.	58	40°09′N	44°02′E
Öland, i., Swe. (û-länd′)	52	57°03′N	17°15′E
Old Harbor, Ak., U.S. (här′bẽr)	40	57°18′N	153°20′W
Olean, N.Y., U.S. (ō-lė-ăn′)	39	42°05′N	78°25′W
Olëkminsk, Russia (ō-lyĕk-mėnsk′)	57	60°39′N	120°40′E
Olenëk, r., Russia (ō-lyĕ-nyôk′)	57	68°00′N	113°00′E
Ol′ga, Russia (ôl′gȧ)	57	43°48′N	135°44′E
Olhão, Port. (ōl-youɴ′)	54	37°02′N	7°54′W
Ólimbos, mtn., Cyp.	67a	34°56′N	32°52′E
Olongapo, Phil.	74	14°49′S	120°17′E
Olot, Spain (ō-lōt′)	54	42°09′N	2°30′E
Olt, r., Rom.	55	44°09′N	24°40′E
Olympia, Wa., U.S. (ō-lĭm′pĭ-ȧ)	38	47°02′N	122°52′W
Ólympos, mtn., Grc.	54	40°05′N	22°21′E
Olyutorskiy, Mys, c., Russia (ŭl-yo͞o′tŏr-skė)	57	59°49′N	167°16′E
Omaha, Ne., U.S. (ō′mȧ-hä)	39	41°18′N	95°57′W
Oman, nation, Asia	68	20°00′N	57°45′E
Oman, Gulf of, b., Asia	68	24°24′N	58°58′E
Omdurman, Sudan	85	15°45′N	32°30′E
Omsk, Russia (ômsk)	56	55°12′N	73°19′E
Öndörhaan, Mong.	77	47°20′N	110°40′E
Onega, Russia (ō-nyĕ′gȧ)	56	63°50′N	38°08′E
Onega, Lake *see* Onezhskoye Ozero, l., Russia	51	62°02′N	34°35′E
Onezhskoye Ozero, Russia (ō-năsh′skō-yĕ ô′zĕ-rō)	51	62°02′N	34°35′E
Ongiin Hiid, Mong.	76	46°00′N	102°46′E
Ongole, India	73	15°36′N	80°03′E
Onon, r., Asia (ô′nōn)	57	49°00′N	112°00′E
Onslow, Austl. (ŏnz′lō)	80	21°53′S	115°00′E
Ontario, Lake, l., N.A.	36	43°35′N	79°05′W
Oodnadatta, Austl. (o͞od′nȧ-dȧ′tȧ)	80	27°38′S	135°40′E
Ooldea Station, Austl. (o͞ol-dā′ä)	80	30°35′S	132°08′E
Ophir, Ak., U.S. (ō′fẽr)	40	63°10′N	156°28′W
Ophir, Mount, mtn., Malay.	67b	2°22′N	102°37′E
Oradea, Rom. (ō-räd′yä)	50	47°02′N	21°55′E
Oral, Kaz.	59	51°14′N	51°22′E
Orange, Austl. (ŏr′ĕnj)	81	33°15′S	149°08′E
Orange, r., Afr.	85	29°15′S	17°30′E
Orangeville, S. Afr.	87c	27°05′S	28°13′E
Orani, Phil. (ō-rä′nė)	75a	14°47′N	120°32′E
Ordos Desert, des., China	76	39°12′N	108°10′E
Ordu, Tur. (ōr′dȯ)	55	41°00′N	37°50′E
Oregon, state, U.S.	38	43°40′N	121°50′W
Orekhovo-Zuyevo, Russia (ōr-yĕ′ᴋō-vō zȯ′yĕ-vō)	56	55°46′N	39°00′E
Orël, Russia (ôr-yôl′)	56	52°59′N	36°05′E
Ore Mountains *see* Erzgebirge, mts., Eur.	52	50°29′N	12°40′E
Orenburg, Russia (ô′rĕn-bo͞org)	56	51°50′N	55°05′E
Orhon, r., Mong.	76	48°33′N	103°07′E
Orinoco, r., Ven. (ō-rĭ-nō′kō)	46	8°32′N	63°13′W
Orion, Phil. (ō-rė-ōn′)	75a	14°37′N	120°34′E
Orissa, state, India (ŏ-rĭs′ȧ)	69	25°09′N	83°50′E
Oristano, Italy (ō-rės-tä′nō)	54	39°53′N	8°38′E
Orizaba, Mex. (ō-rė-zä′bä)	43	18°52′N	97°05′E
Orizaba, Pico de, vol., Mex.	42	19°04′N	97°14′W
Orkney, S. Afr. (ôrk′nĭ)	87c	26°58′S	26°39′E
Orkney Islands, is., Scot., U.K.	52	59°01′N	2°08′W
Orlando, Fl., U.S. (ôr-lăn′dō)	39	28°32′N	81°22′W
Orléans, Fr. (ōr-lā-äɴ′)	50	47°55′N	1°56′E
Orsk, Russia (ôrsk)	56	51°15′N	58°50′E
Ortegal, Cabo, c., Spain (kȧ′bō-ôr-tå-gäl′)	54	43°46′N	8°15′W
Ortigueira, Spain (ôr-tė-gā′ė-rä)	54	43°40′N	7°50′W
Orūmīyeh, Iran	68	37°30′N	45°15′E
Orūmīyeh, Daryacheh-ye, l., Iran	68	38°01′N	45°17′E
Ōsaka, Japan (ō′sä-kä)	77	34°40′N	135°27′E
Osh, Kyrg. (ôsh)	59	40°33′N	72°48′E
Oshkosh, Wi., U.S.	39	44°01′N	88°35′W
Osijek, Cro. (ōs′ĭ-yĕk)	55	45°33′N	18°48′E
Öskemen, Kaz.	59	49°58′N	82°38′E
Oslo, Nor. (ōs′lō)	50	59°56′N	10°41′E
Osmaniye, Tur.	55	37°10′N	36°30′E
Ostrava, Czech Rep.	50	49°51′N	18°18′E
Oswego, N.Y., U.S.	39	43°25′N	76°30′W
Otaru, Japan (ō′tȧ-rȯ)	77	43°07′N	141°00′E
Otranto, Strait of, strt., Eur.	52	40°30′N	18°45′E
Ottawa, Can. (ŏt′ȧ-wȧ)	36	45°25′N	75°43′W
Ottumwa, Ia., U.S. (ô-tŭm′wȧ)	39	41°00′N	92°26′W
Ouagadougou, Burkina (wä′gȧ-do͞o′go͞o)	85	12°22′N	1°31′W
Oulu, Fin. (ō′lȯ)	50	64°58′N	25°43′E
Outer Brass, i., V.I.U.S. (bräs)	43c	18°24′N	64°58′W
Oviedo, Spain (ō-vė-ā′dhō)	50	43°22′N	5°50′W
Owensboro, Ky., U.S. (ō′ĕnz-bŭr-ō)	39	37°45′N	87°05′W
Owen Stanley Range, mts., Pap. N. Gui. (stăn′lė)	75	9°00′S	147°30′E
Oymyakon, Russia (oi-myŭ-kôn′)	57	63°14′N	142°58′E
Ozamiz, Phil. (ō-zä′mėz)	75	8°06′N	123°43′E
Ozark Plateau, plat., U.S.	39	36°37′N	93°56′W
Ozieri, Italy	54	40°38′N	8°53′E
Ozurgeti, Geor.	58	41°56′N	42°00′E

ng-sing; ŋ-baŋk; ɴ-nasalized n; nŏd; cŏmmit; ōld; ôbey; ôrder; oi-boil; fo͞od; ȯ-as oo in foot; ou-out; s-soft; sh-dish; th-thin; pūre; ůnite; ûrn; stŭd; circŭs; ü-as in French tu; ′-indeterminate vowel.

ăt; fin*a*l; rāte; senăte; ärm; ȧsk; sof*a*; fāre; ch-choose; dh-as th in other; bē; ĕvent; bĕt; recĕnt; cratĕr; g-gō; gh-guttural g; bĭt; *ĭ*-short neutral; rīde; κ-guttural k as ch in German ich;

PLACE (Pronunciation)	PAGE	LAT.	LONG.
Praha *see* Prague, Czech Rep.	50	50°05′N	14°26′E
Prescott, Az., U.S. (prĕs′kŏt)	38	34°30′N	112°30′W
Pretoria, S. Afr. (pre̊-tō′rĭ-ȧ)	85	25°43′S	28°16′E
Pretoria North, S. Afr. (prĕ-tô′rĭ-ȧ no͞ord)	87c	25°41′S	28°11′E
Pribilof Islands, is., Ak., U.S. (prĭ′bĭ-lof)	40	57°00′N	169°20′W
Prilep, Mac. (prē′lĕp)	55	41°20′N	21°35′E
Prince Edward Islands, is., S. Afr.	82	46°36′S	37°57′E
Prince of Wales, i., Ak., U.S.	40	55°47′N	132°50′W
Prince of Wales, Cape, c., Ak., U.S. (wālz)	40	65°48′N	169°08′W
Prince William Sound, strt., Ak., U.S. (wĭl′yăm)	40	60°40′N	147°10′W
Priština, Serb. (prēsh′tĭ-nä)	55	42°39′N	21°12′E
Prizren, Serb. (prē′zrĕn)	55	42°11′N	20°45′E
Progreso, Mex. (pro̊-grā′sō)	42	21°14′N	89°39′W
Prokhladnyy, Russia	58	43°46′N	44°00′E
Prome, Mya.	74	18°46′N	95°15′E
Providence, R.I., U.S.	39	41°50′N	71°23′W
Provideniya, Russia (pro̊-vĭ-dă′nĭ-yȧ)	40	64°30′N	172°54′W
Provo, Ut., U.S. (prō′vō)	38	40°15′N	111°40′W
Prudhoe Bay, b., Ak., U.S.	40	70°40′N	147°25′W
Prut, r., Eur. (pro͞ot)	52	48°05′N	27°07′E
Przemyśl, Pol. (pzhĕ′mĭsh′l)	50	49°47′N	22°45′E
Przheval′sk, Kyrg. (p′r-zhĭ-välsk′)	59	42°29′N	78°24′E
Pskov, Russia (pskôf)	56	57°48′N	28°19′E
Puebla, Mex. (pwā′blä)	42	19°02′N	98°11′W
Pueblo, Co., U.S. (pwā′blō)	38	38°15′N	104°36′W
Puerto Barrios, Guat. (pwĕ′r-to̊ bär′re̊-ôs)	42	15°43′N	88°36′W
Puerto Cortés, Hond. (pwĕ′r-to̊ ko̊r-tās′)	42	15°48′N	87°57′W
Puertollano, Spain (pwĕ-to̊l-yä′nō)	54	38°41′N	4°05′W
Puerto Peñasco, Mex. (pwĕ′r-to̊ pĕn-yä′s-ko̊)	42	31°39′N	113°15′W
Puerto Plata, Dom. Rep. (pwĕ′r-to̊ plä′tä)	43	19°50′N	70°40′W
Puerto Princesa, Phil. (pwĕr-to̊ pre̊n-sā′sä)	74	9°45′N	118°41′E
Puerto Rico, dep., N.A. (pwĕr′to̊ rē′kō)	43	18°16′N	66°50′W
Puerto Rico Trench, deep,	43	19°45′N	66°30′W
Pukin, r., Malay.	67b	2°53′N	102°54′E
Pula, Cro. (po͞o′lä)	54	44°52′N	13°55′E
Pulicat, r., India	73	13°58′N	79°52′E
Pulog, Mount, mtn., Phil. (po͞o′lo̊g)	75a	16°38′N	120°53′E
Puma Yumco, l., China (po͞o-mä yo͞om-tswo)	72	28°30′N	90°10′E
Punakha, Bhu. (po͞o-nŭk′ŭ)	69	27°45′N	89°59′E
Pune, India	69	18°38′N	73°53′E
Punjab, state, India (pŭn′jäb′)	69	31°00′N	75°30′E
Puntarenas, C.R. (po̅nt-ä-rā′näs)	43	9°59′N	84°49′W
Puri, India (po̅′re̊)	69	19°52′N	85°51′E
Purús, r., S.A. (po͞o-ro͞o′s)	46	6°45′S	64°34′W
Pusan, Kor., S.	77	35°08′N	129°05′E
Putorana, Gory, mts., Russia	57	68°45′N	93°15′E
Puttalam, Sri L.	73	8°02′N	79°44′E
Putung, Tanjung, c., Indon.	74	3°35′S	111°50′E
Pyinmana, Mya. (pye̊n-mä′nŭ)	69	19°47′N	96°15′E
P′yŏngyang, Kor., N.	77	39°03′N	125°48′E
Pyramids, hist., Egypt	87b	29°53′N	31°10′E
Pyrenees, mts., Eur. (pĭr-e-nēz′)	52	43°00′N	0°05′E
Pýrgos, Grc.	55	37°51′N	21°28′E

Q

PLACE (Pronunciation)	PAGE	LAT.	LONG.
Qal′at Bishah, Sau. Ar.	68	20°01′N	42°30′E
Qamdo, China (chyäm-dwō)	76	31°06′N	96°30′E
Qandala, Som.	71	11°28′N	49°52′E
Qaraghandy (Karaganda), Kaz.	59	49°42′N	73°18′E
Qarqan	76	38°02′N	85°16′E
Qarqan, r., China	76	38°55′N	87°15′E
Qarqaraly, Kaz.	59	49°18′N	75°28′E
Qasr el Boukhari, Alg.	54	35°50′N	2°48′E
Qatar, nation, Asia (kä′tȧr)	68	25°00′N	52°45′E
Qausuittuq (Resolute), Can.	37	74°41′N	95°00′W
Qāyen, Iran	68	33°45′N	59°08′E
Qazvīn, Iran	68	36°10′N	49°59′E
Qeshm, Iran	68	26°51′N	56°10′E
Qeshm, i., Iran	68	26°52′N	56°15′E
Qezel Owzan, r., Iran	68	36°30′N	49°00′E
Qezi′ot, Isr.	67a	30°53′N	34°28′E
Qiblīyah, Jabal al Jalālat al, mts., Egypt	67a	28°49′N	32°21′E
Qilian Shan, mts., China (chyē-lĭĕn shän)	76	38°43′N	98°00′E
Qindao, China (chyĭn-dou)	77	36°05′N	120°10′E
Qinghai, prov., China (chyĭŋ-hī)	76	36°14′N	95°30′E
Qinghai Hu *see* Koko Nor, l., China	76	37°26′N	98°30′E
Qingyang, China (chyĭŋ-yäŋ)	76	36°02′N	107°42′E
Qinhuangdao, China (chyĭn-huaŋ-dou)	77	39°57′N	119°34′E
Qin Ling, mts., China (chyĭn lĭŋ)	76	33°25′N	108°58′E
Qiqian, China (chyē-chyĕn)	77	52°23′N	121°04′E
Qiqihar, China	77	47°18′N	124°00′E
Qiryat Gat, Isr.	67a	31°38′N	34°36′E
Qiryat Shemona, Isr.	67a	33°12′N	35°34′E
Qitai, China (chyē-tī)	76	44°07′N	89°04′E
Qogir Feng *see* K2, mtn., Asia	69	36°06′N	76°38′E
Qom, Iran	68	34°28′N	50°53′E
Qongyrat, Kaz.	59	47°25′N	75°10′E
Qostanay, Kaz.	59	53°10′N	63°39′E
Quang Ngai, Viet. (kwäng n′gä′e̊)	74	15°05′N	108°58′E
Quanzhou, China (chyuän-jō)	77	24°58′N	118°40′E
Qūchān, Iran	71	37°06′N	58°30′E
Québec, Can. (kwe̊-bĕk′) (kå-bĕk′)	36	46°49′N	71°13′W
Québec, prov., Can.	36	51°07′N	70°25′W
Queen Charlotte Islands, is., Can. (kwēn shär′lŏt)	36	53°30′N	132°25′W
Queen Elizabeth Islands, is., Can. (e̊-lĭz′ȧ-bĕth)	37	78°20′N	110°00′W
Queen Maud Land, reg., Ant.	82	75°00′S	10°00′E
Queen Maud Mountains, mts., Ant.	82	85°00′S	179°00′W
Queensland, state, Austl. (kwēnz′lănd)	81	22°45′S	141°01′E
Querétaro, Mex. (kå-rā′tä-rō)	42	20°37′N	100°25′W
Quetta, Pak. (kwĕt′ä)	69	30°19′N	67°01′E
Quezaltenango, Guat. (kå-zäl′tå-näŋ′gō)	42	14°50′N	91°30′W
Quezon City, Phil. (kā-zōn)	74	14°40′N	121°02′E
Quilon, India (kwe̊-lōn′)	73	8°58′N	76°16′E
Quilpie, Austl. (kwĭl′pe̊)	81	26°34′S	149°20′E
Quincy, Il., U.S.	39	39°55′N	91°23′W
Qui Nhon, Viet. (kwĭnyo̊n)	74	13°51′N	109°03′E
Quintana Roo, state, Mex. (rô′o̊)	42	19°30′N	88°30′W
Quito, Ec. (kē′tō)	46	0°17′S	78°32′W
Qurayyah, Wādī, r., Egypt	67a	30°08′N	34°27′E
Qusmuryn köli, l., Kaz.	59	52°30′N	64°15′E
Quxian, China (chyo͞o-shyĕn)	77	28°58′N	118°58′E
Qyzylorda, Kaz.	59	44°58′N	65°45′E

R

PLACE (Pronunciation)	PAGE	LAT.	LONG.
Raba, Indon.	74	8°32′S	118°49′E
Rabat, Mor. (rȧ-bät′)	85	33°59′N	6°47′W
Rabaul, Pap. N. Gui. (rä′boul)	75	4°15′S	152°19′E
Rābigh, Sau. Ar.	71	22°48′N	39°01′E
Rachado, Cape, c., Malay.	67b	2°26′N	101°29′E
Racine, Wi., U.S. (rȧ-sēn′)	39	42°43′N	87°49′W
Rădăuţi, Rom.	55	47°53′N	25°55′E
Rādhanpur, India	72	23°57′N	71°38′E
Radium, S. Afr. (rā′dĭ-ŭm)	87c	25°06′S	28°18′E
Radwah, Jabal, mtn., Sau. Ar.	68	24°44′N	38°14′E
Rafah, Pak. (rä′fä)	67a	31°14′N	34°12′E
Rafsanjān, Iran	68	30°45′N	56°30′E
Ragay, Phil. (rä-gī′)	75a	13°49′N	122°45′E
Ragay Gulf, b., Phil.	75a	13°44′N	122°38′E
Ragusa, Italy (rä-go͞o′sä)	54	36°58′N	14°41′E
Rāichūr, India (rä′e̊-cho͞or′)	69	16°23′N	77°18′E
Raigarh, India (rī′gŭr)	69	21°57′N	83°32′E
Rainbow City, Pan.	42a	9°20′N	79°53′W
Rainier, Mount, mtn., Wa., U.S. (rā-nēr′)	38	46°52′N	121°46′W
Raipur, India (rä′jŭ-bo͞o-re̊′)	72	21°25′N	81°37′E
Rājahmundry, India (räj-ŭ-mŭn′dre̊)	69	17°03′N	81°51′E
Rajang, r., Malay.	74	2°10′N	113°30′E
Rājapālaiyam, India	73	9°30′N	77°33′E
Rājasthān, state, India (rä′jŭs-tän)	69	26°00′N	72°00′E
Rājkot, India (räj′ko̊t)	69	22°20′N	70°48′E
Rājpur, India	72a	22°24′N	88°25′E
Rājshāhi, Bngl.	69	24°26′S	88°39′E
Raleigh, N.C., U.S.	39	35°45′N	78°39′W
Ramanāthapuram, India	73	9°13′N	78°52′E
Ramlat as Sab′atayn, reg., Asia	68	16°08′N	45°15′E
Ramm, Jabal, mtn., Jord.	67a	29°37′N	35°32′E
Râmnicu Sărat, Rom.	55	45°24′N	27°06′E
Rampart, Ak., U.S. (răm′pȧrt)	40	65°28′N	150°18′W
Rāmpur, India (räm′po͞or)	69	28°53′N	79°03′E
Ramree Island, i., Mya. (räm′rē′)	74	19°01′N	93°23′E
Ramu, r., Pap. N. Gui. (rä′mo͞o)	75	5°35′S	145°16′E
Rānchī, India	69	23°21′N	85°20′E
Ranger, Tx., U.S. (rān′jẽr)	38	32°26′N	98°41′W
Rangia, India	72	26°32′N	91°39′E
Rangoon (Yangon), Mya. (rȧŋ-go͞on′)	69	16°46′N	96°09′E
Rangpur, Bngl. (rŭng′po͞or)	69	25°48′N	89°19′E
Rangsang, i., Indon. (räng′säng′)	67b	0°53′N	103°05′E
Rānīganj, India (rä-nē-gŭnj′)	72	23°40′N	87°08′E
Rantau, Malay.	67b	2°35′N	101°58′E
Rantekombola, Bulu, mtn., Indon.	74	3°22′S	119°50′E
Rapid City, S.D., U.S.	38	44°06′N	103°14′W
Rarotonga, Cook Is. (rä′ro̊-tŏŋ′gȧ)	2	20°40′S	163°00′W
Ra′s an Naqb, Jord.	67a	30°00′N	35°29′E
Ras Dashen Terara, mtn., Eth. (räs dä-shän′)	85	12°49′N	38°14′E
Rashayya, Leb.	67a	33°30′N	35°50′E
Rashīd, Egypt (rȧ-shēd′) (ro̊-zĕt′ȧ)	70	31°22′N	30°25′E
Rashîd, Masabb, mth., Egypt	87b	31°30′N	29°58′E
Rasht, Iran	68	37°13′N	49°45′E
Ratangarh, India (rŭ-tŭn′gŭr)	72	28°10′N	74°30′E
Rat Islands, is., Ak., U.S. (răt)	40a	51°35′N	176°48′E
Ratlām, India	72	23°19′N	75°05′E
Ratnāgiri, India	73	17°04′N	73°24′E
Raton, N.M., U.S. (rȧ-tōn′)	38	36°52′N	104°26′W
Raurkela, India	69	22°15′N	84°53′E
Ravenna, Italy (rä-vĕn′nä)	54	44°27′N	12°13′E
Ravensthorpe, Austl. (rā′vĕns-thôrp)	80	33°30′S	120°20′E
Rāwalpindi, Pak. (rä-wŭl-pe̊n′de̊)	69	33°40′N	73°10′E
Rawlina, Austl. (rôr-lēnȧ)	80	31°13′S	125°45′E
Rawlins, Wy., U.S. (rô′lĭnz)	38	41°46′N	107°15′W
Raya, Bukit, mtn., Indon.	74	0°45′S	112°11′E
Ray Mountains, mts., Ak., U.S.	40	65°40′N	151°45′W
Razdan, Arm.	58	40°30′N	44°46′E
Razgrad, Blg.	55	43°32′N	26°32′E
Reading, Pa., U.S.	39	40°20′N	75°55′W
Recife, Braz. (rå-sē′fĕ)	46	8°09′S	34°59′W
Red, r., Asia	74	21°00′N	103°00′E
Red, r., U.S.	39	31°40′N	92°55′W
Red Sea, sea,	68	23°15′N	37°00′E
Reggio di Calabria, Italy (rĕ′jo̊ dē kä-lä′bre̊-ä)	55	38°07′N	15°42′E
Reggio nell′ Emilia, Italy	54	44°43′N	10°34′E
Regina, Can. (re̊-jī′nȧ)	36	50°25′N	104°39′W
Rehovot, Isr.	67a	31°53′N	34°49′E
Reims, Fr. (răɴs)	50	49°16′N	4°00′E
Reitz, S. Afr.	87c	27°48′S	28°25′E
Rema, Jabal, mtn., Yemen	68	14°13′N	44°38′E
Rembau, Malay.	67b	2°36′N	102°06′E
Rempang, i., Indon.	67b	0°51′N	104°04′E
Rengam, Malay. (rĕn′gäm′)	67b	1°53′N	103°24′E
Renmark, Austl. (rĕn′märk)	81	34°10′S	140°50′E
Rennes, Fr. (rĕn)	50	48°07′N	1°02′W
Reno, Nv., U.S. (rē′nō)	38	39°32′N	119°49′W
Requena, Spain (rå-kā′nä)	54	39°29′N	1°03′W
Resolute *see* Qausuittuq, Can.	37	74°41′N	95°00′W
Réunion, dep., Afr. (rā-ü-nyôn′)	3	21°06′S	55°36′E
Reus, Spain (rā′o͞os)	54	41°08′N	1°05′E
Revillagigedo, Islas, is., Mex. (e̊′s-läs-rē-ve̊l-yä-he̊′gĕ-do̊)	42	18°45′N	111°00′W
Rewa, India (rā′wä)	69	24°41′N	81°11′E
Rewāri, India	72	28°19′N	76°39′E
Rey, Iran	71	35°35′N	51°25′E
Reykjanes, c., Ice. (rā′kyȧ-nĕs)	52	63°37′N	24°33′W
Reykjavík, Ice. (rā′kyȧ-vēk)	50	64°09′N	21°39′W
Rhine, r., Eur.	52	50°34′N	7°21′E
Rhode Island, state, U.S. (rōd ī′lȧnd)	39	41°35′N	71°40′W
Rhodes *see* Ródos, i., Grc.	55	36°00′N	28°29′E
Rhodesia *see* Zimbabwe, nation, Afr.	85	17°50′S	29°30′E
Rhodope Mountains, mts., Eur. (rô′do̊-pĕ)	52	42°00′N	24°08′E
Rhône, r., Fr. (rōn)	52	44°30′N	4°45′E
Riau, prov., Indon.	67b	0°56′N	101°25′E
Riau, Kepulauan, i., Indon.	74	0°30′N	104°55′E
Riau, Selat, strt., Indon.	67b	0°40′N	104°27′E
Richards Island, i., Can. (rĭch′ẽrds)	40	69°45′N	135°30′W
Richmond, Austl. (rĭch′mŭnd)	81	20°47′S	143°14′E
Richmond, Va., U.S.	39	37°35′N	77°30′W
Rieti, Italy (re̊-ā′tē)	54	42°25′N	12°51′E
Rīga, Lat. (rē′gȧ)	56	56°55′N	24°05′E
Rīgān, Iran	68	28°45′N	58°55′E
Rīgestān, des., Afg.	68	30°53′N	64°42′E
Rijeka, Cro. (rĭ-yĕ′kä)	54	45°22′N	14°24′E
Rimini, Italy (rē′me̊-nē)	54	44°03′N	12°33′E
Rinjani, Gunung, mtn., Indon.	74	8°39′S	116°22′E
Río Abajo, Pan. (re̊′ō-ä-bä′ko̊)	42a	9°01′N	78°30′W
Rio de Janeiro, Braz. (rē′o̅ dā zhä-nå′e̊-ro̅)	46b	22°50′S	43°20′W
Rioni, r., Geor.	58	42°08′N	41°39′E
Ríoverde, Mex. (rē′ō-vĕr′då)	42	21°54′N	99°59′W
Risdon, Austl. (rĭz′dŭn)	81	42°37′S	147°32′E
Rishon le Ziyyon, Isr.	67a	31°57′N	34°48′E
Rishra, India	72a	22°42′N	88°22′E
Riverside, Ca., U.S. (rĭv′ẽr-sīd)	38	33°59′N	117°21′W
Riyadh, Sau. Ar.	68	24°31′N	46°47′E
Rize, Tur. (rē′zĕ)	55	41°00′N	40°30′E
Roanoke, Va., U.S.	39	37°16′N	79°55′W
Rochester, Mn., U.S.	39	44°01′N	92°30′W
Rochester, N.Y., U.S.	39	43°15′N	77°35′W
Rockford, Il., U.S. (rŏk′fẽrd)	39	42°16′N	89°07′W
Rockhampton, Austl. (rŏk-hămp′tŭn)	81	23°26′S	150°29′E
Rock Hill, S.C., U.S. (rŏk′hĭl)	39	34°55′N	81°01′W
Rock Island, Il., U.S.	39	41°31′N	90°37′W
Rock Springs, Wy., U.S.	38	41°35′N	109°13′W
Rocky Mountains, mts., N.A.	37	50°00′N	114°00′W
Ródos, Grc.	55	36°24′N	28°15′E
Ródos, i., Grc.	55	36°00′N	28°29′E
Roebourne, Austl. (rō′bŭrn)	80	20°50′S	117°15′E
Roedtan, S. Afr.	87c	24°37′S	29°08′E
Rojo, Cabo, c., P.R. (rō′hō)	43b	17°55′N	67°14′W
Roma, Austl. (rō′mȧ)	81	26°30′S	148°48′E
Roma *see* Rome, Italy	50	41°52′N	12°37′E
Romania, nation, Eur. (rō-mā′ne̊-ȧ)	50	46°18′N	22°53′E
Romblon, Phil. (rōm-blōn′)	75a	12°34′N	122°16′E
Romblon Island, i., Phil.	75a	12°33′N	122°17′E
Rome (Roma), Italy	50	41°52′N	12°37′E
Rome, Ga., U.S. (rōm)	39	34°14′N	85°10′W
Rompin, Malay.	67b	2°42′N	102°30′E
Rompin, r., Malay.	67b	2°54′N	103°10′E
Ronne Ice Shelf, ice, Ant.	82	77°30′S	38°00′W
Rooiberg, S. Afr.	87c	24°46′S	27°42′E
Roosevelt Island, i., Ant.	82	79°30′S	168°00′W
Rosa, Monte, mtn., Italy (mōn′tā rō′zä)	54	45°56′N	7°51′E
Rosales, Phil. (ro̊-sä′lĕs)	75a	15°54′N	120°38′E
Rosario, Phil.	75a	13°49′N	121°13′W
Roseberg, Or., U.S. (rōz′bûrg)	38	43°13′N	123°30′W
Rosendal, S. Afr. (ro̊-sĕn′täl)	87c	28°32′S	27°56′E
Rosetta *see* Rashīd, Egypt	70	31°22′N	30°25′E
Rossano, Italy (rô-sä′nō)	55	39°34′N	16°38′E
Ross Ice Shelf, ice, Ant.	82	81°30′S	175°00′W
Ross Sea, sea, Ant.	82	76°00′S	178°00′W
Rostov-na-Donu, Russia (ro̊stôv-nå-do̊-no͞o)	56	47°16′N	39°47′E
Roswell, N.M., U.S.	38	33°23′N	104°32′W
Roti, Pulau, i., Indon. (rō′te̊)	74	10°30′S	122°52′E
Rotterdam, Neth. (rôt′ẽr-däm′)	50	51°55′N	4°27′E
Rouen, Fr. (ro͞o-äɴ′)	50	49°25′N	1°05′E
Roxas, Phil. (rô-xäs)	74	11°30′N	122°47′E
Ruapehu, vol., N.Z. (ro̅-ȧ-pā′ho͞o)	81a	39°15′S	175°37′E
Rub′ al Khali *see* Ar Rub′ al Khālī, des., Asia	68	20°00′N	51°00′E
Rubtsovsk, Russia	56	51°31′N	81°17′E
Rudolf, Lake, l., Afr. (ro͞o′dŏlf)	85	3°30′N	36°05′E
Rügen, i., Ger. (rü′ghĕn)	52	54°28′N	13°47′E
Rummah, Wādī ar, val., Sau. Ar.	68	26°17′N	41°45′E

ng-sing; ŋ-baŋk; ɴ-nasalized n; nŏd; cŏmmit; ōld; o̊bey; ôrder; oi-boil; fo͞od; o̅-as oo in foot; ou-out; s-soft; sh-dish; th-thin; pūre; ůnite; ûrn; stŭd; circŭs; ü-as in French tu; ′-indeterminate vowel.

ăt; finál; rāte; senåte; ärm; åsk; sofá; fåre; ch-choose; dh-as th in other; bē; ĕvent; bĕt; recĕnt; cratĕr; g-gō; gh-guttural g; bĭt; ī-short neutral; rīde; ᴋ-guttural k as ch in German ich;

āt; fināl; rāte; senāte; ärm; àsk; sofà; fâre; ch-choose; dh-as th in other; bē; ĕvent; bĕt; recĕnt; cratĕr; g-gō; gh-guttural g; bĭt; ĭ-short neutral; rīde; к-guttural k as ch in German ich;

PRONOUNCING INDEX Tilichiki – Vanderbijlpark 133

PLACE (Pronunciation)	PAGE	LAT.	LONG.
Tilichiki, Russia (tyĭ-le-chĭ-kė)	57	60°49'N	166°14'E
Timişoara, Rom.	55	45°44'N	21°21'E
Timor, i., Asia (tė-mōr')	75	10°08'S	125°00'E
Timor Sea, sea,	80	12°40'S	125°00'E
Timsāh, l., Egypt (tĭm'sä)	87b	30°34'N	32°22'E
Tinah, Khalij at, b., Egypt	67a	31°06'N	32°42'E
Tinggi, i., Malay.	67b	2°16'N	104°16'E
Tínos, i., Grc.	55	37°45'N	25°12'E
Tinsukia, India (tin-sōō'kĭ-á)	68	27°18'N	95°29'W
Tioman, i., Malay.	67b	2°50'N	104°15'E
Tiranë, Alb. (tė-rä'nä)	50	41°48'N	19°50'E
Tire, Tur. (tē'rĕ)	55	38°05'N	27°48'E
Tiruchchirāppalli, India (tĭr'ŏ-chĭ-rä'pá-lĭ)	69	10°49'N	78°48'E
Tirunelveli, India	73	8°53'N	77°43'E
Tiruppur, India	73	11°11'N	77°08'E
Tista, r., Asia	72	26°00'N	89°30'E
Tisza, r., Eur. (tē'sä)	52	47°30'N	21°00'E
Titāgarh, India	72a	22°44'N	88°23'E
Tivoli, Italy (tē'vô-lė)	54	41°38'N	12°48'E
Tizard Bank and Reef, rf., Asia (tĭz'ȧrd)	74	10°51'N	113°20'E
Tkvarcheli, Geor.	58	42°15'N	41°41'E
Tlaxcala, Mex. (tläs-kä'lä)	42	19°16'N	98°14'W
Tobago, i., Trin. (tô-bā'gō)	43	11°15'N	60°30'W
Togian, Kepulauan, is., Indon.	74	0°20'S	122°00'E
Togo, nation, Afr. (tō'gō)	85	8°00'N	0°52'E
Tohor, Tanjong, c., Malay.	67b	1°53'N	102°29'E
Tokat, Tur. (tô-kät')	68	40°20'N	36°30'E
Tokelau, dep., Oc. (tō-kė-lä'ó)	2	8°00'S	176°00'W
Tokmak, Kyrg. (tôk'mák)	59	42°44'N	75°41'E
Tokuno, i., Japan (tô-kōō'nō)	77	27°42'N	129°25'E
Tokushima, Japan (tō'kô'shē-mä)	77	34°06'N	134°31'E
Tōkyō, Japan	77	35°42'N	139°46'E
Toledo, Spain (tô-lĕ'dô)	54	39°53'N	4°02'W
Toledo, Oh., U.S.	39	41°40'N	83°35'W
Tolo, Teluk, b., Indon. (tō'lō)	74	2°00'S	122°06'E
Tolosa, Spain (tô-lō'sä)	54	43°10'N	2°05'W
Toluca, Mex. (tô-lōō'kȧ)	42	19°17'N	99°40'W
Toluca, Nevado de, mtn., Mex. (nĕ-vä-dô-dĕ-tô-lōō'kä)	42	19°09'N	99°42'W
Tomanivi, mtn., Fiji	78g	17°37'S	178°01'E
Tombouctou, Mali	85	16°46'N	3°01'W
Tommot, Russia (tŏm-mŏt')	57	59°13'N	126°22'E
Tomsk, Russia (tŏmsk)	56	56°29'N	84°57'E
Tondano, Indon. (tôn-dä'nō)	75	1°15'N	124°50'E
Tonga, nation, Oc. (tŏŋ'gȧ)	88	18°50'S	175°20'W
Tonga Trench, deep,	88	23°00'S	172°30'W
Tongbei, China (tôŋ-bā)	77	48°00'N	126°48'E
Tongguan, China (tôŋ-gūän)	77	34°48'N	110°25'E
Tonghua, China (tôŋ-hwä)	77	41°43'N	125°50'E
Tongjiang, China (tôŋ-jyäŋ)	77	47°38'N	132°54'E
Tongren, China (tôŋ-rŭn)	76	27°45'N	109°12'E
Tongtian, r., China (tôŋ-tĭĕn)	76	33°00'N	97°00'E
Tonk, India (Tōŋk)	69	26°13'N	75°45'E
Tonkin, Gulf of, b., Asia (tôn-kăn')	74	20°30'N	108°10'E
Tonle Sap, l., Camb. (tôn'lā säp')	74	13°03'N	102°49'E
Tonopah, Nv., U.S. (tō-nō-pä')	38	38°04'N	117°15'W
Toowoomba, Austl. (tò wōōm'bȧ)	81	27°32'S	152°10'E
Topeka, Ks., U.S. (tô-pē'kȧ)	39	39°02'N	95°41'W
Topolobampo, Mex. (tō-pō-lô-bä'm-pô)	42	25°45'N	109°00'W
Torbat-e Ḩeydarīyeh, Iran	71	35°16'N	59°13'E
Torbat-e Jām, Iran	71	35°14'N	60°36'E
Torino *see* Turin, Italy	50	45°05'N	7°44'E
Torneälven, r., Eur.	52	67°00'N	22°30'E
Tornio, Fin. (tôr'nĭ-ô)	50	65°55'N	24°09'E
Toronto, Can. (tô-rŏn'tō)	36	43°40'N	79°23'W
Toros Dağları, mts., Tur. (tô'rŭs)	68	37°00'N	32°40'E
Torrens, Lake, l., Austl. (tŏr-ĕns)	80	30°07'S	137°40'E
Torreón, Mex. (tôr-rā-ōn')	42	25°32'N	103°26'W
Torres Islands, is., Vanuatu (tôr'rĕs) (tôr'ĕz)	81	13°18'N	165°59'E
Torres Strait, strt., Austl. (tôr'rĕs)	81	10°30'S	141°30'E
Torrijos, Phil. (tôr-rē'hōs)	75a	13°19'N	122°06'E
Tórshavn, Far. Is. (tôrs-houn')	50	62°00'N	6°55'W
Tortola, i., Br. Vir. Is. (tôr-tō'lä)	43b	18°34'N	64°40'W
Tortosa, Spain (tôr-tō'sä)	50	40°59'N	0°33'E
Toruń, Pol.	50	53°02'N	18°35'E
Tosya, Tur. (tôz'yä)	55	41°00'N	34°00'E
Totonicapán, Guat. (tôtō-nē-kä'pän)	42	14°55'N	91°20'W
Tottori, Japan (tô'tô-rė)	77	35°30'N	134°15'E
Toubkal, Jebel, mtn., Mor.	85	31°15'N	7°46'W
Touil, Oued, r., Alg. (tōō-ēl')	54	34°42'N	2°16'E
Toulon, Fr. (tōō-lôN')	50	43°09'N	5°54'E
Toulouse, Fr. (tōō-lōōz')	50	43°37'N	1°27'E
Toungoo, Mya. (tô-ôŋ-gōō')	74	19°00'N	96°29'E
Tours, Fr. (tōōr)	50	47°23'N	0°39'E
Townsville, Austl. (tounz'vĭl)	81	19°18'S	146°50'E
Towuti, Danau, l., Indon. (tô-wōō'tė)	74	3°00'S	121°45'E
Toxkan, r., China	76	40°34'N	77°15'E
Toyama, Japan (tō'yä-mä)	77	36°42'N	137°14'E
Tozeur, Tun. (tô-zûr')	54	33°59'N	8°11'E
Trabzon, Tur. (tráb'zôn)	68	41°00'N	39°45'E
Trangan, Pulau, i., Indon. (träŋ'gän)	75	6°52'S	133°30'E
Transylvania, hist. reg., Rom. (trăn-sĭl-vā'nĭ-ȧ)	50	46°30'N	22°35'E
Trapani, Italy	54	38°01'N	12°31'E
Trás-os-Montes, hist. reg., Port. (träzh'ōzh mōn'tāzh)	54	41°33'N	7°13'W
Trento, Italy (trĕn'tô)	54	46°04'N	11°07'E
Trenton, N.J., U.S.	39	40°13'N	74°46'W
Treviso, Italy (trĕ-vē'sō)	54	45°39'N	12°15'E
Trichardt, S. Afr. (trĭ-kärt')	87c	26°32'N	29°16'E
Trieste, Italy (trė-ĕs'tä)	50	45°39'N	13°48'E
Tríkala, Grc.	55	39°33'N	21°49'E
Trikora, Puncak, mtn., Indon.	75	4°15'S	138°45'E
Trincomalee, Sri L. (trĭŋ-kô-mȧ-lē')	73	8°39'N	81°12'E
Trinidad, Cuba (trē-nē-dhädh')	43	21°50'N	80°00'W
Trinidad, Co., U.S. (trĭn'ĭdăd)	38	37°11'N	104°31'W
Trinidad, r., Pan.	42a	8°55'N	80°01'W
Trinidad and Tobago, nation, N.A. (trĭn'ĭ-dăd) (tô-bā'gō)	43	11°00'N	61°00'W
Trinity, is., Ak., U.S.	40	56°25'N	153°15'W
Trípoli, Grc.	55	37°32'N	22°32'E
Tripoli (Tarābulus), Libya	85	32°50'N	13°13'E
Tripura, state, India	69	24°00'N	92°00'E
Tristan da Cunha Islands, is., St. Hel. (três-tän'dä kōōn'yä)	2	35°30'S	12°15'W
Trobriand Islands, is., Pap. N. Gui. (trō-brē-änd')	75	8°25'S	151°45'E
Troitsko-Pechorsk, Russia (trô'ĭtsk-ô-pyĕ-chôrsk')	56	62°18'N	56°07'E
Trondheim, Nor. (trôn'hâm)	50	63°25'N	11°35'E
Troy, N.Y., U.S.	39	42°45'N	73°45'W
Troy, hist., Tur.	68	39°59'N	26°14'E
Trstenik, Serb. (t'r'stĕ-nĕk)	55	43°36'N	21°00'E
Trucial States *see* United Arab Emirates, nation, Asia	68	24°00'N	54°00'E
Trujillo, Peru	46	8°08'S	79°00'W
Trujillo, Spain (trōō-κĕ'l-yô)	54	39°27'N	5°50'W
Truk *see* Chuuk, is., Micron.	78c	7°25'N	151°47'E
Tsaidam Basin, basin, China (tsī-däm)	76	37°19'N	94°08'E
Tsast Bogd, mtn., Mong.	76	46°44'N	92°34'E
Tskhinvali, Geor.	58	42°13'N	43°56'E
Tsugaru Kaikyō, strt., Japan	77	41°25'N	140°20'E
Tsu Shima, is., Japan (tsōō shē'mä)	77	34°28'N	129°30'E
Tsushima Strait, strt., Asia	77	34°00'N	129°00'E
Tuamoto, Îles, Fr. Poly. (tōō-ä-mō'tōō)	89	19°00'S	141°20'W
Tucson, Az., U.S. (tōō-sŏn')	38	32°15'N	111°00'W
Tudela, Spain (tōō-dhā'lä)	54	42°03'N	1°37'W
Tuguegarao, Phil. (tōō-gā-gä-rä'ō)	74	17°37'N	121°44'E
Tuinplaas, S. Afr.	87c	24°54'S	28°46'E
Tukangbesi, Kepulauan, is., Indon.	75	6°00'S	124°15'E
Tulaghi, Sol. Is.	78e	9°06'S	160°09'E
Tulancingo, Mex. (tōō-län-sĭŋ'gō)	42	20°04'N	98°24'W
Tulangbawang, r., Indon.	74	4°17'S	105°00'E
Tulcea, Rom. (tòl'chȧ)	55	45°10'N	28°47'E
Tulik Volcano, vol., Ak., U.S. (tò'lĭk)	40a	53°28'N	168°10'W
Tūlkarm, W.B. (tōōl kärm)	67a	32°19'N	35°02'E
Tulsa, Ok., U.S. (tŭl'sȧ)	39	36°08'N	95°58'W
Tulun, Russia (tò-lōōn')	57	54°29'N	100°43'E
Tumkūr, India	73	13°22'N	77°05'E
Tungabhadra Reservoir, res., India	73	15°26'N	75°57'E
Tuni, India	73	17°29'N	82°38'E
Tunis, Tun. (tū'nĭs)	85	36°59'N	10°06'E
Tunis, Golfe de, b., Tun.	54	37°06'N	10°43'E
Tunisia, nation, Afr. (tu-nĭzh'ĕ-á)	85	35°00'N	10°11'E
Tura, Russia (tòr'ȧ)	57	64°08'N	99°58'E
Turfan Depression, depr., China	76	42°16'N	90°00'E
Turin, Italy	50	45°05'N	7°44'E
Turkestan, hist. reg., Asia	56	43°27'N	62°14'E
Turkey, nation, Asia	51	38°45'N	32°00'E
Türkistan, Kaz.	59	44°00'N	68°00'E
Turkmenbashy, Turkmen.	59	40°00'N	52°50'E
Turkmenistan, nation, Asia	56	40°46'N	56°01'E
Turks, is., T./C. Is. (tûrks)	43	21°40'N	71°45'W
Turku, Fin. (tòrgokò)	50	60°28'N	22°12'E
Turneffe, i., Belize	42	17°25'N	87°43'W
Turnu Măgurele, Rom.	55	43°54'N	24°49'E
Turpan, China (tōō-är-pän)	76	43°06'N	88°41'E
Turtkul', Uzb. (tòrt-kòl')	59	41°28'N	61°02'E
Turukhansk, Russia (tōō-rōō-κänsk')	56	66°03'N	88°39'E
Tuscaloosa, Al., U.S. (tŭs-kȧ-lōō'sȧ)	39	33°10'N	87°35'W
Tuticorin, India (tōō-tė-kô-rĭn')	73	8°51'N	78°09'E
Tutrakan, Blg.	55	44°02'N	26°36'E
Tutuila, i., Am. Sam.	78a	14°18'S	170°42'W
Tuvalu, nation, Oc.	3	5°20'S	174°00'E
Tuwayq, Jabal, mts., Sau. Ar.	68	20°45'N	46°30'E
Túxpan, Mex.	42	20°57'N	97°26'W
Tuxtla Gutiérrez, Mex. (tòs'tlä gōō-tyâr'rĕs)	42	16°44'N	93°08'W
Tuzla, Bos. (tòz'lȧ)	55	44°33'N	18°46'E
Tver', Russia	56	56°52'N	35°57'E
Tweeling, S. Afr. (twē'lĭng)	87c	27°34'S	28°31'E
Twin Falls, Id., U.S. (fôls)	38	42°33'N	114°29'W
Tyler, Tx., U.S.	39	32°21'N	95°19'W
Tyndinskiy, Russia	57	55°22'N	124°45'E
Tyre *see* Şūr, Leb.	67a	33°16'N	35°13'E
Tyrrhenian Sea, sea, Italy (tĭr-rē'nĭ-ȧn)	52	40°10'N	12°15'E
Tyukalinsk, Russia (tyò-kȧ-lĭnsk')	56	56°03'N	71°43'E
Tyumen', Russia (tyōō-mĕn')	56	57°02'N	65°28'E

U

PLACE (Pronunciation)	PAGE	LAT.	LONG.
Ubangi, r., Afr. (ōō-bäŋ'gė)	85	3°00'N	18°00'E
Ubon Ratchathani, Thai. (ōō'bŭn rä'chätá-nē)	74	15°15'N	104°52'E
Udaipur, India (ò-dŭ'ē-pōōr)	72	24°41'N	73°41'E
Udine, Italy (ōō'dē-nå)	54	46°05'N	13°14'E
Udon Thani, Thai.	74	17°31'N	102°51'E
Udskaya Guba, b., Russia	57	55°00'N	136°30'E
Ufa, Russia (ò'fa)	56	54°45'N	55°57'E
Uganda, nation, Afr. (ōō-gän'dä) (û-gän'dȧ)	85	2°00'N	32°28'E
Ugashik Lake, l., Ak., U.S. (ōō'gá-shĕk)	40	57°36'N	157°10'W
Uglegorsk, Russia (ōō-glĕ-gôrsk)	57	49°00'N	142°31'E
Uiju, Kor., N. (ò'ējōō)	77	40°09'N	124°33'E
Ujjain, India (ōō-jŭēn)	69	23°18'N	75°37'E
Ujungpandang, Indon.	74	5°08'S	119°28'E
Ukraine, nation, Eur.	56	49°15'N	30°15'E
Ulaangom, Mong.	76	50°23'N	92°14'E
Ulan Bator (Ulaanbaatar), Mong.	76	47°56'N	107°00'E
Ulan-Ude, Russia (ōō'län ōō'dȧ)	57	51°59'N	107°41'E
Ulcinj, Serb. (ōōl'tsēn')	55	41°56'N	19°15'E
Ulhās, r., India	73b	19°13'N	73°03'E
Ulhāsnagar, India	72	19°10'N	73°07'E
Uliastay, Mong.	76	47°49'N	97°00'E
Ulmer, Mount, mtn., Ant. (ŭl'mûr')	82	77°30'S	86°00'W
Ulubāria, India	72a	22°27'N	88°09'E
Ulukışla, Tur. (ōō-lōō-kēsh'lȧ)	55	36°40'N	34°30'E
Ulungur, r., China (ōō-lōōn-gûr)	76	46°31'N	88°00'E
Uluru (Ayers Rock), mtn., Austl.	80	25°23'S	131°05'E
Ulverstone, Austl. (ŭl'vēr-stŭn)	81	41°20'S	146°22'E
Ul'yanovsk, Russia (ōō-lyä'nôfsk)	56	54°20'N	48°24'E
Umberpāda, India	73b	19°28'N	73°04'E
Umeälven, r., Swe.	52	64°57'N	18°51'E
Umnak Pass, Ak., U.S.	40a	53°10'N	168°04'W
Unalakleet, Ak., U.S. (ū-nȧ-lăk'lēt)	40	63°50'N	160°42'W
Unalaska, Ak., U.S. (ū-nȧ-lăs'kȧ)	40a	53°30'N	166°20'W
Unayzah, Sau. Ar.	68	25°50'N	44°02'E
Unimak, i., Ak., U.S. (ōō-nē-măk')	40	54°30'N	163°35'W
Unimak Pass, Ak., U.S.	40a	54°22'N	165°22'W
Unisan, Phil. (ōō-nē'sän)	75a	13°50'N	121°59'E
United Arab Emirates, nation, Asia	68	24°00'N	54°00'E
United Kingdom, nation, Eur.	50	56°30'N	1°40'W
United States, nation, N.A.	38	38°00'N	110°00'W
Ünye, Tur. (ün'yĕ)	55	41°00'N	37°10'E
Upolu, i., Samoa	78a	13°55'S	171°45'W
Upper Kapuas Mountains, mts., Asia	74	1°45'N	112°06'E
Upper Volta *see* Burkina Faso, nation, Afr.	85	13°00'N	2°00'W
Uppsala, Swe. (ōōp'sȧ-lä)	50	59°53'N	17°39'E
Ural, r., (ò-räl'') (ū-rôl)	56	48°00'N	51°00'E
Urals, mts., Russia	56	56°28'N	58°13'E
Uran, India (ōō-rän')	73b	18°53'N	72°46'E
Urdaneta, Phil. (ōōr-dä-nā'tä)	75a	15°59'N	120°34'E
Uruguay, nation, S.A. (ōō-rōō-gwī') (ū'rōō-gwā)	46	32°45'S	56°00'W
Uruguay, r., S.A. (ōō-rōō-gwī')	46	27°05'S	55°15'W
Ürümqi, China (ů-rům-chyē)	76	43°49'N	87°43'E
Urup, i., Russia (ò'ròp')	77	46°00'N	150°00'E
Ürzhar, Kaz.	59	47°28'N	82°00'E
Uşak, Tur. (ōō'shȧk)	55	38°45'N	29°15'E
Ussuriysk, Russia	57	43°48'N	132°09'E
Ust'-Bol'sheretsk, Russia	57	52°41'N	157°00'E
Ust'-Kamchatsk, Russia	57	56°13'N	162°18'E
Ust'-Kulom, Russia (kò'lŭm)	56	61°38'N	54°00'E
Ust'-Maya, Russia (mä'yȧ)	57	60°33'N	134°43'E
Ust' Olenëk, Russia	57	72°52'N	120°15'E
Ust' Port, Russia (òst'pôrt')	56	69°20'N	83°41'E
Ust'-Tsil'ma, Russia (tsĭl'mȧ)	56	65°25'N	52°10'E
Ust'-Tyrma, Russia (tur'mȧ)	57	50°27'N	131°17'E
Ust-Urt, Plateau, plat., Asia	56	44°03'N	54°58'E
Usu, China (ů-sōō)	76	44°28'N	84°07'E
Utah, state, U.S. (ū'tô)	38	39°25'N	112°40'W
Utan, India	73b	19°17'N	72°43'E
Utica, N.Y., U.S.	39	43°05'N	75°10'W
Utrera, Spain (ōō-trā'rä)	54	37°12'N	5°48'W
Utsunomiya, Japan (ōōt'sò-nô-mē-yȧ')	77	36°35'N	139°52'E
Uttaradit, Thai.	74	17°47'N	100°10'E
Uttaranchal, state, India	69	29°30'N	78°30'E
Uttarpara-Kotrung, India	72a	22°40'N	88°21'E
Uttar Pradesh, state, India (òt-tär-prä-dēsh)	69	27°00'N	80°00'E
Utuado, P.R. (ōō-tōō-ä'dhō)	43b	18°16'N	66°40'W
Uvs Nuur, l., Asia	76	50°29'N	93°32'E
Uzbekistan, nation, Asia	56	42°42'N	60°00'E

V

PLACE (Pronunciation)	PAGE	LAT.	LONG.
Vaal, r., S. Afr. (väl)	85	28°15'S	24°30'E
Vaaldam, res., S. Afr.	87c	26°58'S	28°37'E
Vaalplaas, S. Afr.	87c	25°39'S	28°56'E
Vaalwater, S. Afr.	87c	24°17'S	28°08'E
Vaasa, Fin. (vä'sȧ)	50	63°06'N	21°39'E
Vaigai, r., India	73	10°20'N	78°13'E
Valdepeñas, Spain (väl-dā-pān'yäs)	54	38°46'N	3°22'W
Valdez, Ak., U.S. (văl'dēz)	40	61°10'N	146°18'W
Valdosta, Ga., U.S. (văl-dŏs'tȧ)	39	30°50'N	83°18'W
Valencia, Spain	50	39°26'N	0°23'W
Valentine, Ne., U.S. (vă lăn-tė-nyē')	38	42°52'N	100°34'W
Valladolid, Mex. (väl-yä-dhō-lēdh')	42	20°39'N	88°13'W
Valladolid, Spain (väl-yä-dhô-lēdh')	50	41°41'N	4°41'W
Vallejo, Ca., U.S. (vä-yā'hō) (vȧ-lā'hō)	38	38°06'N	122°15'W
Valles, Mex.	42	21°59'N	99°02'W
Valletta, Malta (väl-lĕt'ä)	54	35°50'N	14°29'E
Valley City, N.D., U.S.	38	46°55'N	97°59'W
Valls, Spain (väls)	54	41°15'N	1°15'E
Valona *see* Vlorë, Alb.	55	40°28'N	19°31'E
Valparaíso, Chile (väl'pä-rä-ē'sō)	46	33°02'S	71°32'W
Vals, r., S. Afr.	87c	27°32'S	26°51'E
Vals, Tanjung, c., Indon.	75	8°30'S	137°15'E
Van, Tur. (vän)	68	38°04'N	43°10'E
Vancouver, Can. (văn-kōō'vēr)	36	49°16'N	123°06'W
Vancouver, Wa., U.S.	38	45°37'N	122°40'W
Vancouver Island, i., Can.	36	49°50'N	125°05'W
Vanderbijlpark, S. Afr.	87c	26°43'S	27°50'E

ng-sing; ŋ-baŋk; N-nasalized n; nŏd; cŏmmit; ōld; ôbey; ôrder; oi-boil; fōōd; ò-as oo in foot; ou-out; s-soft; sh-dish; th-thin; pūre; ůnite; ûrn; stŭd; circŭs; ü-as in French tu; '-indeterminate vowel.

PLACE (Pronunciation)	PAGE	LAT.	LONG.
Vanegas, Mex. (vä-nĕ′gäs)	42	23°54′N	100°54′W
Vänern, l., Swe.	52	58°52′N	13°17′E
Vangani, India	73b	19°07′N	73°15′E
Van Rees, Pegunungan, mts., Indon.	75	2°30′S	138°45′E
Vanua Levu, i., Fiji	78g	16°33′S	179°15′E
Vanuatu, nation, Oc.	81	16°02′S	169°15′E
Vārānasi (Benares), India	69	25°25′N	83°00′E
Varangerfjorden, b., Nor.	53	70°05′N	30°20′E
Varaždin, Cro. (vä′räzh′dĕn)	55	46°17′N	16°20′E
Varna, Blg. (vär′nȧ)	50	43°14′N	27°58′E
Vasa, India	73b	19°20′N	72°47′E
Vasto, Italy (väs′tô)	54	42°06′N	12°42′E
Vatican City, nation, Eur.	50	41°54′N	12°22′E
Vättern, l., Swe.	52	58°15′N	14°24′E
Vaygach, i., Russia (vī-gȧch′)	56	70°00′N	59°00′E
Vehār Lake, l., India	73b	19°11′N	72°52′E
Velebit, mts., Serb. (vā′lĕ-bĕt)	55	44°25′N	15°23′E
Velika Kapela, mts., Serb. (vĕ′lė-kä kä-pĕ′lä)	55	45°03′N	15°20′E
Velika Morava, r., Serb. (mô′rä-vä)	55	44°00′N	21°30′E
Velikiye Luki, Russia (vyē-lē′-kyė lōō′ke)	56	56°19′N	30°32′E
Velikiy Ustyug, Russia (vå-lē′kĭ ōōs-tyôg′)	56	60°45′N	46°38′E
Veliko Tŭrnovo, Blg.	55	43°06′N	25°38′E
Vellore, India (vĕl-lōr′)	69	12°57′N	79°09′E
Vel′sk, Russia (vĕlsk)	56	61°00′N	42°18′E
Venezia *see* Venice, Italy	50	45°25′N	12°18′E
Venezuela, nation, S.A. (vĕn-ė-zwē′lȧ)	46	8°00′N	65°00′W
Veniaminof, Mount, mtn., Ak., U.S.	40	56°12′N	159°20′W
Venice, Italy	50	45°25′N	12°18′E
Venice, Gulf of, b., Italy	54	45°23′N	13°00′E
Ventersburg, S. Afr. (vĕn-tērs′bûrg)	87c	28°06′S	27°10′E
Ventersdorp, S. Afr. (vĕn-tērs′dôrp)	87c	26°20′S	26°48′E
Veracruz, Mex.	42	19°13′N	96°07′W
Veracruz, state, Mex. (vā-rä-krōōz′)	42	20°30′N	97°15′W
Verāval, India (vĕr′vŭ-väl)	69	20°59′N	70°49′E
Verde, i., Phil. (vĕr′då)	75a	13°34′N	121°11′E
Verde Island Passage, strt., Phil. (vĕr′dē)	75a	13°36′N	120°39′E
Vereeniging, S. Afr. (vĕ-rā′nĭ-gĭng)	87c	26°40′S	27°56′E
Verena, S. Afr. (vĕr-ĕn ȧ)	87c	25°30′S	29°02′E
Verkhne-Kamchatsk, Russia (vyĕrk′nyĕ kȧm-chatsk′)	57	54°42′N	158°41′E
Verkhne Ural′sk, Russia (ö-ralsk′)	56	53°53′N	59°13′E
Verkhoyansk, Russia (vyĕr-kô-yänsk′)	57	67°43′N	133°33′E
Verkhoyanskiy Khrebet, mts., Russia (vyĕr-kô-yänskĭ)	57	67°45′N	128°00′E
Vermont, state, U.S. (vẽr-mŏnt′)	39	43°50′N	72°50′W
Verona, Italy (vā-rō′nä)	54	45°28′N	11°02′E
Vert, Cap, c., Sen.	86	14°43′N	17°30′W
Vestfjord, b., Nor.	52	67°33′N	12°59′E
Vesuvio, vol., Italy (vĕ-sōō′vyä)	52	40°35′N	14°26′E
Vet, r., S. Afr. (vĕt)	87c	28°25′S	26°37′E
Viana do Castelo, Port. (dö käs-tā′lö)	54	41°41′N	8°45′W
Viangchan, Laos	74	18°07′N	102°33′E
Vicenza, Italy (vė-chĕnt′sä)	54	45°33′N	11°33′E
Vicksburg, Ms., U.S.	39	32°20′N	90°50′W
Victoria, Can. (vĭk-tō′rĭ-ȧ)	36	48°26′N	123°23′W
Victoria, Phil. (vėk-tô-ryä)	75a	15°34′N	120°41′E
Victoria, state, Austl.	81	36°46′S	143°15′E
Victoria, l., Afr.	85	0°50′S	32°50′E
Victoria, Mount, mtn., Mya.	69	21°26′N	93°59′E
Victoria, Mount, mtn., Pap. N. Gui.	75	9°35′S	147°45′E
Victoria Falls, wtfl., Afr.	85	17°55′S	25°51′E
Victoria Island, i., Can.	37	70°13′N	107°45′W
Victoria Land, reg., Ant.	82	75°00′S	160°00′E
Victoria River Downs, Austl. (vĭc-tôr′ĭȧ)	80	16°30′S	131°10′E
Vidin, Blg. (vĭ′dĕn)	55	44°00′N	22°53′E
Viedma, Arg. (vyād′mä)	46	40°55′S	63°03′W
Vienna (Wien), Aus.	50	48°13′N	16°22′E
Vientiane *see* Viangchan, Laos	74	18°07′N	102°33′E
Vieques, P.R. (vyā′kås)	43b	18°09′N	65°27′W
Vieques, i., P.R. (vyā′kås)	43b	18°05′N	65°28′W
Vierfontein, S. Afr. (vēr′fôn-tān)	87c	27°06′S	26°45′E
Vietnam, nation, Asia (vyĕt′näm′)	74	18°00′N	107°00′E
Vigan, Phil. (vēgän)	74	17°36′N	120°22′E
Vigo, Spain (vē′gō)	50	42°18′N	8°42′W
Vijayawāda, India	69	16°31′N	80°37′E
Vila Real, Port. (rā-äl′)	54	41°18′N	7°48′W
Viljoenskroon, S. Afr.	87c	27°13′S	26°58′E
Villahermosa, Mex. (vēl′yä-ĕr-mō′sä)	42	17°59′N	92°56′W
Villaldama, Mex. (vēl-yäl-dä′mä)	42	26°30′N	100°26′W
Villarrobledo, Spain (vēl-yär-rô-blā′dhō)	54	39°15′N	2°37′W
Villena, Spain (vē-lyā′nȧ)	54	38°37′N	0°52′W
Villiers, S. Afr. (vĭl′ĭ-ērs)	87c	27°03′S	28°38′E
Villupuram, India	73	11°59′N	79°33′E
Vilnius, Lith. (vĭl′nė-ös)	56	54°40′N	25°26′E
Vilyuy, r., Russia (vēl′yĭ)	57	63°00′N	121°00′E
Vilyuysk, Russia (vė-lyōō′isk′)	57	63°41′N	121°47′E
Vincennes, In., U.S. (vĭn-zĕnz′)	39	38°40′N	87°30′W
Vindhya Range, mts., India (vĭnd′yä)	69	22°30′N	75°50′E
Vinh, Viet. (vėn′y′)	74	18°38′N	105°42′E
Vinnytsia, Ukr.	56	49°13′N	28°31′E
Vinson Massif, mtn., Ant.	82	77°40′S	87°00′W
Virginia, S. Afr.	87c	28°07′S	26°54′E
Virginia, Mn., U.S. (vẽr-jĭn′yȧ)	39	47°32′N	92°36′W
Virginia, state, U.S.	39	37°00′N	80°45′W
Virgin Islands, is., N.A. (vûr′jĭn)	43	18°15′N	64°00′W
Vis, i., Serb.	55	43°00′N	16°10′E
Visby, Swe. (vĭs′bü)	50	57°39′N	18°19′E
Viscount Melville Sound, strt., Can.	37	74°00′N	110°00′W
Vishākhapatnam, India	69	17°48′N	83°21′E
Vistula *see* Wisła, r., Pol.	52	52°30′N	20°00′E
Viterbo, Italy (vė-tĕr′bō)	54	42°24′N	12°08′E
Viti Levu, i., Fiji	78g	18°00′S	178°00′E
Vitim, Russia (vē′tėm)	57	59°22′N	112°43′E
Vitim, r., Russia (vē′tėm)	57	54°00′N	115°00′E
Vitoria, Spain (vē-tô-ryä)	54	42°43′N	2°43′W
Vizianagaram, India	69	18°10′N	83°29′E
Vladimir, Russia (vlȧ-dyē′mēr)	56	56°08′N	40°24′E
Vladivostok, Russia (vlȧ-dė-vôs-tôk′)	57	43°06′N	131°47′E
Vlorë, Alb.	55	40°27′N	19°30′E
Volga, r., Russia (vôl′gä)	56	47°30′N	46°20′E
Volgograd, Russia (vŏl-gō-gră′t)	56	48°40′N	42°20′E
Volgogradskoye, res., Russia (vôl-gô-grad′skô-yĕ)	56	51°10′N	45°10′E
Vologda, Russia (vô′lôg-dȧ)	56	59°12′N	39°52′E
Volta, Lake, res., Ghana (vôl′tȧ)	85	7°10′N	0°30′W
Vorkuta, Russia (vôr-kōō′tȧ)	56	67°28′N	63°40′E
Voronezh, Russia (vô-rô′nyĕzh)	56	51°39′N	39°11′E
Vrangelya (Wrangel), i., Russia	56	71°25′N	178°30′W
Vratsa, Blg. (vrät′tsȧ)	55	43°12′N	23°31′E
Vrede, S. Afr. (vrĭ′dĕ)(vrēd)	87c	27°25′S	29°11′E
Vredefort, S. Afr. (vrĭ′dĕ-fôrt) (vrēd′fōrt)	87c	27°00′S	27°21′E
Vršac, Serb. (v′r′shȧts)	55	45°08′N	21°18′E
Vyborg, Russia (vwē′bôrk)	56	60°43′N	28°46′E
Vyshniy Volochëk, Russia (vėsh′nyĭ vôl-ô-chĕk′)	56	57°34′N	34°35′E
Vytegra, Russia (vû′tĕg-rȧ)	56	61°00′N	36°20′E

W

PLACE (Pronunciation)	PAGE	LAT.	LONG.
Wabamun, Grc.	55	39°23′N	22°56′E
Waco, Tx., U.S. (wā′kō)	38	31°35′N	97°06′W
Wagga Wagga, Austl. (wôg′ȧ wôg′ă)	81	35°10′S	147°30′E
Waha, Libya	70	28°16′N	19°54′E
Waigeo, Pulau, i., Indon. (wä-ė-gā′ô)	75	0°07′N	131°00′E
Wainganga, r., India (wä-ēn-gŭŋ′gä)	69	20°30′N	80°15′E
Waingapu, Indon.	74	9°32′S	120°00′E
Wainwright, Ak., U.S. (wān-rīt)	40	74°40′N	159°00′W
Wakayama, Japan (wä-kä′yä-mä)	77	34°14′N	135°11′E
Wake, i., Oc. (wāk)	3	19°25′N	167°00′E
Wakkanai, Japan (wä′kä-nä′ė)	77	45°19′N	141°43′E
Wales, Ak., U.S. (wālz)	40	65°35′N	168°14′W
Wales, state, U.K.	50	52°12′N	3°40′W
Walgett, Austl. (wôl′gĕt)	81	30°00′S	148°10′E
Wallaroo, Austl. (wŏl-ȧ-rōō)	80	33°52′S	137°45′E
Walla Walla, Wa., U.S. (wŏl′ȧ wŏl′ȧ)	38	46°03′N	118°20′W
Wallis and Futuna Islands, dep., Oc.	88	13°00′S	176°10′E
Walvis Bay, Nmb. (wôl′vĭs)	85	22°50′S	14°30′E
Wanda Shan, mts., China (wän-dä shän)	77	45°54′N	131°45′E
Wanxian, China (wän-shyĕn)	76	30°48′N	108°22′E
Warangal, India (wŭ′rȧŋ-gȧl)	69	18°03′N	79°45′E
Wardān, Wādī, r., Egypt	67a	29°22′N	33°00′E
Warden, S. Afr. (wôr′dĕn)	87c	27°52′N	28°59′E
Wardha, India (wŭr′dä)	69	20°46′N	78°42′E
Warmbad, S. Afr.	87c	24°52′S	28°18′E
Warrnambool, Austl. (wôr′năm-bōōl)	81	38°20′S	142°28′E
Warsaw, Pol.	50	52°15′N	21°05′E
Warszawa *see* Warsaw, Pol.	50	52°15′N	21°05′E
Warwick, Austl. (wŏr′ĭk)	81	28°05′S	152°10′E
Washington, D.C., U.S. (wŏsh′ĭng-tŭn)	39	38°50′N	77°00′W
Washington, state, U.S.	38	47°30′N	121°10′W
Washington, Mount, mtn., N.H., U.S.	39	44°15′N	71°15′W
Water, i., V.I.U.S. (wô′tẽr)	43c	18°20′N	64°57′W
Waterberge, mts., S. Afr. (wôrtẽr′bûrg)	87c	24°25′S	27°53′E
Waterloo, Ia., U.S.	39	42°30′N	92°22′W
Watertown, N.Y., U.S.	39	44°00′N	75°55′W
Watertown, S.D., U.S.	38	44°53′N	97°07′W
Waukegan, Il., U.S. (wŏ-kē′găn)	39	42°22′N	87°51′W
Wausau, Wi., U.S. (wô′sô)	39	44°58′N	89°40′W
Wazīrbad, Pak.	72	32°39′N	74°11′E
Weddell Sea, sea, Ant. (wĕd′ĕl)	82	73°00′S	45°00′W
Wei, r., China (wā)	76	34°00′N	108°10′E
Weichang, China (wā-chäŋ)	77	41°50′N	118°00′E
Weifang, China	77	36°43′N	119°08′E
Weihai, China (wa′hăī′)	77	37°30′N	122°05′E
Weipa, Austl.	81	12°25′S	141°54′E
Weixi, China (wā-shyĕ)	76	27°27′N	99°30′E
Wellington, N.Z.	81a	41°15′S	174°45′E
Welverdiend, S. Afr. (vĕl-vẽr-dēnd′)	87c	26°23′S	27°16′E
Wenquan, China (wŭn-chyůän)	77	47°10′N	120°00′E
Wenshan, China	76	23°20′N	104°15′E
Wensu, China (wĕn-sö)	76	41°45′N	80°30′E
Wentworth, Austl. (wĕnt′wûrth)	81	34°03′S	141°53′E
Wenzhou, China (wŭn-jō)	77	28°00′N	120°40′E
Weser, r., Ger. (vā′zĕr)	52	51°00′N	10°30′E
Wesselsbron, S. Afr. (wĕs′ĕl-brŏn)	87c	27°51′S	26°22′E
West, Mount, mtn., Pan.	42a	9°10′N	79°52′W
West Bengal, state, India (bĕn-gôl′)	69	23°30′N	87°30′E
Western Australia, state, Austl. (ôs-trā′lĭ-ȧ)	80	24°15′S	121°30′E
Western Ghāts, mts., India	69	17°35′N	74°00′E
Western Sahara, dep., Afr. (sȧ-hä′rȧ)	85	23°05′N	15°33′W
Western Samoa *see* Samoa, nation, Oc.	2	14°30′S	172°00′W
Western Siberian Lowland, depr., Russia	56	63°37′N	72°45′E
West Indies, is., (ĭn′dēz)	43	19°00′N	78°30′W
Westleigh, S. Afr. (wĕst-lē)	87c	27°39′S	27°18′E
Westonaria, S. Afr.	87c	26°19′S	27°38′E
West Palm Beach, Fl., U.S. (päm bēch)	39	26°44′N	80°04′W
West Virginia, state, U.S. (wĕst vẽr-jĭn′ĭ-ȧ)	39	39°00′N	80°50′W
West Wyalong, Austl. (wīălông)	81	34°00′S	147°20′E
Wetar, Pulau, i., Indon. (wĕt′ȧr)	75	7°34′S	126°00′E
Wewak, Pap. N. Gui. (wå-wäk′)	75	3°19′S	143°30′E
Whitehorse, Can. (whīt′hôrs)	36	60°39′N	135°01′W
White Pass, p., N.A.	40	59°35′N	135°03′W
White Russia *see* Belarus, nation, Eur.	56	53°30′N	25°33′E
White Sea, sea, Russia	56	66°00′N	40°00′E
Whitney, Mount, mtn., Ca., U.S.	38	36°34′N	118°18′W
Whyalla, Austl. (hwī-ăl′ȧ)	80	33°00′S	137°32′E
Wichita, Ks., U.S. (wĭch′ĭ-tô)	38	37°42′N	97°21′W
Wichita Falls, Tx., U.S. (fôls)	38	33°54′N	98°29′W
Wien *see* Vienna, Aus.	50	48°13′N	16°22′E
Wilcannia, Austl. (wĭl-cān-ĭȧ)	81	31°30′S	143°30′E
Wilge, r., S. Afr. (wĭl′jė)	87c	25°38′S	29°09′E
Wilge, r., S. Afr.	87c	27°27′S	28°46′E
Wilhelm, Mount, mtn., Pap. N. Gui.	75	5°58′S	144°58′E
Wilkes-Barre, Pa., U.S. (wĭlks′băr-ė)	39	41°15′N	75°50′W
Wilkes Land, reg., Ant.	82	71°00′S	126°00′E
Willemstad, Neth. Ant.	43	12°12′N	68°58′W
William Creek, Austl. (wĭl′yăm)	80	28°45′S	136°20′E
Williston, N.D., U.S. (wĭl′ĭs-tŭn)	38	48°08′N	103°38′W
Willow, Ak., U.S.	40	61°50′N	150°00′W
Wilmington, De., U.S.	39	39°45′N	75°33′W
Wilmington, N.C., U.S.	39	34°12′N	77°56′W
Wilno *see* Vilnius, Lith.	56	54°40′N	25°26′E
Wilpoort, S. Afr.	87c	26°57′S	26°17′E
Wiluna, Austl. (wĭ-lōō′nȧ)	80	26°35′S	120°25′E
Winburg, S. Afr. (wĭm-bûrg)	87c	28°31′S	27°02′E
Windhoek, Nmb. (vĭnt′hök)	85	22°05′S	17°10′E
Windorah, Austl. (wĭn-dō′rȧ)	81	25°15′S	142°50′E
Windward Islands, is., N.A. (wind′wẽrd)	43	12°45′N	61°40′W
Windward Passage, strt., N.A.	43	19°30′N	74°20′W
Winnemucca, Nv., U.S. (wĭn-ė-mŭk′ȧ)	38	40°59′N	117°43′W
Winnipeg, Can. (wĭn′ĭ-pĕg)	36	49°53′N	97°09′W
Winnipeg, Lake, l., Can.	36	52°00′N	97°00′W
Winona, Mn., U.S.	39	44°03′N	91°40′W
Winston-Salem, N.C., U.S. (wĭn stŭn-sā′lĕm)	39	36°05′N	80°15′W
Winton, Austl. (wĭn-tŭn)	81	22°17′S	143°08′E
Wisconsin, state, U.S. (wĭs-kŏn′sĭn)	39	44°30′N	91°00′W
Wisła, r., Pol. (vės′wȧ)	52	52°30′N	20°00′E
Witbank, S. Afr. (wĭt-băŋk)	87c	25°53′S	29°14′E
Witu Islands, is., Pap. N. Gui.	75	4°45′S	149°50′E
Witwatersrand, mtn., S. Afr. (wĭt-wôr′tẽrs-rănd)	87c	25°55′S	26°27′E
Wollongong, Austl. (wöl′ŭn-gŏng)	81	34°26′S	151°05′E
Wolwehoek, S. Afr.	87c	26°55′S	27°50′E
Wŏnsan, Kor., N. (wŭn′sän′)	77	39°08′N	127°24′E
Wonthaggi, Austl. (wônt-hăg′ė)	81	38°45′S	145°42′E
Woodlark Island, i., Pap. N. Gui. (wöd′lärk)	75	9°07′S	152°00′E
Woods, Lake of the, l., N.A.	36	49°25′N	93°25′W
Woomera, Austl. (wōōm′ẽrȧ)	80	31°15′S	136°43′E
Worcester, Ma., U.S. (wös′tẽr)	39	42°16′N	71°49′W
Wowoni, Pulau, i., Indon. (wō-wō′nė)	75	4°05′S	123°45′E
Wrangell, Cape, c., Ak., U.S.	40a	52°55′N	172°30′E
Wrangell, Mount, mtn., Ak., U.S.	40	61°58′N	143°50′W
Wrangell Mountains, mts., Ak., U.S.	40	62°28′N	142°40′W
Wrangell-Saint Elias National Park, rec., Ak., U.S.	40	61°00′N	142°00′W
Wu, r., China (wōō′)	76	27°30′N	107°00′E
Wuchang, China (wōō-chäŋ)	77	30°32′N	114°25′E
Wuhan, China	77	30°30′N	114°15′E
Wushi, China (wōō-shr)	76	41°13′N	79°08′E
Wuxi, China (wōō-shyē)	77	31°36′N	120°17′E
Wuxing, China (wōō-shyĭŋ)	77	30°38′N	120°10′E
Wuzhou, China (wōō-jō)	77	23°32′N	111°25′E
Wyndham, Austl. (wĭnd′ăm)	80	15°30′S	128°15′E
Wyoming, state, U.S.	38	42°50′N	108°30′W

X

PLACE (Pronunciation)	PAGE	LAT.	LONG.
Xalapa, Mex.	42	19°32′N	96°53′W
Xánthi, Grc.	55	41°08′N	24°53′E
Xàtiva, Spain	54	38°58′N	0°31′W
Xiamen, China	77	24°30′N	118°10′E
Xi′an, China (shyē-än)	76	34°20′N	109°00′E
Xiang, r., China (shyäŋ)	77	27°30′N	112°30′E
Xiangtan, China (shyäŋ-tän)	77	27°55′N	112°45′E
Xiapu, China (shyä-pōō)	77	27°00′N	120°00′E
Xining, China (shyē-nĭŋ)	76	36°52′N	101°36′E
Xinjiang (Sinkiang), prov., China (shyĭn-jyäŋ)	76	40°15′N	82°15′E
Xinyang, China (shyĭn-yäŋ)	77	32°08′N	114°04′E
Xizang (Tibet), prov., China (shyē-dzäŋ)	76	31°15′N	87°30′E
Xudat, Azer.	58	41°38′N	48°42′E
Xuddur, Som.	87a	3°55′N	43°45′E
Xuzhou, China	77	34°17′N	117°10′E

ăt; finăl; rāte; senåte; ärm; åsk; sofȧ; fâre; ch-choose; dh-as th in other; bē; ėvent; bĕt; recĕnt; cratẽr; g-gō; gh-guttural g; bĭt; ĭ-short neutral; rīde; κ-guttural k as ch in German ich;

PLACE (Pronunciation)	PAGE	LAT.	LONG.

Y

Ya'an, China (yä-än) .... 76 30°00′N 103°20′E
Yablonovyy Khrebet, mts., Russia
  (yá-blŏ-nô-vê′) .... 57 51°15′N 111°30′E
Yakima, Wa., U.S. (yăk′ĭmá) .... 38 46°35′N 120°30′W
Yaku, i., Japan (yä′kōō) .... 77 30°15′N 130°41′E
Yakutat, Ak., U.S. (yák′ō-tát) .... 40 59°32′N 139°35′W
Yakutsk, Russia (yá-kŏtsk′) .... 57 62°13′N 129°49′E
Yalong, r., China (yä-lŏŋ) .... 76 32°29′N 98°41′E
Yalu, r., Asia .... 77 41°20′N 126°35′E
Yalutorovsk, Russia (yä-lōō-tô′rôfsk) ... 56 56°42′N 66°32′E
Yamagata, Japan (yä-má′gä-tä) .... 77 38°12′N 140°24′E
Yamal, Poluostrov, pen., Russia
  (yä-mäl′) .... 56 71°15′N 70°00′E
Yambol, Blg. (yám′bôl) .... 55 42°28′N 26°31′E
Yamdena, i., Indon. .... 75 7°23′S 130°30′E
Yamethin, Mya. (yŭnŭm′) .... 69 20°14′N 96°27′E
Yamoussoukro, C. Iv. .... 85 6°49′N 5°17′W
Yamsk, Russia (yämsk) .... 57 59°41′N 154°09′E
Yamuna, r., India .... 69 25°30′N 80°30′E
Yamzho Yumco, l., China
  (yäm–jwo yōōm-tswo) .... 76 29°11′N 91°26′E
Yana, r., Russia (yä′ná) .... 57 71°00′N 136°00′E
Yanac, Austl. (yăn′ák) .... 81 36°10′S 141°30′E
Yanam, India (yŭnŭm′) .... 69 16°45′N 82°15′E
Yan'an, China (yän-än) .... 76 36°46′N 109°15′E
Yanbu', Sau. Ar. .... 68 23°57′N 38°02′E
Yangon see Rangoon, Mya. .... 69 16°46′N 96°09′E
Yangtze (Chang), r., China (yäng′tse)
  (chän) .... 77 30°30′N 117°25′E
Yangzhou, China (yän-jō) .... 77 32°24′N 119°24′E
Yanji, China (yän-jyē) .... 77 42°55′N 129°35′E
Yankton, S.D., U.S. (yănk′tŭn) .... 38 42°51′N 97°24′W
Yantai, China .... 77 37°32′N 121°22′E
Yanzhou, China (yän-jō) .... 77 35°35′N 116°50′E
Yaoundé, Cam. .... 85 3°52′N 11°31′E
Yap, i., Micron. (yäp) .... 3 11°00′N 138°00′E
Yapen, Pulau, i., Indon. .... 75 1°30′S 136°15′E
Yaque del Norte, r., Dom. Rep.
  (yä′kå dĕl nôr′tå) .... 43 19°40′N 71°25′W
Yaqui, r., Mex. (yä′kē) .... 42 28°15′N 109°40′W
Yaraka, Austl. (yä-räk′á) .... 81 24°50′S 144°08′E
Yaransk, Russia (yä-ränsk′) .... 56 57°18′N 48°05′E
Yarkand see Shache, China .... 76 38°15′N 77°15′E
Yaroslavl′, Russia (yä-rŏ-släv′′l) ... 56 57°37′N 39°54′E
Yartsevo, Russia .... 57 60°13′N 89°52′E
Yasawa Group, is., Fiji .... 78g 17°00′S 177°23′E
Yazd, Iran .... 68 31°59′N 54°03′E
Ye, Mya. (yā) .... 74 15°13′N 97°52′E
Yekaterinburg, Russia .... 56 56°51′N 60°36′E
Yelets, Russia (yĕ-lyĕts′) .... 56 52°35′N 38°28′E
Yelizavety, Mys, c., Russia
  (yĕ-lyĕ-sá-vyĕ′tĭ) .... 57 54°28′N 142°59′E
Yellow see Huang, r., China .... 77 35°06′N 113°39′E
Yellow Sea, sea, Asia .... 77 35°20′N 122°15′E
Yellowstone, r., U.S. .... 38 46°00′N 108°00′W
Yellowstone National Park, rec., U.S.
  (yĕl′ô-stōn) .... 38 44°45′N 110°35′W
Yemen, nation, Asia (yĕm′ĕn) .... 68 15°00′N 47°00′E

Yenangyaung, Mya. (yä′năn-d oung) ... 69 20°27′N 94°59′E
Yencheng, China .... 76 37°30′N 79°26′E
Yengisar, China (yŭn-gĕ-sär) .... 76 39°01′N 75°29′E
Yenisey, r., Russia (yĕ-nĕ-sĕ′ĕ) .... 56 71°00′N 82°00′E
Yeniseysk, Russia (yĕ-nĭĕsä′ĭsk) .... 57 58°27′N 90°28′E
Yerevan, Arm. (yĕ-rĕ-vän′) .... 51 40°10′N 44°30′E
Yevlax, Azer. .... 58 40°36′N 47°09′E
Yibin, China (yĕ-bĭn) .... 76 28°50′N 104°40′E
Yichang, China (yĕ-chän) .... 77 30°38′N 111°22′E
Yilan, China (yĕ-län) .... 77 46°10′N 129°40′E
Yinchuan, China (yĭn-chŭän) .... 76 38°22′N 106°22′E
Yingkou, China (yĭn-kō) .... 77 40°35′N 122°10′E
Yining, China (yĕ-nĭŋ) .... 76 43°58′N 80°40′E
Yishan, China (yĕ-shän) .... 76 24°32′N 108°42′E
Yitong, China (yĕ-tôŋ) .... 77 43°15′N 125°10′E
Yogyakarta, Indon. (yŏg-yä-kär′tá) ... 74 7°50′S 110°20′E
Yokohama, Japan (yō′kô-hä′mạ) .... 77 35°37′N 139°40′E
Yongshun, China .... 76 29°05′N 109°58′E
York, Austl. .... 80 32°00′S 117°00′E
York, Pa., U.S. .... 39 40°00′N 76°40′W
York, Cape, c., Austl. .... 81 10°45′S 142°35′E
York, Kap, c., Grnld. .... 37 75°30′N 73°00′W
Yos Sudarsa, Pulau, i., Indon. ...... 75 7°20′S 138°30′E
Yozgat, Tur. (yôz′gåd) .... 68 39°50′N 34°50′E
Yrghyz, Kaz. .... 59 48°30′N 61°17′E
Yrghyz, r., Kaz. .... 52 49°30′N 60°32′E
Ystädeh-ye Moqor, Åb-e, l., Afg. .... 72 32°35′N 68°00′E
Yu'alliq, Jabal, mts., Egypt .... 67a 30°12′N 33°42′E
Yuan, r., China (yüän) .... 77 28°50′N 110°50′E
Yucatán, state, Mex. (yōō-kä-tän′) ... 42 20°45′N 89°00′W
Yucatán Channel, strt., N.A. .... 42 22°30′N 87°00′W
Yucatán Peninsula, pen., N.A. .... 36 19°30′N 89°00′W
Yueyang, China (yüĕ-yän) .... 77 29°25′N 113°05′E
Yugoslavia see Serbia and
  Montenegro, nation, Eur.
  (yōō-gô-slä-vī-á) .... 50 44°00′N 21°00′E
Yukon, ter., Can. (yōō′kŏn) .... 36 63°16′N 135°30′W
Yukon, r., N.A. .... 36 64°00′N 159°30′W
Yukutat Bay, b., Ak., U.S. (yōō-kū tät′) ... 40 59°34′N 140°50′W
Yulin, China .... 76 38°18′N 109°45′E
Yuma, Az., U.S. (yōō′mä) .... 38 32°40′N 114°40′W
Yumen, China (yōō-mŭn) .... 76 40°14′N 96°56′E
Yunnan, prov., China (yun′nän′) ..... 76 24°23′N 101°03′E
Yunnan Plat, plat., China (yò-nän) ... 76 26°03′N 101°00′E
Yunxian, China (yón shyĕn) .... 77 32°50′N 110°55′E
Yü Shan, mtn., Tai. .... 77 23°38′N 121°00′E
Yutian, China (yōō-tĭĕn) (kŭ-r-yä) ... 76 36°55′N 81°39′E
Yuzhno-Sakhalinsk, Russia
  (yōōzh′nô-sä-kä-lĭnsk′) .... 57 47°11′N 143°04′E

Zagreb, Cro. (zä′grĕb) .... 50 45°50′N 15°58′E
Zagros Mountains, mts., Iran .... 68 33°30′N 46°30′E
Zähedän, Iran (zä′hå-dän) .... 68 29°37′N 60°31′E
Zahlah, Leb. (zä′lä′) .... 67a 33°50′N 35°54′E
Zaire see Congo, Democratic
  Republic of the, nation, Afr. ..... 85 1°00′S 22°15′E
Zákynthos, i., Grc. .... 55 37°45′N 20°32′E
Zambezi, r., Afr. (zám-bä′zĕ) .... 85 16°00′S 29°45′E
Zambia, nation, Afr. (zăm′bĕ-ä) ..... 85 14°23′S 24°15′E
Zamboanga, Phil. (säm-bô-aŋ′gä) ..... 74 6°58′N 122°02′E
Zamora, Mex. (sä-mō′rä) .... 42 19°59′N 102°16′W
Zamora, Spain (thä-mō′rä) .... 54 41°32′N 5°43′W
Zanjän, Iran .... 68 36°26′N 48°24′E
Zanzibar, i., Tan. .... 85 6°20′S 39°37′E
Zaporizhzhia, Ukr. .... 56 47°50′N 35°10′E
Zaragoza, Spain (thä-rä-gō′thä) ..... 50 41°39′N 0°53′W
Zaranj, Afg. .... 71 31°06′N 61°53′E
Zarqä′, r., Jord. .... 67a 32°13′N 35°43′E
Zashiversk, Russia (zá′shĭ-vĕrsk′) ... 57 67°08′N 144°02′E
Zäyandeh, r., Iran .... 68 32°15′N 51°00′E
Zaysan, Kaz. (zĭ′sán) .... 59 47°43′N 84°44′E
Zebediela, S. Afr. .... 87c 24°19′S 29°21′E
Zefat, Isr. .... 67a 32°58′N 35°30′E
Zemlya Frantsa-Iosifa (Franz Josef
  Land), is., Russia .... 56 81°32′N 40°00′E
Zemun, Serb. (zĕ′mōōn) (sĕm′lĭn) ..... 55 44°50′N 20°25′E
Zeya, Russia (zá′yá) .... 57 53°43′N 127°29′E
Zhambyl, Kaz. .... 59 42°51′N 71°29′E
Zhangaqazaly, Kaz. .... 59 45°47′N 62°00′E
Zhangbei, China (jän-bā) .... 77 41°12′N 114°50′E
Zhangjiakou, China .... 77 40°45′N 114°58′E
Zhangye, China (jän-yu) .... 76 38°46′N 101°00′E
Zhangzhou, China (jän-jō) .... 77 24°35′N 117°45′E
Zhanjiang, China (jän-jyän) .... 77 21°20′N 110°28′E
Zhaotong, China (jou-tôŋ) .... 76 27°18′N 103°50′E
Zharkent, Kaz. .... 59 44°02′N 79°58′E
Zhaysang köli, l., Kaz. .... 59 48°16′N 84°05′E
Zhejiang, prov., China (jü-jyän) .... 77 29°30′N 120°00′E
Zhelaniya, Mys, c., Russia
  (zhĕ′lä-nĭ-yá) .... 56 75°43′N 69°10′E
Zhengzhou, China (jüŋ-jō) .... 77 34°46′N 113°42′E
Zhenjiang, China (jün-jyän) .... 77 32°13′N 119°24′E
Zhetiqara, Kaz. .... 59 52°12′N 61°18′E
Zhigalovo, Russia (zhĕ-gä′lô-vô) .... 57 54°52′N 105°05′E
Zhigansk, Russia (zhĕ-gánsk′) .... 57 66°45′N 123°20′E
Zhongwei, China (jôŋ-wä) .... 76 37°32′N 105°10′E
Zhongxian, China (jôŋ shyĕn) .... 76 30°20′N 108°00′E
Zhoushan Qundao, is., China
  (jō-shän-chyòn-dou) .... 77 30°00′N 123°00′E
Zhytomyr, Ukr. .... 56 50°15′N 28°40′E
Zile, Tur. (zĕ-lĕ′) .... 55 40°20′N 35°50′E
Zimbabwe, nation, Afr. (rǒ-dē′zhĭ-à) ... 85 17°50′S 29°30′E
Zin, r., Isr. .... 67a 30°50′N 35°12′E
Zlatoust, Russia (zlä-tô-ôst′) ...... 56 55°13′N 59°39′E
Zonguldak, Tur. (zôn′gōōl′dák) ...... 68 41°25′N 31°50′E
Zugdidi, Geor. .... 58 42°30′N 41°53′E
Zunyi, China .... 76 27°58′N 106°40′E
Zürich, Switz. (tsü′rĭk) .... 50 47°22′N 8°32′E
Zuwayzā, Jord. .... 67a 31°42′N 35°55′E
Zyryanka, Russia (zĕ-ryän′ká) .... 57 65°45′N 151°15′E
Zyryanovsk, Kaz. .... 59 49°43′N 84°20′E

Z

Za, r., Mor. .... 54 34°19′N 2°23′W
Zacatecas, Mex. (sä-kä-tā′käs) ...... 42 22°44′N 102°32′W
Zacatecas, state, Mex. .... 42 24°00′N 102°45′W
Zadar, Cro. (zä′där) .... 50 44°08′N 15°16′E

---

ng-sing; ŋ-baŋk; N-nasalized n; nŏd; cŏmmit; ōld; ôbey; ôrder; oi-boil; fōōd; ò-as oo in foot; ou-out; s-soft; sh-dish; th-thin; pūre; ûnite; ûrn; stŭd; circŭs; ü-as in French tu; ′-indeterminate vowel.

SUBJECT INDEX

Listed below are major topics covered by the thematic maps, graphs and/or statistics.
Page citations are for world, continent and country maps and for world tables.

SOURCES

The following sources have been consulted during the process of creating and updating the thematic maps and statistics for the 21st Edition.

Air Carrier Traffic at Canadian Airports, Statistics Canada

Annual Coal Report, U.S. Dept. of Energy, Energy Information Administration

Armed Conflicts Report, Project Ploughshares

Atlas of Canada, Natural Resources Canada

Canadian Minerals Yearbook, Statistics Canada

Census of Canada, Statistics Canada

Census of Population, U.S. Census Bureau

Chromium Industry Directory, International Chromium Development Association

Coal Fields of the Conterminous United States, U.S. Geological Survey

Coal Quality and Resources of the Former Soviet Union, U.S. Geological Survey

Coal-Bearing Regions and Structural Sedimentary Basins of China and Adjacent Seas, U.S. Geological Survey

Commercial Service Airports in the United States with Percent Boardings Change, Federal Aviation Administration (FAA)

Completed Peacekeeping Operations, Center for Defense Information

Conventional Arms Transfers to Developing Nations, Library of Congress, Congressional Research Service

Current Status of the World's Major Episodes of Political Violence: Hot Wars and Hot Spots, Center for Systemic Peace

Dependencies and Areas of Special Sovereignty, U.S. Dept. of State, Bureau of Intelligence and Research

Earth's Seasons—Equinoxes, Solstices, Perihelion, and Aphelion, U.S. Naval Observatory

EarthTrends: The Environmental Information Portal, World Resources Institute and World Conservation Monitoring Centre 2003. Available at http://earthtrends.wri.org/ Washington, D.C.: World Resources Institute

Economic Census, U.S. Census Bureau

Employment, Hours, and Earnings from the Current Employment Statistics Survey, U.S. Dept. of Labor, Bureau of Labor Statistics

Energy Statistics Yearbook, United Nations Dept. of Economic and Social Affairs

Epidemiological Fact Sheets by Country, Joint United Nations Program on HIV/AIDS (UNAIDS), World Health Organization, United Nations Children's Fund (UNICEF)

Estimated Water Use in the United States, U.S. Geological Survey

Estimates of Health Personnel, World Health Organization

FAO Food Balance Sheet, Food and Agriculture Organization of the United Nations (FAO)

FAO Statistical Databases (FAOSTAT), Food and Agriculture Organization of the United Nations (FAO)

Fishstat Plus, Food and Agriculture Organization of the United Nations (FAO)

Geothermal Resources Council Bulletin, Geothermal Resources Bulletin

Geothermal Resources in China, Bob Lawrence and Associates, Inc.

Global Alcohol Database, World Health Organization

Global Forest Resources Assessment, Food and Agriculture Organization of the United Nations (FAO), Forest Resources Assessment Programme

Great Lakes Factsheet Number 1, U.S. Environmental Protection Agency

The Hop Atlas, Joh. Barth & Sohn GmbH & Co. KG

Human Development Report 2003, United Nations Development Programme, © 2003 by United Nations Development Programme. Used by permission of Oxford University Press, Inc.

Installed Generating Capacity, International Geothermal Association

International Database, U.S. Census Bureau

International Energy Annual, U.S. Dept. of Energy, Energy Information Administration

International Journal on Hydropower and Dams, International Commission on Large Dams

International Petroleum Encyclopedia, PennWell Publishing Co.

International Sugar and Sweetener Report, F.O. Licht, Licht Interactive Data

International Trade Statistics, World Trade Organization

International Water Power and Dam Construction Yearbook, Wilmington Publishing

Iron and Steel Statistics, U.S. Geological Survey, Thomas D. Kelly and Michael D. Fenton

Lakes at a Glance, LakeNet

Land Scan Global Population Database, U.S. Dept. of Energy, Oak Ridge National Laboratory (© 2003 UT-Battelle, LLC. All rights reserved. Notice: These data were produced by UT-Battelle, LLC under Contract No. DE-AC05-00OR22725 with the Department of Energy. The Government has certain rights in this data. Neither UT-Battelle, LLC nor the United States Department of Energy, nor any of their employees, makes any warranty, express or implied, or assumes any legal liability or responsibility for the accuracy, completeness, or usefulness of any data, apparatus, product, or process disclosed, or represents that its use would not infringe privately owned rights.)

Largest Rivers in the United States, U.S. Geological Survey

Lengths of the Major Rivers, U.S. Geological Survey

Likely Nuclear Arsenals Under the Strategic Offensive Reductions Treaty, Center for Defense Information

Major Episodes of Political Violence, Center for Systemic Peace

Maps of Nuclear Power Reactors, International Nuclear Safety Center

Mineral Commodity Summaries, U.S. Geological Survey, Bureau of Mines

Mineral Industry Surveys, U.S. Geological Survey, Bureau of Mines

Minerals Yearbook, U.S. Geological Survey, Bureau of Mines

National Priorities List, U.S. Environmental Protection Agency

National Tobacco Information Online System (NATIONS), U.S. Dept. of Health and Human Services, Centers for Disease Control and Prevention (CDC)

Natural Gas Annual, U.S. Dept. of Energy, Energy Information Administration

New and Recent Conflicts of the World, The History Guy

Nuclear Power Reactors in the World, International Atomic Energy Agency

Oil and Gas Journal DataBook, PennWell Publishing Co.

Oil and Gas Resources of the World, Oilfield Publications, Ltd.

Petroleum Supply Annual, U.S. Dept. of Energy, Energy Information Administration

Population of Capital Cities and Cities of 100,000 and More Inhabitants, United Nations Dept. of Economic and Social Affairs

Preliminary Estimate of the Mineral Production of Canada, Natural Resources Canada

Red List of Threatened Species, International Union for Conservation and Natural Resources

Significant Earthquakes of the World, U.S. Geological Survey

State of Food Insecurity in the World, Food and Agriculture Organization of the United Nations (FAO)

State of the World's Children, United Nations Children's Fund (UNICEF)

Statistical Abstract of the United States, U.S. Census Bureau

Statistics on Asylum-Seekers, Refugees and Others of Concern to UNHCR, United Nations High Commissioner for Refugees (UNHCR)

Survey of Energy Resources, World Energy Council

Tables of Nuclear Weapons Stockpiles, Natural Resources Defense Council

TeleGeography Research, PriMetrica, Inc. (www.primetrica.com)

Tobacco Atlas, World Health Organization

Tobacco Control Country Profiles, World Health Organization

Transportation in Canada, Minister of Public Works and Government Services, Transport Canada

UNESCO Statistical Tables, United Nations Educational, Scientific and Cultural Organization (UNESCO)

United Nations Commodity Trade Statistics (COMTRADE), United Nations Dept. of Economic and Social Affairs

United Nations Peacekeeping in the Service of Peace, United Nations Dept. of Peacekeeping Operations

United Nations Peacekeeping Operations, United Nations Dept. of Peacekeeping Operations

Uranium: Resources, Production and Demand, United Nations Organization for Economic Co-operation and Development (OECD)

Volcanoes of the World, Smithsonian National Museum of Natural History

Water Account for Australia, Australian Bureau of Statistics

Women in National Parliaments, Inter-Parliamentary Union

Women's Suffrage, Inter-Parliamentary Union

The World at War, Center for Defense Information, The Defense Monitor

The World at War, Federation of American Scientists, Military Analysis Network

World Conflict List, National Defense Council Foundation

World Contraceptive Use, United Nations Dept. of Economic and Social Affairs

The World Factbook, U.S. Dept. of State, Central Intelligence Agency (CIA)

World Facts and Maps, Rand McNally

World Lakes Database, International Lake Environment Committee

World Population Prospects, United Nations Dept. of Economic and Social Affairs

World Urbanization Prospects, United Nations Dept. of Economic and Social Affairs

World Water Resources and Their Use, State Hydrological Institute of Russia/UNESCO

The World's Nuclear Arsenal, Center for Defense Information

Special Acknowledgements

The American Geographical Society, for permission to use the Miller cylindrical projection.

The Association of American Geographers, for permission to use R. Murphy's landforms map.

The McGraw-Hill Book Company, for permission to use G. Trewartha's climatic regions map.

The University of Chicago Press, for permission to use Goode's Homolosine equal-area projection.